AF844930

Cancer Drug Discovery and Development

Series Editor

Beverly A. Teicher
Genzyme Corporation, Framington, MA, USA

For further volumes:
http://www.springer.com/series/7625

Lawrence Panasci • Raquel Aloyz
Moulay Alaoui-Jamali
Editors

Advances in DNA Repair in Cancer Therapy

Humana Press

Editors
Lawrence Panasci
Jewish General Hospital
Montreal, QC, Canada

Moulay Alaoui-Jamali
Department of Medicine and Oncology
McGill University
Montreal, QC, Canada

Raquel Aloyz
Department of Medicine and Oncology
McGill University
Montreal, QC, Canada

ISBN 978-1-4614-4740-5 ISBN 978-1-4614-4741-2 (eBook)
DOI 10.1007/978-1-4614-4741-2
Springer New York Heidelberg Dordrecht London

Library of Congress Control Number: 2012952084

© Springer Science+Business Media, LLC 2013
This work is subject to copyright. All rights are reserved by the Publisher, whether the whole or part of the material is concerned, specifically the rights of translation, reprinting, reuse of illustrations, recitation, broadcasting, reproduction on microfilms or in any other physical way, and transmission or information storage and retrieval, electronic adaptation, computer software, or by similar or dissimilar methodology now known or hereafter developed. Exempted from this legal reservation are brief excerpts in connection with reviews or scholarly analysis or material supplied specifically for the purpose of being entered and executed on a computer system, for exclusive use by the purchaser of the work. Duplication of this publication or parts thereof is permitted only under the provisions of the Copyright Law of the Publisher's location, in its current version, and permission for use must always be obtained from Springer. Permissions for use may be obtained through RightsLink at the Copyright Clearance Center. Violations are liable to prosecution under the respective Copyright Law.
The use of general descriptive names, registered names, trademarks, service marks, etc. in this publication does not imply, even in the absence of a specific statement, that such names are exempt from the relevant protective laws and regulations and therefore free for general use.
While the advice and information in this book are believed to be true and accurate at the date of publication, neither the authors nor the editors nor the publisher can accept any legal responsibility for any errors or omissions that may be made. The publisher makes no warranty, express or implied, with respect to the material contained herein.

Printed on acid-free paper

Humana Press is a brand of Springer
Springer is part of Springer Science+Business Media (www.springer.com)

Preface

The field of DNA repair has been constantly evolving from the initial description of RecA as a key factor in homologous recombinational repair (HRR) in *Escherichia coli*. The Rec A protein plays a central role in the process. DNA damage induces a complicated response in *E. coli* called the SOS response. This process involves many enzymes. The RecA homologue in mammalian cells is Rad51. Rad51 along with several Rad51 paralogues and other proteins is involved in error-free homologous recombinational repair (HRR) of complicated DNA damage which can result from many different anticancer drugs, in particular DNA cross-linking agents, and ionizing radiation.

An alternative DNA repair system is non-homologous endjoining (NHEJ). NHEJ includes both classical NHEJ and alternative NHEJ (alt-NHEJ). Alt-NHEJ is promoted by PARP-1 and is more error-prone than classical NHEJ. Classical NHEJ has several components including Ku70, Ku80, DNA-PKcs, XRCC4, ligase IV, and Artemis. DNA-PK is a complex composed of the catalytic DNA-PKcs and the regulatory Ku subunits, Ku70 and Ku80. The DSB repair process is initiated by the association of DNA ends with the Ku70/80 dimer which acts as a scaffold to assemble other NHEJ proteins including DNA-PKcs. Once the DNA-PK complex is located on DNA ends, the serine/threonine kinase activity of DNA-PK is activated resulting in autophosphorylation and phosphorylation of several proteins including p53. DNA-PK autophosphorylation results in release of the DNA-PK complex and further processing/ligation by other NHEJ proteins.

There are several components involved in recognition of DNA damage including PARP, MRN, and H2AX. A major cellular defense for DNA damage produced by anticancer agents and ionizing radiation are signaling pathways that are largely regulated by the phosphoinositol 3-kinase-like serine/threonine protein kinases such as ataxia telangiectasia mutated (ATM) and ATM- and Rad3-related (ATR). This recognition of DNA damage then activates the DNA repair systems. Damage caused by anticancer drugs and ionizing radiation can result in double strand breaks (DSBs) which can be repaired by either DNA repair system mentioned above. The choice of HRR or NHEJ depends on several factors including the phase of the cell cycle (NHEJ is implicated in the repair of DSBs in all phases of the cell cycle unlike HRR which is most active during S and G2 because HRR mediates DSB repair using homologous DNA sequences which are found on sister chromosomes). However, the choice of HRR or NHEJ may also be dictated by as yet unknown factors linked to the recognition of the DNA lesions produced by various anticancer drugs and ionizing radiation.

A large body of evidence has been generated that suggests that alterations in the expression/function of many DNA repair genes are associated with resistance to anticancer drugs/ionizing radiation that result in DNA damage. Furthermore inhibition or interference with many of these DNA repair-related genes can sensitize cells to DNA damage. This along with the fact that DNA repair is often coupled with cell cycle checkpoint arrest which can be impaired in cancer cells with various genetic defects has resulted in a stimulus to examine compounds that inhibit DNA repair vis-à-vis anticancer therapy efficacy.

What is clear is that inhibition of ATR, ATM, PARP, NHEJ, or HRR can sensitize cancer cells to anticancer drugs and radiotherapy. The only class of DNA repair inhibitors in clinical trials is the PARP inhibitors.

Recently inhibitors of ATR have been proven to be very efficacious in anticancer therapy in vitro and that there is synthetic lethality when ATR is inhibited in cancer cells with ATM/p53 defects.

While there are many specific inhibitors of ATM and DNA-PKcs, a key component of NHEJ, none of the currently available agents are in clinical trial because of rapid hepatic inactivation of all these inhibitors. Hopefully, within the next 5 years, new DNA-PK inhibitors with better pharmacokinetics will be synthesized and thus allow for testing of the principle that direct transient of DNA repair pathway can improve the therapeutic index of anticancer therapy.

In addition, inhibition of the Rad51 HRR pathway is just beginning, and this may prove very promising. Ultimately, the possibility that transient inhibition of DNA repair will increase the efficacy of anticancer therapy should be put to the test in the next 5–10 years.

Montreal, QC, Canada Lawrence Panasci
Montreal, QC, Canada Raquel Aloyz
Montreal, QC, Canada Moulay Alaoui-Jamali

Contents

Repair of DNA Interstrand Cross-links Produced by Cancer Chemotherapeutic Drugs

Anderson T. Wang, Peter J. McHugh, and John A. Hartley

1 DNA Interstrand Cross-linking Drugs

It has been clear for over 50 years that bifunctional reactivity is an essential prerequisite for the potent cytotoxic and antitumour activity of agents such as the nitrogen mustards [1]. DNA was later identified as a target for these drugs [2, 3], and the covalent modification of DNA almost certainly accounts for the antitumour activity of these drugs [1]. The fact that a bifunctional covalent reaction with DNA (cross-linking) is essential for the toxicity of these agents is evident from studies employing monofunctional analogues; for drugs such as the nitrogen mustard's mechlorethamine and melphalan, their monofunctional counterparts are many orders of magnitude less toxic [4, 5]. Cross-links can be formed on the same strand of DNA (intrastrand), between the two complementary strands of DNA (interstrand), or between a base on DNA and a reactive group on a protein (DNA–protein). For the bifunctional alkylating drugs (e.g. the nitrogen mustard class and mitomycin C), it is clear that the interstrand cross-link (ICL), although accounting for only a small proportion of the total DNA adducts, is the critical cytotoxic lesion [6, 7]. For the platinum drugs (e.g. cisplatin and carboplatin) the majority (>80%) of DNA adducts are intrastrand cross-links, although the <5% of ICLs are critical cytotoxic lesions [8].

Drug-induced ICLs, which are generally irreversible, prevent the separation of the two strands of DNA which is essential for cellular processes such as replication and transcription. Since both DNA strands are involved, ICLs pose problems for the cellular DNA repair machinery and it is clear that there is a

A.T. Wang • P.J. McHugh
Weatherall Institute of Molecular Medicine, John Radcliffe Hospital, University of Oxford, Oxford, UK

J.A. Hartley (✉)
UCL Cancer Institute, Faculty of Medical Sciences, University College London, London, UK
e-mail: john.hartley@ucl.ac.uk

L. Panasci et al. (eds.), *Advances in DNA Repair in Cancer Therapy*, Cancer Drug Discovery and Development 72, DOI 10.1007/978-1-4614-4741-2_1,
© Springer Science+Business Media, LLC 2013

co-ordination of ICL-induced cellular responses, including cell cycle arrest, DNA damage repair and cell death [9]. Different human tumour types differ in their inherent sensitivity to DNA cross-linking agents, and this appears to be the result, at least in part, of their differing abilities to repair specific types of drug-induced DNA damage [10, 11]. Increased repair of ICLs is clearly also a critical mechanism of clinical acquired resistance to agents such as the nitrogen mustards, chlorambucil and melphalan. This has been shown in chronic lymphocytic leukaemia [12], and in multiple myeloma [11]. More recently, this has also been demonstrated for platinum drugs in ovarian cancer [13]. In addition, the capacity to repair ICLs appears to decline with age in normal cells, which may be a factor in the poor tolerance of chemotherapy in the elderly [14].

Although there are many ICL anticancer drugs approved for clinical use, relatively few of these agents have been widely employed in the most detailed mechanistic studies of ICL repair. In fact, one of the cross-linking agents most commonly employed in such studies is not used in cancer treatment. Of the anticancer agents, the original nitrogen mustard mechlorethamine [chemically, 2-chloro-*N*-(2-chloroethyl)-*N*-methylethanamine] is by far the best characterised. This agent cross-links preferentially between opposed guanines in the sequence 5′-GNC-3′/3′-CNG-5′ [15], and this cross-linking represents only a small fraction of the total DNA lesions that this drug produces (<5%), the remainder being monofunctional alkylations at guanine N7 and adenine N3 [1]. Cisplatin also cross-links between guanine N7 positions in the DNA major groove, but in this case in the sequence 5′-GC-3′/5′-CG-3′ [16]. Another anticancer agent commonly used in mechanistic studies is mitomycin C. This natural product molecule requires metabolic reduction in order to generate the reactive species, which produces cross-links in the DNA minor groove through reaction with the N2 position of guanines, cross-linking the opposed guanines in the sequence 5′-GC-3′/5′-CG-3′ (up to 13% of total adducts are ICLs) [17].

The non-anticancer compounds that have been studied in great detail are the psoralens [18], particularly 8-methoxypsoralen. Following 405 nm visible radiation, the formation of DNA monoadducts is favoured, whereas ultraviolet A (UVA) (365 nm) is required to convert these to abundant ICLs (up to 40% of the total adducts). The basis of the activity of psoralens is UVA-induced reactivity at 5′-AT-3′/5′-TA-3′ base pairs to form ICLs. Because of the more complex, multi-ringed structure of the psoralens, they form asymmetric cross links that bear a furan-ringed side and a pyrone-ringed side.

One property of the ICLs produced by drugs such as the nitrogen mustards and platinum drugs is that they significantly distort the structure of the DNA [16, 19]. Pyrrolobenzodiazepine dimer-based drugs such as SJG-136 (SG2000) have been rationally designed as highly efficient minor groove ICL agents [20, 21]. Interstrand cross-linking is primarily between the two guanine N2 positions in the sequence 5′-purine-GATC-pyrimidine-3′, and an important property is that the ICL produces minimal distortion of the normal DNA structure. Because of the high specificity and efficiency of ICL formation, substrates containing single SJG-136 ICLs are proving useful in detailed mechanistic studies of ICL repair [22, 23].

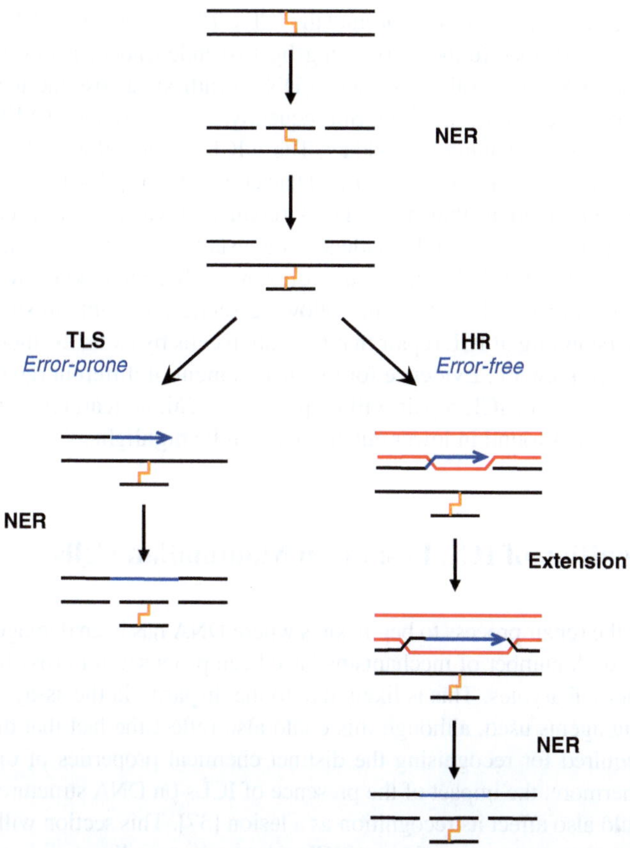

Fig. 1 A basic model for ICL repair based on understanding from bacteria and yeast. The NER pathway recognises the lesion and makes incisions around the lesion, unhooking the interstrand cross-link. This substrate can then be processed by two different pathways. The first is an error-prone process involving TLS, and the second is an error-free process dependent on HR. Both pathways result in DNA synthesis across the lesion, followed by a second round of NER to fully remove the cross-link

2 Basic Model of ICL Repair Based on Bacterial and Yeast Studies

Given the physical constraints ICL lesions impose on the DNA double helix, their repair requires the co-ordination of multiple repair pathways. A general model of ICL repair has been proposed based on studies in bacteria and yeast [24, 25]. Early genetic studies involving epistasis analysis of bacterial or yeast mutants sensitive to a range of DNA damaging agents, identified three major groups of genes involved in ICL repair, corresponding to the nucleotide excision repair (NER), homologous recombination (HR) and the translesion synthesis (TLS) DNA synthesis pathways [26–29]. Collectively, a model has emerged in which two parallel, but non-redundant, pathways are implicated in the repair of ICLs (Fig. 1). Both of these pathways require the pivotal, initial action of the

NER machinery, making incisions around the ICLs. This "unhooked" ICL-repair intermediate then undergoes further processing by two independent pathways involving either HR or TLS. Both pathways lead to DNA synthesis across the tethered ICLs, filling the gap created by the incisions. Subsequently, a second round of NER is required to fully remove the remaining ICL adduct. These ICL repair pathways have both been partially reconstituted *in vitro* using purified bacterial proteins [30–36].

It has become evident that many DNA repair pathways are well-conserved in higher eukaryotes. The available evidence suggests that the models outlined above for bacteria and yeast ICL repair are relevant to higher eukaryotes, although several key differences do exist. The following section attempts to summarise the current understanding of ICL repair in mammalian cells by using the model depicted in Fig. 1 as a framework. Evidence for the involvement of different repair pathways in the distinct steps of ICL repair will be provided. Unique features of mammalian cell ICL repair not found in lower eukaryotes will be highlighted.

3 Recognition of ICL Lesions in Mammalian Cells

In order for the repair process to begin, sites where DNA has been damaged must first be recognised. A number of mechanisms have been proposed for cross-link recognition in higher eukaryotes. This is likely due to the disparity in the assay systems and cross-linking agents used, although this could also reflect the fact that different systems are required for recognising the distinct chemical properties of different ICL types. Furthermore, the impact of the presence of ICLs on DNA structure and helical density would also affect its recognition as a lesion [37]. This section will summarise the literature on damage recognition of ICLs in vertebrate cells.

3.1 NER and ICL Lesion Recognition

The NER pathway appears to be essential for the repair of ICLs in both bacteria and yeast. Therefore, it has been postulated that NER also functions in vertebrate ICL repair. In the context of ICL repair, given that the two strands are covalently linked by the presence of the cross-link, no base pairs could be "flipped out"; therefore binding to the undamaged single-stranded DNA is impossible in principle. However, it is likely that distorting ICL lesions would result in a degree of unwinding that could provide an entry site for XPC binding adjacent to the ICL lesion [38]. Evidence of the involvement of XPC in the damage recognition of psoralen ICLs exists whereby both XPC-hHR23B and XPA-RPA can bind to triplex forming oligonucleotide with a psoralen cross link [39]. XPC proteins were found to be recruited rapidly to sites of a laser-induced damage "stripe" containing psoralen ICLs in G1 phase human cells [40]. However, a conflicting report on the involvement of XPC proteins in recognising cisplatin adducts exists, whereby XPC cells were not more sensitive to cisplatin than the wild-type cells [41]. Furthermore, the mechanism of XPC recognition of mitomycin C ICLs remains unclear, given the modest distortion they induce without significant unwinding of the DNA. It is a possibility that the recognition of ICLs by XPC requires an interacting partner, such as the high mobility group protein B1 (HMGB1) that has recently been demonstrated to play a role in facilitating XPC in the recognition of psoralen-triplex ICLs [42].

3.2 Transcription-Coupled Mechanism of ICL Recognition

Transcription-coupled nucleotide excision repair (TC-NER) has been proposed to play an important role in the repair of ICLs, especially during G1 phase, as an ICL represents an absolute block to the RNA polymerases [43]. TC-NER has been described to be linked to ICL repair in the early 1990s when it was demonstrated that the repair of ICLs was more efficient in transcribed regions of an active gene [44–46]. Furthermore, host-cell reactivation experiments using expression plasmids containing ICLs placed in between the promoter and downstream reporter gene showed reduced ICL repair efficiency in cells defective in TC-NER [47–49]. However, it is important to note that the host-cell reactivation system is heavily transcriptionally biased as the readout of the assay is dependent on transcription. Therefore, although TC-NER is capable of repairing ICLs, the actual importance of TC-NER in contributing toward the repair of ICLs in cells remains to be determined although genetic studies also provide some evidence of TC-NER's involvement in ICL repair as both CSA and CSB defective cells were found to be sensitive to cisplatin [41].

3.3 Recognition of ICLs by Replication Forks

It has been proposed by a number of groups that the repair of ICLs in mammalian cells is replication-dependent during S-phase [9, 25, 50–53]. The replication-dependent repair of ICLs was first evident in observation that psoralen/UVA treatment only induced cell cycle arrest when synchronised human skin fibroblasts pass through S-phase, regardless of where in the cell cycle the cross-linking agent was initially administered [54]. This implies that the recognition of ICLs occurs exclusively in S-phase, and the replication fork arresting at the site of an ICL triggers the cellular repair response. It was also noticed that the repair of ICLs during S-phase results in generation of DNA double strand breaks (DSBs), which is not evident in stationary yeast or CHO cells [5, 55]. This leads to the proposal that replication-dependent repair of ICLs during S-phase involves the DSB repair pathway.

It has been proposed more recently that the replication-dependent repair of ICLs occurs when two forks converge on a single ICL (Fig. 2) [56, 57]. The convergence of two replication forks was observed by electron microscopy when ICL-containing plasmid substrates were replicated in the presence of *Xenopus* egg extract. The

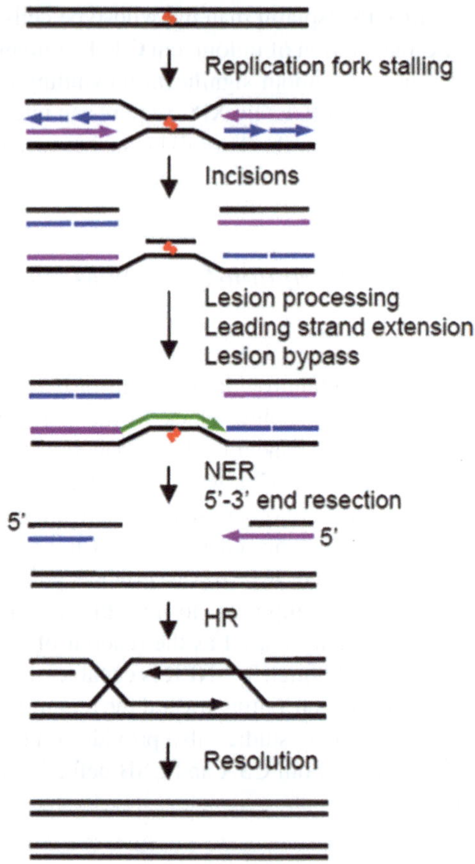

Fig. 2 Converging fork model of ICL repair. Based upon data from in vitro studies using *Xenopus* egg extracts with a plasmid-based system that favours converging replication forks [56, 57]. Replication fork stalling at the site of ICL is likely to provide a signal for monoubiquitination and activation of FANCD2-FANCI which orchestrates the repair of ICLs. The initiation of repair is thought to involve dual incisions around the ICL on one DNA strand. The "unhooked" ICL repair intermediate undergoes further processing, before the leading strand extends and TLS polymerase bypasses the remaining adduct that allow the restoration of a DNA template. Following a second round of incision, most likely involving NER that fully removes the ICL lesion, the HR machinery can utilise the DSB ends to re-establish the replication fork and complete DNA synthesis

repair of ICLs in such a context was found to be entirely replication-dependent. Using this system, it was observed that the converging replication forks initially stall 20–40 nucleotides from the lesion before one of the leading strands advances to within one nucleotide from the ICL. Subsequent dual incisions of the ICL result in the uncoupling of the two sister chromatids and lesion bypass DNA synthesis. The authors proposed that the double fork collision model of ICL repair is advantageous to cells as lesion bypass can readily occur from a nascent leading strand, preventing prolonged lag time between the incision and HR steps before the DNA synthesis is completed (Fig. 2). The absolute replication-dependence of ICL repair remains controversial as another similar *in vitro* study observed that replication-independent repair of ICL occurs [58]. Furthermore, the possibility of two forks arriving at the ICL is likely to be low in the *in vivo* setting, and is excluded in the situation when there are two ICLs formed in between neighbouring origins of firing. It has also been shown that ICL-induced checkpoint signalling would inhibit origin firing and slow fork elongation, limiting the possibility of two forks converging on an ICL [59]. Moreover, a two-sided DSB would be generated following incision of the two forks. This would be a potential substrate for non homologous end joining

(NHEJ) although it has been shown that NHEJ has a limited role in the repair of DSBs induced by ICLs [5]. ICLs have also been shown to induce sister chromatid exchanges (SCEs), which do not commonly occur in the context of repair of two-sided DSBs [53].

4 Unhooking of ICLs in Mammalian Cells

Following the recognition of ICLs as lesions, the repair machinery acts to make incisions on either side of the ICL to "unhook" the lesion. This represents a pivotal step, regardless of the mechanism of unhooking, as this relieves the torsional stress an ICL imposes on the DNA helix and permits processing of the repair intermediates by downstream pathways. A number of nucleases have been suggested to play a role in the unhooking step of ICLs.

Given the role of XPF-ERCC1 in making 5′ incision during NER, which is found to be essential in ICL repair in yeast and bacteria, this structure-specific endonuclease has long been implicated in the unhooking of ICLs. However, the extreme sensitivity of many *XPF* and *ERCC1* defective cell lines to cross-linking agents, compared to cells bearing mutations in other components of the NER apparatus, supports a role of XPF-ERCC1 in ICL repair processes other than NER [5, 60–62]. Purified XPF-ERCC1 proteins were able to make incisions on ICL placed on a duplex with splayed arm structure [63]. Incisions were observed on both the 5′ side and 3′ side of the ICL. The 3′ incision was stimulated when the cross-link was moved further away from the splayed arms. The inability of XPF-ERCC1 proteins to make incisions around the same psoralen ICL placed on a linear DNA duplex suggests that the splayed arm structure mimicking a stalled replication fork provides the substrate for XPF-ERCC1 recognition, further supporting the replication-dependent model of ICL repair. However, XPF-ERCC1 incision on DNA substrates containing a site-specific SJG-136 ICL has been found to be lesion specific [23].

Another structure-specific endonuclease related to XPF at the sequence level, Mus81-Eme1, has also been implicated in taking part in the unhooking step of ICLs. Mouse embryonic stem (ES) cells disrupted of MUS81 or EME1 were found to be mildly sensitive to mitomycin C [64, 65]. Based on the observation that *mus81*$^{-/-}$ MEFs showed suppression of DSBs [65], whereas *ercc1*$^{-/-}$ MEFs accumulate DSBs [62], several authors have proposed that ICL repair is initiation by MUS81-dependent incision, possibly on the leading strand template of the replication fork, followed by a second XPF-ERCC1 dependent incision 5′ to the ICL, where the net result is unhooking [62, 65, 66]. However, the observation that XPF-ERCC1 depleted human cells accumulated more MUS81-dependent DSBs argues against an initiating role for Mus81 in the unhooking [23]. Given that MUS81-dependent DSBs occur as a late response to ICLs, it is suggested that MUS81-dependent incision occurs when the XPF-ERCC1 incision fails or in the situation of a converging fork.

Another structure-specific endonuclease, the SLX1 and SLX4 heterodimer, is implicated in ICL repair as cells depleted of SLX4 has been shown to be hypersensitive to cross-linking agents, and SLX4 has recently been identified to be a factor mutated in the complementation group P form of Fanconi anaemia (FA) (see below). However, given the very mild sensitivity of cells depleted of SLX1, the catalytic subunit of the heterodimer, it is unlikely that the Holliday junction resolvase SLX1–SLX4 acts to incise ICLs. The involvement of SLX4 in ICL repair may come from its interaction with XPF-ERCC1 and MUS81-EME1 [67–70]. It is postulated that SLX4 acts as a scaffold protein for recruiting these structure specific endonucleases to the sites of ICLs.

Most recently, a nuclease that is associated with the FA pathway, FAN1, has been identified as a factor that is required for resistance to mitomycin C [71–74]. It has not yet been demonstrated whether purified FAN1 is involved in the unhooking reaction. However, given that FAN1 depleted cells are also defective in HR repair as directly measured by the GFP reporter assay, it has been suggested that FAN1 is involved in the HR repair of DSBs induced by cross-linking agents [72, 73]. Consistently, FAN1 depleted cells also show increased chromosome aberration and radial chromosome formation.

5 Post-incision/excision Gap-Filling

Regardless of the mechanism which leads to unhooking of an ICL, the repair intermediate generated requires further processing in order to restore an undamaged template before the removal of the remaining ICL on the opposite strand. In the canonical NER reaction, the incision step is followed by the removal/excision of the released oligonucleotide to create a gap on the incised strand opposite the lesion before DNA synthesis could take place. In the context of ICLs, the incised oligonucleotide remains attached to the non-incised strand via the cross-link. Several studies have demonstrated that the downstream DNA repair synthesis opposite the remaining ICL lesion would be more efficient if the tethered oligonucleotide is displaced and/or removed [31, 75, 76]. This would involve the concerted action of a helicase and/or exonuclease.

The SNM1/PSO2 gene was initially identified in two independent screens in budding yeast [28, 77, 78]. Cells mutated for *PSO2* showed high sensitivity to cross-linking agents, but not to other forms of DNA damage, including that produced by monofunctional alkylating agent, IR or UV light. Three human orthologues of yeast Pso2 were identified: SNM1A, SNM1B (also known as Apollo) and SNM1C (Artemis) [79–81]. Structurally, SNM1A has been suggested to be the human homologue of Pso2 as there is greater degree of homology in the overall domain structure between Pso2 and mammalian SNM1A [82], and only complementing human SNM1A (hSNM1A) proteins in yeast *pso2* mutant cells would partially rescue the hypersensitivity to HN2 [83]. Purified hSNM1A protein demonstrates 5′ exonuclease activity on both double- and single-stranded DNA [83, 84], and it has been demonstrated recently that hSNM1A is capable of removing the tethered oligonucleotide from a 5′ incision initiated by XPF-ERCC1 pausing at the cross link [23].

Intriguingly, hSNM1A has been shown to interact with the mono-ubiquitinated form of PCNA [85], suggesting that the processing of hSNM1A might be associated with downstream TLS reactions.

TLS polymerases have been shown to play an essential role for gap-filling following unhooking and subsequent processing of ICLs [75]. The monoubiquitination of PCNA at lysine 164 by Rad6–Rad18 complex following replication fork stalling appears to be pivotal and allows the switch from replicative polymerases to TLS polymerase [86, 87]. The involvement of Rad18 in ICL repair in vertebrate cells is evident from the hypersensitivity of Rad18 and Rev3 cells to mitomycin C, mechlorethamine and cisplatin in chicken DT40 and mouse cells [88]. Recently, a number of TLS polymerases have been shown to be involved in the repair of ICLs, including Pol ζ [88–90], Rev1 [91–93], Pol η [48, 49, 94], Pol κ [76], Pol ι [75] and Pol ν [95–97]. However, the role and the regulation of the polymerases in performing lesion bypass have remained poorly defined. Mechanistic studies using *Xenopus* egg extracts showed that Pol ζ is not involved in inserting a nucleotide opposite a cisplatin ICL but is required for extension beyond the inserted nucleotide [57]. A detailed analysis of *in vitro* lesion bypass activities of translesion polymerases has shown that the Y-family polymerases, Pol κ, Pol ι and Pol η are capable of inserting a nucleotide opposite a cisplatin lesion, whereas Polζ alone or in complex with Rev1 is much less efficient [98]. Interestingly, Pol ζ complex failed to extend insertion products by Rev1 *in vitro*, whereas Pol η could perform translesion nucleotide insertion and a further extension of two nucleotides downstream [98]. The contrasting results of the poor *in vitro* efficiency of Pol ζ in lesion bypass in comparison to the more important role in TLS demonstrated in the *in vivo* context highlights the importance of the yet-to-be fully elucidated regulation of lesion bypass events during ICL repair.

5.1 HR in Resolving ICL-Induced DSBs

The presence of an ICL represents a blockage to replication fork progression that leads to replisome instability. If a single replication fork encounters an ICL, the collapse of the replication fork will result in one-ended DSBs, and the resolution of such DSBs and the re-establishment of replication forks require the break-induced repair the component of the HR pathway. In the situation when two replication forks converge on an ICL, dual incisions around the ICL result in the formation of two-ended DSBs, which are utilised by the HR machinery to complete replication and repair (Fig. 2). In higher eukaryotes, the involvement of HR in ICL repair has been demonstrated by the hypersensitivity of cells defective in HR to cross-linking agents. CHO cells defective in XRCC2 and XRCC3 showed hypersensitivities to mechlorethamine and mitomycin C, and are defective in the resolution of DSBs generated [5, 99]. Similar results were also seen in chicken DT40 cells in which knockout of five Rad51 paralogues were all sensitive to mitomycin C and cisplatin [100]. In contrast, Ku knockout chicken DT40 cells defective in NHEJ were resistant to cross-linking agents [101]. It was also observed that Rad54 and Rad54B protect cells from mitomycin C in mouse ES cells [102]. Consistently, mouse cells defective in Rad51 interacting partners, BRCA1 and BRCA2 are extremely sensitive to a range of cross-linking agents [103–105]. The increase in sensitivity of these cells to cross-linking agents is associated with a suppression in ICL-induced Rad51 foci, indicating defective HR [103]. Consistently, *in vitro* repair of ICL-containing plasmids using *Xenopus* egg extracts has provided support for the involvement of Rad51-mediated HR in resolving ICL-induced DSBs, and in this system almost all ICL repair is HR-dependent [106]. The involvement of HR in ICL repair is further supported by the induction of SCEs by many cross-linking agents [107]. It has also been observed that the presence of a psoralen ICL on one homologous duplex stimulates short tract gene conversion events with associated crossovers [108].

A number of studies have implicated XPF-ERCC1 in HR during repair of ICLs. Yeast Rad1/Rad10 is involved in crossover generation through the processing of SSA recombination intermediates, removing terminal non-homologous sequences [109]. This function appears to be evolutionarily conserved in mammalian cells. In response to ionising radiation, XPF-ERCC1 is required for trimming 3′ ends extending beyond the annealed regions of microhomology during MMEJ [110]. In CHO cells, XPF-ERCC1 is needed for accurate resolution of recombination intermediates following strand invasion, involving 5′ incision at a single-stranded to double-stranded transition structure, which would otherwise block branch migration [111–113]. It was also found that in addition to Rad51 paralogues (RAD51D, XRCC2 and XRCC3), proteins from the FA network and REV3, XPF-ERCC1 were required for the homology-directed repair of psoralen ICLs involving DSBs in host-cell reactivation assays [114]. The role of XPF-ERCC1 has been suggested to be late during HR, as XPF-ERCC1 deficiency does not impair Rad51 foci formation but results in defective HR completion as measured in a reporter construct [115]. However, whether such a reporter system provides a direct measurement of HR specifically induced by ICLs remains questionable, given that XPF-ERCC1 deficient cells are already defective in HR within the constructs used in the absence of cross-linking agent treatment [115].

The role of HR in ICL repair is further supported by its interaction with the FA pathway. One FA factor, BRCA2, also known as FANCD1, is an HR factor. The following section will discuss the involvement of the FA pathway in ICL repair.

6 Fanconi Anaemia Pathway and ICL Repair

FA is an inherited recessive disease, caused by mutations in at least 15 different genes [116]. Mutations of FA genes are autosomal recessive with an exception of FANCB, which is X-linked [117]. In addition to developmental abnormalities and bone marrow failure, FA patients are also predisposed to cancer development, especially leukaemia and carcinomas [118]. Cells derived from these patients show genomic instability with increased radial formations. In addition, the hallmark of FA is cellular hypersensitivity to cross-linking agents, a trait that has been used as a

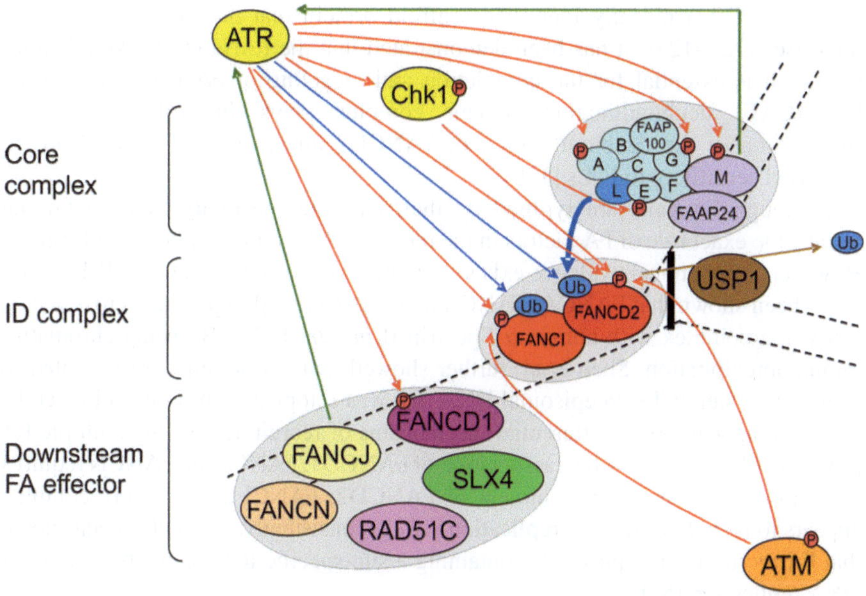

Fig. 3 A schematic representation of the FA pathway and its post-translational modifications. FA pathway becomes activated following detection of ICLs (*black solid line*) by FANCM-FAAP24 at sites of stalled replication fork (*dashed lines*). FANCM directly interacts with FANCF to recruit the FA core complex. The E3 ligase activity of FANCL subunit within the core complex monoubiquitinates FANCD2 and FANCI (ID complex), which becomes activated and associates to damaged chromatin. The activated ID complex recruits downstream FA effectors and possibly other factors (not shown) to facilitate repair of ICLs. The deubiquitination of the ID complex is essential for the completion of the FA pathway and ICL repair. The checkpoint kinases, ATM, ATR and Chk1 regulate the activity of the FA pathway by phosphorylating a number of FA factors (*red arrows*). ATR also regulates the monoubiquitination of the ID complex (*blue arrows*). The FA pathway, in turn, regulates the ATR pathway whereby FANCM has been found to be essential for activating the ATR pathway (*green arrow*). Adapted from [119], and [116]

diagnostic marker for FA patients [118]. The 15 gene products of FA have been found to be implicated in a common pathway involved in both DNA repair and checkpoint signalling, especially those involved in ICL repair (Fig. 3) [116, 119]. The monoubiquitination of FANCD2 and FANCI proteins at lysine 561 and lysine 523, respectively, appears to be the central event of the pathway [120, 121]. This occurs constitutively during S-phase and responds to the presence of DNA damage. Along with other accessory factors including FAAP24 and FAAP100, eight of the FA proteins (FANCA, B, C, F, G, L and M) form the FA core complex. FANCL is the catalytic E3 ligase that directly ubiquitinates FANCD2-FANCI [122]. There appears to be additional role of FANCM outside of the core complex, as FANCM deficiency only partially inactivates FANCD2 monoubiquination [123]. The remainder of the FA proteins are downstream effectors of the activated FANCD2-FANCI proteins. These include FANCD1/BRCA2, FANCN/PALB2, FANCJ/BRIP1 and RAD51C that have roles in facilitating homologous recombination (HR) [124], and

SLX4/FANCP, a recently identified scaffold protein that interacts with multiple nucleases [125–127]. It has been demonstrated that the FANCD2-FANCI deubiquitination is essential for the completion and recycling of FA pathway, and the USP1/UAF1 deubiquitinating enzyme complex plays this role [128, 129]. Inactivation of the deubiquitination of FANCD2 due to loss of USP1 leads to increased sensitivity to ICLs [130].

Although FA cells are typified by their extreme sensitivity to cross-linking agents, the exact role of FA factors in the repair of ICLs remains poorly understood. However, biochemical evidence exists showing a role of FA proteins in ICL repair. It has been shown that FANCA, FANCC and FANCG bind to psoralen cross-linked DNA in a complex with human α-Spectrin II *in vitro* [131]. By using chromatin-immunoprecipitation, Shen et al. further showed that FA proteins are recruited to sites of psoralen ICLs on episomal DNA that was ectopically introduced into cells, and there appears to be differential mechanisms of recruitment of the multiple FA proteins [132]. The cross-link association of FANCD1, FANCJ and FANCN requires replication, whereas the recruitment of FANCD2, FANCI and the core complex appears to be independent of replication [132]. Consistently, it has also been shown that FA recruitment to plasmids containing a site-specific ICL occurs regardless of DNA replication [58].

In vitro data using a site-specific ICL in plasmid DNA with *Xenopus* egg extracts have shown that the FANCD2-FANCI monoubiquitination is essential for the replication-dependent repair of ICLs [56, 57]. Inactivation of FANCD2-FANCI affected both the nucleolytic incision step and insertion of nucleotide opposite the tethered ICL during TLS [56]. It has been shown that monoubiquitination of FANCD2-FANCI is required for its nuclear foci formation at sites of ssDNA gaps and DSBs for orchestrating repair [120]. The newly identified nuclease, FAN1, has been shown to be recruited to sites of ICLs by FANCD2 [72–74, 133]. Interestingly, it has recently been shown that the monoubiquitination of FANCD2 following mitomycin C treatment occurs independently of XPF-ERCC1 [66]. However, chromatin association and deubiquitination of FANCD2 requires the presence of XPF-ERCC1, providing a link between the nucleolytic processing step of ICL repair and the FA pathway [66].

A clinical feature of FA patient cells is the increased radial structures generated between non-homologous chromosomes following treatment with cross-linking agents [134]. This is a result of reduced HR activity, accompanied by increased non-allelic homologous recombination, or use of f regions of microhomology as templates for replication fork restart, or increased NHEJ activities. It has been demonstrated recently that Ku deletion renders FANCC or FANCD2 mutant cells more resistant to cross-linking agents and reverses impaired homologous recombination in *Caenorhabditis elegans*, chicken DT-40 cells and human patient derived cell lines. The observation indicates that the FA pathway is involved in channelling the repair of ICL-induced DSBs away from inappropriate engagement with NHEJ and promoting HR during ICL repair [135, 136].

The link between the FA pathway and HR during ICL repair has been provided by the *in vitro* observation that BRCA2-defective cell extract could not fully remove psoralen ICLs present on plasmids, although unhooking of the ICLs was unaffected [137].

The mechanistic interaction between the FA pathway and HR core components appears to take place at several levels and remains poorly understood. The activated FANCD2 appears to be a critical regulator for HR activity. FANCD2 was found to recruit and co-localise with essential HR factors at sites of DNA damage, including BRCA1, BRCA2 and Rad51 [120, 138–140]. The interaction between FANCD2 and BRCA2 appears to require a role of FANCG independent of the core complex [141]. Furthermore, FANCD2 also co-localises with MRN complex following mitomycin C treatment [142]. Conversely, FANCD2 was found to be regulated by the MRN complex, providing a negative feedback loop in controlling FA activation during ICL repair [143].

The FA pathway also directly participates in checkpoint signalling in response to ICLs. It was demonstrated that the FA pathway plays an important role in activating the ATR/Chk1 checkpoint signalling as depleting FANCD2 inhibits phosphorylation of Chk1 and replication of an undamaged plasmid *in trans* [58]. This indicates that FA acts upstream of the ATR/Chk1 signalling. In addition, the FA pathway also appears to be regulated by checkpoint pathways. The monoubiquitination of FANCD2 on K561 in response to ICLs is mediated by ATR [144], and has been found to be dependent on the phosphorylation of FANCD2 on Ser 331 by Chk1 [145].

7 Clinical Relevance and Future Prospects

As our understanding of the complex molecular mechanisms involved in the repair of ICLs in human cells and the critical determinants of cellular sensitivity to damage of this type increases, so does the potential to develop sensitive screens to predict clinical response. Methods that allow the determination of formation and repair of ICLs in patient-derived material have clearly shown that repair of ICLs is an important determinant of inherent sensitivity or acquired resistance in the clinic. For example, using a modification of the single cell gel electrophoresis (comet) assay [146], the formation and repair of melphalan-induced DNA ICLs in plasma cells from melphalan naïve and melphalan-treated patients (i.e. those who had relapsed after a melphalan-conditioned autologous stem cell transplant or oral melphalan therapy) was examined [11]. Although similar levels of cross links were observed in cells from both melphalan naïve and treated patients, marked differences in repair (unhooking) of the cross-links were observed. Whereas cells from naïve patients showed no repair, those from treated patients exhibited efficient repair. A similar result has been found more recently for cisplatin-induced ICLs in ovarian cancer [13]. Cells from newly diagnosed patients were generally deficient at unhooking cisplatin ICLs, whereas cells taken following platinum-treatment (including from the same patients following therapy) showed enhanced ICL unhooking.

It has generally been assumed that the repair of different types of drug-induced ICL will be by a common mechanism or mechanisms. Indeed, resistance to melphalan due to enhanced removal of melphalan-induced ICLs in chronic lymphocytic leukaemia resulted in cross-resistance to other nitrogen mustard drugs [12]. Similarly, disruption of the FA-BRCA pathway in cisplatin sensitive ovarian tumours resulted in sensitivity to mitomycin C [147]. However, this may not always be the case since the clinical acquired resistant myeloma cells described above, which are able to unhook the ICLs produced by melphalan, were not able to unhook the ICLs produced by cisplatin [11]. Conversely, the ovarian cancer cells from platinum-treated patients, which could unhook cisplatin-induced ICLs, could not unhook those produced by melphalan. Although the molecular basis for these differences remains unclear the results have important clinical implications and indicate that inherent sensitivity to DNA ICL agents can be the result of a specific DNA repair defect, and acquired resistance can be ICL agent class specific.

Key proteins involved in this specific process can also be considered as novel therapeutic targets, whose inhibition could increase sensitivity to cross-linking drugs in tumours normally inherently resistant or which have acquired resistance following initial therapy. Conversely, strategies to activate ICL repair in normal tissue, such as bone marrow, could protect from drug toxicity. Since the cellular response to DNA ICLs is co-ordinated to include cell cycle arrest, DNA damage repair and cell death [9], alternative strategies to sensitise tumours to ICL agents can target DNA damage response pathways such as the MAPK pathway [148] or the checkpoint kinase Chk1 [149].

Cisplatin and carboplatin have shown synergistic activity with the antimetabolite gemcitabine *in vitro* [150]. Recently, it has been demonstrated clinically that in patients with platinum-*resistant* ovarian cancer the combination of carboplatin and gemcitabine can be active, with the *in vivo* demonstration using the comet assay of a significant reduction in the repair (unhooking) of carboplatin-induced cross-links following the addition of gemcitabine [151]. Similar data have been observed with the clinical combination of gemcitabine and the cross-linking agent treosulfan [152]. Fludarabine has also been shown to inhibit the repair of cisplatin- and oxaliplatin-induced DNA cross-linking [153, 154]. The combination of fludarabine and the novel minor groove interstrand cross-linking agent SJG-136 has also been found to be synergistic in B-CLL cells, even in samples derived from patients with fludarabine resistance [155]. The combination resulted in higher levels of SJG-136-induced ICLs and it was demonstrated that fludarabine suppressed transcription of ERCC1 and inhibited SJG-136-induced ERCC1 transcription when given in combination, offering a mechanistic rationale for the synergistic interaction.

It is also becoming established that a significant proportion of tumours are defective in homologous recombination repair (e.g. BRCA1- and BRCA2-deficient) which confers sensitivity to cross-linking agents such as cisplatin and inhibitors of poly(ADP-ribose)polymerase (PARP), an enzyme required for repair of endogenous DNA damage [156]. The combination of PARP inhibitor and platinum drug can be more effective than either drug alone [157].

Another indirect mechanism to inhibit repair of some ICLs is through targeting extracellular signalling pathways. For example, inhibition of the EGFR (Her1) pathway by gefitinib resulted in synergy with cisplatin in human breast cancer MCF7 cells and was shown to inhibit the unhooking of the ICLs produced by cisplatin [158]. Similarly, incubation of breast cancer cell lines with the anti-HER2 antibody trastuzumab (Herceptin) delayed the repair of ICLs produced by cisplatin, with no effect on the repair of intrastrand cross-links [159]. In each of these examples the effect is dependent on the cross-linking agent used but clearly shows that rational combinations of DNA ICL agents and targeted therapies can be found.

The modified comet assay, in addition to having an important role in mechanistic studies of ICL formation and repair *in vitro*, *in vivo* and in clinical material, is becoming established clinically as a pharmacodynamic endpoint for ICL drugs in early phase clinical trials of novel agents (e.g. SJG-136) [160], or novel targeting strategies involving ICL. For example, in a clinical trial of the prodrug CB1954 and the synthetic nicotinamide cofactor analogue, EP0152R, ICLs resulting from NQO2-reduced CB1954 were detected in biopsies demonstrating successful prodrug activation [161]. The DNA damage produced by DNA ICL agents, including mechlorethamine and cisplatin, can also result in phosphorylation of the histone protein H2AX resulting in γ-H2AX foci [162, 163]. The peak γ-H2AX response was detected 2–3 h after the peak of DNA ICL formation, and detection of γ-H2AX foci by immunofluorescence confocal microscopy was found to be up to ten times more sensitive than detection of cross links using the comet assay [163]. In addition, mechlorethamine or cisplatin-induced γ-H2AX foci persisted longer in cells that were defective in either ICL unhooking or homologous recombination. Together these data raised the possibility that the measurement of γ-H2AX foci could be used as a sensitive pharmacodynamic marker of DNA damage by ICL agents in the clinic and recently this assay has been utilised as a pharmacodynamic endpoint in Phase I studies of the novel agent SJG-136 [164].

Novel cross-linking agents continue to be developed in an attempt to produce more selective, less toxic drugs. Agents that produce cross-links in the minor groove of DNA are of particular interest. For example, the sequence selective ICLs produced by the pyrrolobenzodiazepine dimer SJG-136 in the minor groove of DNA produce minimal distortion of the normal DNA structure and, as a result, it appears to evade some of the recognition and repair mechanisms used for the processing of the distorting cross-links produced in the major groove by conventional drugs. SJG-136-induced ICLs persist in tumour cells able to efficiently unhook ICLs produced by drugs such as cisplatin and melphalan. This highly potent agent has significant antitumour activity in animal models [21, 165] and is currently in clinical development.

It is possible to generate pyrrolobenzodiazepine dimer-based ICL agents with increased potency compared to SJG-136, down to sub-picomolar GI_{50} values *in vitro* [20, 166, 167]. It remains to be established whether agents of this type will find a role in the clinic since their use as unmodified single agents may be limited *in vivo* by issues of delivery and toxicity. The fact that these molecules have multiple potential sites of attachment and that they can be inactivated with appropriate modification suggests that PBDs may also have a potential role in other strategies aimed at targeting and releasing highly cytotoxic agents directly at a tumour site, where production of potent and difficult to repair ICLs is desirable. Examples of such approaches are antibody-directed or gene-directed prodrug therapies (ADEPT and GDEPT) and hypoxia-activated prodrugs which are currently using prodrugs based on more conventional cross-linking agents [168, 169]. One important application is as the "warhead" component of antibody drug conjugates, an area that is fast emerging as one of the principal approaches in the field of monoclonal antibody cancer therapeutics [170].

References

1. Hartley JA (2001) Alkylating agents. In: Souhami RL, Tannock RF, Hohenberger P, Horiot J-C (eds) Oxford textbook of oncology, vol 1. Oxford University Press, Oxford, pp 639–654
2. Kohn KW, Spears CL et al (1966) Inter-strand crosslinking of DNA by nitrogen mustard. J Mol Biol 19(2):266–288
3. Lawley PD, Brookes P (1967) Interstrand crosslinking of DNA by difunctional alkylating agents. J Mol Biol 25:143–160
4. Clingen PH, De Silva IU et al (2005) The XPF-ERCC1 endonuclease and homologous recombination contribute to the repair of minor groove DNA interstrand crosslinks in mammalian cells produced by the pyrrolo[2,1-c][1,4]benzodiazepine dimer SJG-136. Nucleic Acids Res 33(10):3283–3291
5. De Silva IU, McHugh PJ et al (2000) Defining the roles of nucleotide excision repair and recombination in the repair of DNA interstrand cross-links in mammalian cells. Mol Cell Biol 20(21):7980–7990
6. O'Connor PM, Kohn KW (1990) Comparative pharmacokinetics of DNA lesion formation and removal following treatment of L1210 cells with nitrogen mustards. Cancer Commun 2(12):387–394
7. Sunters A, Springer CJ et al (1992) The cytotoxicity, DNA crosslinking ability and DNA sequence selectivity of the aniline mustards melphalan, chlorambucil and 4-[bis(2-chloroethyl)amino] benzoic acid. Biochem Pharmacol 44(1):59–64
8. Trimmer EE, Essigmann JM (1999) Cisplatin. Essays Biochem 34:191–211
9. Osawa T, Davies D et al (2011) Mechanism of cell death resulting from DNA interstrand cross-linking in mammalian cells. Cell Death Dis 2:e187
10. Koberle B, Masters JR et al (1999) Defective repair of cisplatin-induced DNA damage caused by reduced XPA protein in testicular germ cell tumours. Curr Biol 9(5):273–276
11. Spanswick VJ, Craddock C et al (2002) Repair of DNA interstrand crosslinks as a mechanism of clinical resistance to melphalan in multiple myeloma. Blood 100(1):224–229
12. Torres-Garcia SJ, Cousineau L et al (1989) Correlation of resistance to nitrogen mustards in chronic lymphocytic leukemia with enhanced removal of melphalan-induced DNA cross-links. Biochem Pharmacol 38(18):3122–3123
13. Wynne P, Newton C et al (2007) Enhanced repair of DNA interstrand crosslinking in ovarian cancer cells from patients following treatment with platinum-based chemotherapy. Br J Cancer 97(7):927–933
14. Rudd GN, Hartley JA, Souhami RL (1995) Persistence of cisplatin-induced DNA interstrand crosslinks in peripheral blood mononuclear cells from elderly and young individuals. Cancer Chemother Pharmacol 35:323–326
15. Ojwang JO, Grueneberg DA et al (1989) Synthesis of a duplex oligonucleotide containing a nitrogen mustard interstrand DNA-DNA cross-link. Cancer Res 49(23):6529–6537
16. Huang H, Zhu L et al (1995) Solution structure of a cisplatin-induced DNA interstrand cross-link. Science 270(5243):1842–1845
17. Tomasz M (1995) Mitomycin C: small, fast and deadly (but very selective). Chem Biol 2(9):575–579
18. Ben-Hur E, Song PS (1984) The photochemistry and photobiology of furocoumarins (psoralens). Adv Radiat Biol 11:131–171
19. Rink SM, Hopkins PB (1995) A mechlorethamine-induced DNA interstrand cross-link bends duplex DNA. Biochemistry 34(4):1439–1445
20. Hartley JA (2011) The development of pyrrolobenzodiazepines as antitumour agents. Expert Opin Investig Drugs 20(6):733–744
21. Hartley JA, Spanswick VJ et al (2004) SJG-136 (NSC 694501), a novel rationally designed DNA minor groove interstrand cross-linking agent with potent and broad spectrum antitumor activity: part 1: cellular pharmacology, in vitro and initial in vivo antitumor activity. Cancer Res 64(18):6693–6699

22. Kiakos K, Hartley JM et al (2010) Measurement of DNA interstrand crosslinking in naked DNA using gel-based methods. Methods Mol Biol 613:283–302
23. Wang AT, Sengerova B et al (2011) Human SNM1A and XPF-ERCC1 collaborate to initiate DNA interstrand cross-link repair. Genes Dev 25(17):1859–1870
24. Cole RS (1973) Repair of DNA containing interstrand crosslinks in *Escherichia coli*: sequential excision and recombination. Proc Natl Acad Sci USA 70(4):1064–1068
25. McHugh PJ, Spanswick VJ et al (2001) Repair of DNA interstrand crosslinks: molecular mechanisms and clinical relevance. Lancet Oncol 2(8):483–490
26. Berardini M, Mackay W et al (1997) Evidence for a recombination-independent pathway for the repair of DNA interstrand cross-links based on a site-specific study with nitrogen mustard. Biochemistry 36(12):3506–3513
27. Cole RS (1971) Inactivation of *Escherichia coli*, F' episomes at transfer, and bacteriophage lambda by psoralen plus 360-nm light: significance of deoxyribonucleic acid cross-links. J Bacteriol 107(3):846–852
28. Henriques JA, Moustacchi E (1980) Isolation and characterization of pso mutants sensitive to photo-addition of psoralen derivatives in *Saccharomyces cerevisiae*. Genetics 95(2):273–288
29. Lawley PD, Brookes P (1965) Molecular mechanism of the cytotoxic action of difunctional alkylating agents and of resistance to this action. Nature 206(983):480–483
30. Berardini M, Foster PL et al (1999) DNA polymerase II (polB) is involved in a new DNA repair pathway for DNA interstrand cross-links in *Escherichia coli*. J Bacteriol 181(9):2878–2882
31. Kumari A, Minko IG et al (2008) Replication bypass of interstrand cross-link intermediates by *Escherichia coli* DNA polymerase IV. J Biol Chem 283(41):27433–27437
32. Munn M, Rupp W (1991) Interaction of the UvrABC endonuclease with DNA containing a psoralen monoadduct or cross-link. Differential effects of superhelical density and comparison of preincision complexes. J Biol Chem 266(36):24748–24756
33. Pu WT, Kahn R et al (1989) UvrABC incision of N-methylmitomycin A-DNA monoadducts and cross-links. J Biol Chem 264(34):20697–20704
34. Sladek F, Munn M et al (1989) In vitro repair of psoralen-DNA cross-links by RecA, UvrABC, and the 5′- exonuclease of DNA polymerase I. J Biol Chem 264(12):6755–6765
35. Van Houten B, Gamper H et al (1986) Action mechanism of ABC excision nuclease on a DNA substrate containing a psoralen crosslink at a defined position. Proc Natl Acad Sci USA 83(21):8077–8081
36. Zietlow L, Bessho T (2008) DNA polymerase I-mediated translesion synthesis in RecA-independent DNA interstrand cross-link repair in *E. coli*. Biochemistry 47(19):5460–5464
37. Smeaton MB, Hlavin EM et al (2008) Distortion-dependent unhooking of interstrand cross-links in mammalian cell extracts. Biochemistry 47(37):9920–9930
38. Clement FC, Camenisch U et al (2010) Dynamic two-stage mechanism of versatile DNA damage recognition by xeroderma pigmentosum group C protein. Mutat Res 685(1–2):21–28
39. Thoma BS, Wakasugi M et al (2005) Human XPC-hHR23B interacts with XPA-RPA in the recognition of triplex-directed psoralen DNA interstrand crosslinks. Nucleic Acids Res 33(9):2993–3001
40. Muniandy PA, Thapa D et al (2009) Repair of laser-localized DNA interstrand cross-links in G1 phase mammalian cells. J Biol Chem 284(41):27908–27917
41. Furuta T, Ueda T et al (2002) Transcription-coupled nucleotide excision repair as a determinant of cisplatin sensitivity of human cells. Cancer Res 62(17):4899–4902
42. Lange SS, Reddy MC, Vasquez KM (2009) Human HMGB1 directly facilitates interactions between nucleotide excision repair proteins on triplex-directed psoralen interstrand cross-links DNA Repair (Amst) 8:865–872
43. Hlavin EM, Smeaton MB et al (2010) Initiation of DNA interstrand cross-link repair in mammalian cells. Environ Mol Mutagen 51(6):604–624
44. Futscher BW, Pieper RO et al (1992) DNA-damaging and transcription-terminating lesions induced by AF64A in vitro. J Neurochem 58(4):1504–1509

45. Islas AL, Vos JM et al (1991) Differential introduction and repair of psoralen photoadducts to DNA in specific human genes. Cancer Res 51(11):2867–2873
46. Larminat F, Bohr VA (1994) Role of the human ERCC-1 gene in gene-specific repair of cisplatin-induced DNA damage. Nucleic Acids Res 22(15):3005–3010
47. Ahn B, Kang D et al (2004) Repair of mitomycin C cross-linked DNA in mammalian cells measured by a host cell reactivation assay. Mol Cells 18(2):249–255
48. Wang G, Chen Z et al (2001) Detection and determination of oligonucleotide triplex formation-mediated transcription-coupled DNA repair in HeLa nuclear extracts. Nucleic Acids Res 29(8):1801–1807
49. Zheng H, Wang X et al (2003) Nucleotide excision repair- and polymerase eta-mediated error-prone removal of mitomycin C interstrand cross-links. Mol Cell Biol 23(2):754–761
50. Dronkert ML, Kanaar R (2001) Repair of DNA interstrand cross-links. Mutat Res 486:217–247
51. Niedernhofer LJ, Lalai AS et al (2005) Fanconi anemia (cross)linked to DNA repair. Cell 123(7):1191–1198
52. Patel KJ, Joenje H (2007) Fanconi anemia and DNA replication repair. DNA Repair (Amst) 6(7):885–890
53. Thompson LH, Hinz JM (2009) Cellular and molecular consequences of defective Fanconi anemia proteins in replication-coupled DNA repair: mechanistic insights. Mutat Res 668(1–2):54–72
54. Akkari YMN, Bateman RL et al (2000) DNA replication is required to elicit cellular responses to psoralen-induced DNA interstrand cross-links. Mol Cell Biol 20(21):8283–8289
55. McHugh PJ, Sones WR et al (2000) Repair of intermediate structures produced at DNA interstrand cross-links in *Saccharomyces cerevisiae*. Mol Cell Biol 20(10):3425–3433
56. Knipscheer P, Raschle M et al (2009) The Fanconi anemia pathway promotes replication-dependent DNA interstrand cross-link repair. Science 326(5960):1698–1701
57. Räschle M, Knipscheer P et al (2008) Mechanism of replication-coupled DNA interstrand crosslink repair. Cell 134(6):969–980
58. Ben-Yehoyada M, Wang LC et al (2009) Checkpoint signaling from a single DNA interstrand crosslink. Mol Cell 35(5):704–715
59. Lambert S, Froget B et al (2007) Arrested replication fork processing: interplay between checkpoints and recombination. DNA Repair 6(7):1042–1061
60. Andersson BS, Sadeghi T et al (1996) Nucleotide excision repair genes as determinants of cellular sensitivity to cyclophosphamide analogs. Cancer Chemother Pharmacol 38(5):406–416
61. Hoy CA, Thompson LH et al (1985) Defective DNA cross-link removal in Chinese hamster cell mutants hypersensitive to bifunctional alkylating agents. Cancer Res 45(4):1737–1743
62. Niedernhofer LJ (2004) The structure-specific endonuclease Ercc1-Xpf is required to resolve DNA interstrand cross-link-induced double-strand breaks. Mol Cell Biol 24:5776–5787
63. Kuraoka I, Kobertz WR et al (2000) Repair of an interstrand DNA cross-link initiated by ERCC1-XPF repair/recombination nuclease. J Biol Chem 275(34):26632–26636
64. Abraham J, Lemmers B et al (2003) Eme1 is involved in DNA damage processing and maintenance of genomic stability in mammalian cells. EMBO J 22(22):6137–6147
65. Hanada K, Budzowska M et al (2006) The structure-specific endonuclease Mus81-Eme1 promotes conversion of interstrand DNA crosslinks into double-strands breaks. EMBO J 25(20):4921–4932
66. Bhagwat N, Olsen AL et al (2009) XPF-ERCC1 participates in the Fanconi anemia pathway of cross-link repair. Mol Cell Biol 29(24):6427–6437
67. Andersen SL, Bergstralh DT et al (2009) Drosophila MUS312 and the vertebrate ortholog BTBD12 interact with DNA structure-specific endonucleases in DNA repair and recombination. Mol Cell 35(1):128–135
68. Fekairi S, Scaglione S et al (2009) Human SLX4 is a Holliday junction resolvase subunit that binds multiple DNA repair/recombination endonucleases. Cell 138(1):78–89

69. Muñoz IM, Hain K et al (2009) Coordination of structure-specific nucleases by human SLX4/BTBD12 is required for DNA repair. Mol Cell 35(1):116–127
70. Svendsen JM, Smogorzewska A et al (2009) Mammalian BTBD12/SLX4 assembles a Holliday junction resolvase and is required for DNA repair. Cell 138(1):63–77
71. Cannavo E, Gerrits B et al (2007) Characterization of the interactome of the human MutL homologues MLH1, PMS1, and PMS2. J Biol Chem 282(5):2976–2986
72. Kratz K, Schopf B et al (2010) Deficiency of FANCD2-associated nuclease KIAA1018/FAN1 sensitizes cells to interstrand crosslinking agents. Cell 142(1):77–88
73. MacKay C, Declais AC et al (2010) Identification of KIAA1018/FAN1, a DNA repair nuclease recruited to DNA damage by monoubiquitinated FANCD2. Cell 142(1):65–76
74. Smogorzewska A, Desetty R et al (2010) A genetic screen identifies FAN1, a Fanconi anemia-associated nuclease necessary for DNA interstrand crosslink repair. Mol Cell 39(1):36–47
75. Ho TV, Scharer OD (2010) Translesion DNA synthesis polymerases in DNA interstrand crosslink repair. Environ Mol Mutagen 51(6):552–566
76. Minko IG, Harbut MB et al (2008) Role for DNA polymerase kappa in the processing of N2-N2-guanine interstrand cross-links. J Biol Chem 283(25):17075–17082
77. Henriques JA, Moustacchi E (1981) Interactions between mutations for sensitivity to psoralen photoaddition (pso) and to radiation (rad) in *Saccharomyces cerevisiae*. J Bacteriol 148(1):248–256
78. Ruhland A, Kircher M et al (1981) A yeast mutant specifically sensitive to bifunctional alkylation. Mutat Res 91(6):457–462
79. Demuth I, Bradshaw PS et al (2008) Endogenous hSNM1B/Apollo interacts with TRF2 and stimulates ATM in response to ionizing radiation. DNA Repair (Amst) 7(8): 1192–1201
80. Dronkert MLG, de Wit J et al (2000) Disruption of mouse SNM1 causes increased sensitivity to the DNA interstrand cross-linking agent mitomycin C. Mol Cell Biol 20(13):4553–4561
81. Moshous D, Callebaut I et al (2001) Artemis, a novel DNA double-strand break repair/V(D)J recombination protein, is mutated in human severe combined immune deficiency. Cell 105(2):177–186
82. Cattell E, Sengerova B et al (2010) The SNM1/Pso2 family of ICL repair nucleases: from yeast to man. Environ Mol Mutagen 51(6):635–645
83. Hazrati A, Ramis-Castelltort M et al (2008) Human SNM1A suppresses the DNA repair defects of yeast pso2 mutants. DNA Repair (Amst) 7(2):230–238
84. Hejna J, Philip S et al (2007) The hSNM1 protein is a DNA 5′-exonuclease. Nucleic Acids Res 35(18):6115–6123
85. Yang K, Moldovan GL et al (2010) RAD18-dependent recruitment of SNM1A to DNA repair complexes by a ubiquitin-binding zinc finger. J Biol Chem 285(25):19085–19091
86. Kannouche P, Fernandez de Henestrosa AR et al (2003) Localization of DNA polymerases eta and iota to the replication machinery is tightly co-ordinated in human cells. EMBO J 22(5):1223–1233
87. Kannouche PL, Wing J et al (2004) Interaction of human DNA polymerase eta with monoubiquitinated PCNA: a possible mechanism for the polymerase switch in response to DNA damage. Mol Cell 14(4):491–500
88. Nojima K, Hochegger H et al (2005) Multiple repair pathways mediate tolerance to chemotherapeutic cross-linking agents in vertebrate cells. Cancer Res 65(24):11704–11711
89. Gan GN, Wittschieben JP et al (2008) DNA polymerase zeta (pol zeta) in higher eukaryotes. Cell Res 18(1):174–183
90. Sonoda E, Okada T et al (2003) Multiple roles of Rev3, the catalytic subunit of polzeta in maintaining genome stability in vertebrates. EMBO J 22(12):3188–3197
91. Okada T, Sonoda E et al (2005) Multiple roles of vertebrate REV genes in DNA repair and recombination. Mol Cell Biol 25(14):6103–6111
92. Ross A-L, Simpson LJ et al (2005) Vertebrate DNA damage tolerance requires the C-terminus but not BRCT or transferase domains of REV1. Nucleic Acids Res 33(4):1280–1289

93. Shen X, Jun S et al (2006) REV3 and REV1 play major roles in recombination-independent repair of DNA interstrand cross-links mediated by monoubiquitinated proliferating cell nuclear antigen (PCNA). J Biol Chem 281(20):13869–13872
94. Albertella MR, Green CM et al (2005) A role for polymerase η in the cellular tolerance to cisplatin-induced damage. Cancer Res 65(21):9799–9806
95. Moldovan GL, Madhavan MV et al (2010) DNA polymerase POLN participates in cross-link repair and homologous recombination. Mol Cell Biol 30(4):1088–1096
96. Yamanaka K, Minko IG et al (2010) Novel enzymatic function of DNA polymerase nu in translesion DNA synthesis past major groove DNA-peptide and DNA-DNA cross-links. Chem Res Toxicol 23(3):689–695
97. Zietlow L, Smith LA et al (2009) Evidence for the involvement of human DNA polymerase N in the repair of DNA interstrand cross-links. Biochemistry 48(49):11817–11824
98. Ho TV, Guainazzi A et al (2011) Structure-dependent bypass of DNA interstrand crosslinks by translesion synthesis polymerases. Nucleic Acids Res 39(17):7455–7464
99. Jones NJ, Stewart SA et al (1990) Biochemical and genetic analysis of the Chinese hamster mutants irs1 and irs2 and their comparison to cultured ataxia telangiectasia cells. Mutagenesis 5(1):15–23
100. Takata M, Sasaki MS et al (2001) Chromosome instability and defective recombinational repair in knockout mutants of the five Rad51 paralogs. Mol Cell Biol 21(8):2858–2866
101. Takata M, Sasaki MS et al (1998) Homologous recombination and non-homologous end-joining pathways of DNA double-strand break repair have overlapping roles in the maintenance of chromosomal integrity in vertebrate cells. EMBO J 17(18):5497–5508
102. Wesoly J, Agarwal S et al (2006) Differential contributions of mammalian Rad54 paralogs to recombination, DNA damage repair, and meiosis. Mol Cell Biol 26(3):976–989
103. Bhattacharyya A, Ear US et al (2000) The breast cancer susceptibility gene BRCA1 is required for subnuclear assembly of Rad51 and survival following treatment with the DNA cross-linking agent cisplatin. J Biol Chem 275(31):23899–23903
104. Wiegant WW, Overmeer RM et al (2006) Chinese hamster cell mutant, V-C8, a model for analysis of Brca2 function. Mutat Res 600(1–2):79–88
105. Yu VP, Koehler M et al (2000) Gross chromosomal rearrangements and genetic exchange between nonhomologous chromosomes following BRCA2 inactivation. Genes Dev 14(11):1400–1406
106. Long DT, Raschle M et al (2011) Mechanism of RAD51-dependent DNA interstrand cross-link repair. Science 333(6038):84–87
107. Bodell WJ (1990) Molecular dosimetry for sister-chromatid exchange induction and cytotoxicity by monofunctional and bifunctional alkylating agents. Mutat Res 233(1–2): 203–210
108. Liu Y, Nairn RS et al (2008) Processing of triplex-directed psoralen DNA interstrand cross-links by recombination mechanisms. Nucleic Acids Res 36(14):4680–4688
109. Fishman-Lobell J, Haber JE (1992) Removal of nonhomologous DNA ends in double-strand break recombination: the role of the yeast ultraviolet repair gene RAD1. Science 258(5081):480–484
110. Ahmad A, Robinson AR et al (2008) ERCC1-XPF endonuclease facilitates DNA double-strand break repair. Mol Cell Biol 28(16):5082–5092
111. Adair GM, Rolig RL et al (2000) Role of ERCC1 in removal of long non-homologous tails during targeted homologous recombination. EMBO J 19(20):5552–5561
112. Sargent RG, Meservy JL et al (2000) Role of the nucleotide excision repair gene ERCC1 in formation of recombination-dependent rearrangements in mammalian cells. Nucleic Acids Res 28(19):3771–3778
113. Sargent RG, Rolig RL et al (1997) Recombination-dependent deletion formation in mammalian cells deficient in the nucleotide excision repair gene ERCC1. Proc Natl Acad Sci USA 94(24):13122–13127
114. Zhang N, Liu X et al (2007) Double-strand breaks induce homologous recombinational repair of interstrand cross-links via cooperation of MSH2, ERCC1-XPF, REV3, and the Fanconi anemia pathway. DNA Repair (Amst) 6(11):1670–1678

115. Al-Minawi AZ, Lee Y-F et al (2009) The ERCC1/XPF endonuclease is required for completion of homologous recombination at DNA replication forks stalled by inter-strand cross-links. Nucleic Acids Res 37(19):6400–6413
116. Kee Y, D'Andrea AD (2010) Expanded roles of the Fanconi anemia pathway in preserving genomic stability. Genes Dev 24(16):1680–1694
117. Meetei AR, Levitus M et al (2004) X-linked inheritance of Fanconi anemia complementation group B. Nat Genet 36(11):1219–1224
118. Kennedy RD, D'Andrea AD (2006) DNA repair pathways in clinical practice: lessons from pediatric cancer susceptibility syndromes. J Clin Oncol 24(23):3799–3808
119. Wang LC, Gautier J (2010) The Fanconi anemia pathway and ICL repair: implications for cancer therapy. Crit Rev Biochem Mol Biol 45(5):424–439
120. Garcia-Higuera I, Taniguchi T et al (2001) Interaction of the Fanconi anemia proteins and BRCA1 in a common pathway. Mol Cell 7(2):249–262
121. Smogorzewska A, Matsuoka S et al (2007) Identification of the FANCI protein, a monoubiquitinated FANCD2 paralog required for DNA repair. Cell 129(2):289–301
122. Meetei AR, Yan Z et al (2004) FANCL replaces BRCA1 as the likely ubiquitin ligase responsible for FANCD2 monoubiquitination. Cell Cycle 3(2):179–181
123. Kim JM, Kee Y et al (2008) Cell cycle-dependent chromatin loading of the Fanconi anemia core complex by FANCM/FAAP24. Blood 111(10):5215–5222
124. Moldovan G-L, D'Andrea AD (2009) How the Fanconi anemia pathway guards the genome. Annu Rev Genet 43(1):223–249
125. Crossan GP, van der Weyden L et al (2011) Disruption of mouse Slx4, a regulator of structure-specific nucleases, phenocopies Fanconi anemia. Nat Genet 43(2):147–152
126. Kim Y, Lach FP et al (2011) Mutations of the SLX4 gene in Fanconi anemia. Nat Genet 43(2):142–146
127. Stoepker C, Hain K et al (2011) SLX4, a coordinator of structure-specific endonucleases, is mutated in a new Fanconi anemia subtype. Nat Genet 43(2):138–141
128. Cohn MA, Kowal P et al (2007) A UAF1-containing multisubunit protein complex regulates the Fanconi anemia pathway. Mol Cell 28(5):786–797
129. Nijman SM, Huang TT et al (2005) The deubiquitinating enzyme USP1 regulates the Fanconi anemia pathway. Mol Cell 17(3):331–339
130. Kim JM, Parmar K et al (2009) Inactivation of murine Usp1 results in genomic instability and a Fanconi anemia phenotype. Dev Cell 16(2):314–320
131. McMahon LW, Sangerman J et al (2001) Human α spectrin II and the FANCA, FANCC, and FANCG proteins bind to DNA containing psoralen interstrand cross-links, Ä†. Biochemistry 40(24):7025–7034
132. Shen X, Do H et al (2009) Recruitment of fanconi anemia and breast cancer proteins to DNA damage sites is differentially governed by replication. Mol Cell 35(5):716–723
133. Liu T, Ghosal G et al (2010) FAN1 acts with FANCI-FANCD2 to promote DNA interstrand cross-link repair. Science 329(5992):693–696
134. Auerbach AD, Wolman SR (1976) Susceptibility of Fanconi's anaemia fibroblasts to chromosome damage by carcinogens. Nature 261(5560):494–496
135. Adamo A, Collis SJ et al (2010) Preventing nonhomologous end joining suppresses DNA repair defects of Fanconi anemia. Mol Cell 39(1):25–35
136. Pace P, Mosedale G et al (2010) Ku70 corrupts DNA repair in the absence of the Fanconi anemia pathway. Science 329(5988):219–223
137. Cipak L, Watanabe N et al (2006) The role of BRCA2 in replication-coupled DNA interstrand cross-link repair in vitro. Nat Struct Mol Biol 13(8):729–733
138. Hussain S, Wilson JB et al (2004) Direct interaction of FANCD2 with BRCA2 in DNA damage response pathways. Hum Mol Genet 13(12):1241–1248
139. Taniguchi T, Garcia-Higuera I et al (2002) S-phase-specific interaction of the Fanconi anemia protein, FANCD2, with BRCA1 and RAD51. Blood 100(7):2414–2420
140. Wang X, Andreassen PR et al (2004) Functional interaction of monoubiquitinated FANCD2 and BRCA2/FANCD1 in chromatin. Mol Cell Biol 24(13):5850–5862

141. Wilson JB, Yamamoto K et al (2008) FANCG promotes formation of a newly identified protein complex containing BRCA2, FANCD2 and XRCC3. Oncogene 27(26):3641–3652
142. Nakanishi K, Taniguchi T et al (2002) Interaction of FANCD2 and NBS1 in the DNA damage response. Nat Cell Biol 4(12):913–920
143. Roques C, Coulombe Y et al (2009) MRE11-RAD50-NBS1 is a critical regulator of FANCD2 stability and function during DNA double-strand break repair. EMBO J 28(16):2400–2413
144. Andreassen PR, D'Andrea AD et al (2004) ATR couples FANCD2 monoubiquitination to the DNA-damage response. Genes Dev 18(16):1958–1963
145. Zhi G, Wilson JB et al (2009) Fanconi anemia complementation group FANCD2 protein serine 331 phosphorylation is important for fanconi anemia pathway function and BRCA2 interaction. Cancer Res 69(22):8775–8783
146. Spanswick VJ, Hartley JM et al (2010) Measurement of DNA interstrand crosslinking in individual cells using the Single Cell Gel Electrophoresis (Comet) assay. Methods Mol Biol 613:267–282
147. Taniguchi T, Tischkowitz M et al (2003) Disruption of the Fanconi anemia-BRCA pathway in cisplatin-sensitive ovarian tumors. Nat Med 9(5):568–574
148. Reinhardt HC, Aslanian AS et al (2007) p53-deficient cells rely on ATM- and ATR-mediated checkpoint signaling through the p38MAPK/MK2 pathway for survival after DNA damage. Cancer Cell 11(2):175–189
149. Koniaras K, Cuddihy AR et al (2001) Inhibition of Chk1-dependent G2 DNA damage checkpoint radiosensitizes p53 mutant human cells. Oncogene 20(51):7453–7463
150. Bergman AM, Ruiz van Haperen VW et al (1996) Synergistic interaction between cisplatin and gemcitabine in vitro. Clin Cancer Res 2(3):521–530
151. Ledermann JA, Gabra H et al (2010) Inhibition of carboplatin-induced DNA interstrand cross-link repair by gemcitabine in patients receiving these drugs for platinum-resistant ovarian cancer. Clin Cancer Res 16(19):4899–4905
152. Corrie PG, Shaw J et al (2005) Phase I trial combining gemcitabine and treosulfan in advanced cutaneous and uveal melanoma patients. Br J Cancer 92(11):1997–2003
153. Li L, Keating MJ et al (1997) Fludarabine-mediated repair inhibition of cisplatin-induced DNA lesions in human chronic myelogenous leukemia-blast crisis K562 cells: induction of synergistic cytotoxicity independent of reversal of apoptosis resistance. Mol Pharmacol 52(5):798–806
154. Moufarij MA, Sampath D et al (2006) Fludarabine increases oxaliplatin cytotoxicity in normal and chronic lymphocytic leukemia lymphocytes by suppressing interstrand DNA cross-link removal. Blood 108(13):4187–4193
155. Pepper C, Lowe H et al (2007) Fludarabine-mediated suppression of the excision repair enzyme ERCC1 contributes to the cytotoxic synergy with the DNA minor groove crosslinking agent SJG-136 (NSC 694501) in chronic lymphocytic leukaemia cells. Br J Cancer 97:253–259
156. Mukhopadhyay A, Elattar A et al (2010) Development of a functional assay for homologous recombination status in primary cultures of epithelial ovarian tumor and correlation with sensitivity to poly(ADP-ribose) polymerase inhibitors. Clin Cancer Res 16(8):2344–2351
157. Drew Y, Mulligan EA et al (2011) Therapeutic potential of poly(ADP-ribose) polymerase inhibitor AG014699 in human cancers with mutated or methylated BRCA1 or BRCA2. J Natl Cancer Inst 103(4):334–346
158. Friedmann B, Caplin M et al (2004) Modulation of DNA repair in vitro after treatment with chemotherapeutic agents by the epidermal growth factor receptor inhibitor gefitinib (ZD1839). Clin Cancer Res 10(19):6476–6486
159. Boone JJ, Bhosle J et al (2009) Involvement of the HER2 pathway in repair of DNA damage produced by chemotherapeutic agents. Mol Cancer Ther 8(11):3015–3023
160. Puzanov I, Lee W et al (2011) Phase I pharmacokinetic and pharmacodynamic study of SJG-136, a novel DNA sequence selective minor groove cross-linking agent, in advanced solid tumors. Clin Cancer Res 17(11):3794–3802

161. Middleton MR, Knox R et al (2010) Quinone oxidoreductase-2-mediated prodrug cancer therapy. Sci Transl Med 2(40):40ra50
162. Banuelos CA, Banath JP et al (2009) gammaH2AX expression in tumors exposed to cisplatin and fractionated irradiation. Clin Cancer Res 15(10):3344–3353
163. Clingen PH, Wu JY et al (2008) Histone H2AX phosphorylation as a molecular pharmacological marker for DNA interstrand crosslink cancer chemotherapy. Biochem Pharmacol 76(1):19–27
164. Hochhauser D, Meyer T et al (2009) Phase I study of sequence-selective minor groove DNA binding agent SJG-136 in patients with advanced solid tumors. Clin Cancer Res 15(6):2140–2147
165. Alley MC, Hollingshead MG et al (2004) SJG-136 (NSC 694501), a novel rationally designed DNA minor groove interstrand cross-linking agent with potent and broad spectrum antitumor activity: part 2: efficacy evaluations. Cancer Res 64(18):6700–6706
166. Hartley JA, Hamaguchi A et al (2010) SG2285, a novel C2-aryl-substituted pyrrolobenzodiazepine dimer prodrug that cross-links DNA and exerts highly potent antitumor activity. Cancer Res 70(17):6849–6858
167. Hartley JA, Hamaguchi A et al (2012) DNA interstrand cross-linking and in vivo antitumor activity of the extended pyrrolo[2,1-c][1,4]benzodiazepine dimer SG2057. Invest New Drugs 30(3):950–958
168. Both GW (2009) Recent progress in gene-directed enzyme prodrug therapy: an emerging cancer treatment. Curr Opin Mol Ther 11(4):421–432
169. Mayer A, Francis RJ et al (2006) A phase I study of single administration of antibody-directed enzyme prodrug therapy with the recombinant anti-carcinoembryonic antigen antibody-enzyme fusion protein MFECP1 and a bis-iodo phenol mustard prodrug. Clin Cancer Res 12(21):6509–6516
170. Adair JR, Howard et al (2012) Antibody-drug conjugates - a perfect synergy. Expert Opin Biol Ther 12(9):1191–206

Middleton MR, Knox R, [illegible] (2010) Quinone oxidoreductase-2 mediated prodrug cancer therapy. [illegible]

[illegible]

DNA-PK, a Pharmacological Target in Cancer Chemotherapy and Radiotherapy?

Bernard Salles, Patrick Calsou, and Gladys Mirey

1 Introduction

Radiotherapy induces a variety of DNA damage including oxidized base damage, abasic sites, single-strand breaks (SSBs) and double-strand breaks (DSBs). This DNA damage, if unrepaired, triggers cell death through mitotic catastrophe and apoptosis. Amongst these lesions, DSBs are considered to be major actors in cell death [1]. Similarly to ionizing radiations, most untargeted antitumor drugs cause DNA damage that induces death signals in cancer cells as well as in normal cells. DNA lesions trigger a cell response through an interconnected network called the DNA damage response (DDR) that tends to maintain cell viability and genomic stability [2, 3]. The DDR relies on a complex network of proteins that initiate and coordinate DNA repair activity by halting the cell cycle through the activation of checkpoints that block cells at the G1-S transition, the intra-S phase or the G2/M boundary [4]. When DNA repair fails, the DDR plays a key role in the induction of apoptosis. A defective DDR, for instance in the control of cell cycle blockage or in DNA repair processes, is commonly reported in many cancers and some cancer-prone human syndromes arise from defects in specific DDR and DNA repair genes [2].

Within the DDR, DNA damage repair determines the cell response, as illustrated with ionizing radiations [5, 6]. Thus, an excess of DNA lesions or a specific localization of lesions in the genome may overcome the cell repair capacity and trigger cell death.

In the case of DNA damage following radiotherapy, it has been largely documented that cell survival correlates with the number of DSBs in the genome [1].

B. Salles, D.V.M., Ph.D. (✉) • G. Mirey, Ph.D.
INRA, UMR1331, Toxalim (Research Centre in Food Toxicology),
Université de Toulouse III, 180 chemin de Tournefeuille Toulouse, F-31027, France
e-mail: bernard.salles@toulouse.inra.fr

P. Calsou, M.D., Ph.D.
CNRS, IPBS (Institut de Pharmacologie et de Biologie Structurale),
205 route de Narbonne, Toulouse F-31077, France

L. Panasci et al. (eds.), *Advances in DNA Repair in Cancer Therapy*,
Cancer Drug Discovery and Development 72, DOI 10.1007/978-1-4614-4741-2_2,
© Springer Science+Business Media, LLC 2013

Consequently, any increase or decrease in the repair capacity and/or signaling will lead to cell resistance or sensitivity, respectively. Based on such considerations, pharmaceutical companies have undertaken the development of new compounds aimed at modulating DDR processes and/or DNA repair, particularly DSB repair, after chemotherapy and radiotherapy, with the ultimate goal of sensitizing tumor cells to the treatment [7, 8]. For example, inhibitors of the checkpoint kinases Chk1 and Chk2 have recently been shown to sensitize tumor cells to DNA damaging agents [9–12].

In translational research dealing with potential DNA repair proteins as pharmacological targets, the DNA dependent protein kinase (DNA-PK), an heterotrimer comprising the regulatory subunit Ku70/Ku80 bound to the catalytic subunit DNA-PKcs, is of interest since its represents a major actor in DSBs repair.

Cells deficient in Ku70/Ku80 or DNA-PKcs are sensitive to DSBs induced by IR or chemotherapeutic agents [13, 14] supporting the idea that DNA-PK may represent a good target in cancer chemotherapy. We discuss here the rationale of this approach in the field of cancer treatment.

2 Double-Strand Break Induction by Radiotherapy/ Chemotherapy and the Biological Consequences

DNA DSBs are considered the most severe DNA lesions: unrepaired DSBs can induce apoptosis or mitotic cell death or when repaired incorrectly, they can lead to carcinogenesis through mutagenic genome rearrangements [15]. DSBs are produced exogenously by ionizing radiation (IR) or chemicals, but also endogenously during DNA replication fork collapse or physiological processes such as V(D)J recombination and meiotic exchange [1].

Mainly through water molecule radiolysis, IR induces a plethora of DNA damage whose complexity increases with the value of the linear transfer energy (LET) [16, 17]. DNA damage includes DSBs, SSBs, damaged bases, and abasic sites located at a distance from each other when induced by low-LET irradiation. By contrast, high-LET provokes the formation of complex DNA damage within one or two DNA helical turns [18], although this value is under discussion since it has been reported recently that it could cover regions extending over several kilobases of the DNA molecule [19]. The biological consequences of complex DNA damage range from point mutations and loss of genetic material to cell death, due to repair impairment or repair-intermediate persistency. Clustered lesions induce intra- or interchromosomal insertions, and inversions, often in association with large deletions, that appear to promote genome instability that may lead to carcinogenesis [19–21].

Antitumor chemotherapy has evolved from nonspecific cytotoxic agents to targeted therapies, which are directed at unique molecular signatures of cancer cells to produce greater efficacy with less toxicity. However, untargeted drugs are still widely used and most of these compounds induce DNA damage directly or indirectly. Thus, DSBs arise from cell treatments with various anticancer agents such as the following: (1) topoisomerase II inhibitors like anthracyclines and epipodophyllotoxins that trap the DNA–enzyme intermediate complex in a so-called cleavable complex; cellular processing of the cleavable complex converts the protein–DNA cleavable complex into DSBs [22]; (2) topo I inhibitors like camptothecin that block the protein–DNA complex leading to SSBs converted into DSBs when the replication machinery encounters the lesion [23]; (3) cross-linking bifunctional agents like cisplatin or chlorambucil that produce DSBs either during the repair processing of cross links [24, 25] or following replication fork collapse [26]; (4) radiomimetic agents like enediynes that induce a low ratio of SSBs to DSBs (about 5:1 for neocarzinostatin, 2:1 for C-1027) unlike IR, which induces up to 100 SSBs for every DSB [27–31]. Overall, the induction of DNA breaks correlates with toxicity to the cells and cells usually respond to enediyne-induced damage as they do to IR-induced damage [32].

3 DSB Repair Pathways

Whatever the origins of DSBs, such DNA damage can be repaired by two distinct and complementary mechanisms: the Homologous Recombination (HR) or the Nonhomologous End Joining (NHEJ) processes [33, 34]. The activation of one or more of the three related phosphatidylinositol 3-kinase-like kinases (PI3KK) in response to DNA damage is required for the completion of the HR or NHEJ processes. While NHEJ allows fast but possibly γ-H2AX error-prone repair during the entire cell cycle, the slower but high fidelity HR pathway is restricted to the S and G2 phases of the cell cycle. The PI3KK involved in DSB repair are ATM (Ataxia Telangiectasia Mutated) and ATR (ATM and Rad3-related) that are associated with HR and typically activated by DNA breaks or after replication fork collapse. The third one, the DNA protein kinase catalytic subunit, is involved in NHEJ that operates throughout the cell cycle in response to DSBs. Despite competition between HR and NHEJ during the S-G2 phases, it has recently been reported that DNA-PK was able to modulate HR activity through its phosphorylation status (more than 30 phospho-sites have been determined in DNA-PKcs) [35]. These kinases belong to the family of transducer proteins that relay and amplify the damage signal to receptor proteins. A common substrate of the three PI3KKs mentioned above is the histone variant, H2AX, which is phosphorylated on serine 139 and subsequently called γ-H2AX. γ-H2AX has been widely used as a sensitive and early marker of DSBs in various areas including cancer research [36–38]. In line with its high sensitivity of detection (one DSB corresponds to one γ-H2AX focus determined by immunofluorescence) [39], it has been assumed that there is a direct relationship between γ-H2AX labeling and the existence of DSBs. However, since activation of PI3KK could occur in the absence of DSBs under certain circumstances [40–43], γ-H2AX labeling does not account solely for DSB occurrence. For instance, DNA-PK is activated in hypoxic cells independently of DNA breaks by a new mechanism relying on chromatin modifications [44].

4 The Non Homologous End-Joining Pathway

In mammalian cells, NHEJ is the predominant repair pathway for DSB repair which, throughout the cell cycle, ligates the two DNA ends together with minimal end processing [45, 46]. NHEJ consists of at least two genetically and biochemically distinct sub-pathways (Fig. 1): (1) a main canonical end-joining pathway (C-NHEJ) and (2) an alternative NHEJ (A-NHEJ) or backup NHEJ (B-NHEJ) (hereafter referred to as A-NHEJ) [47–49].

Since C-NHEJ is essential both for cell survival after IR treatment and V(D)J recombination, which generates the antibody and T cell receptor diversity required for lymphocyte maturation, cells from RS-SCID (radiosensitive-severe combined immunodeficiency) patients have helped to genetically define the NHEJ components [50]. C-NHEJ is a multi-step process involving several essential factors [51, 52] (Fig. 2).

The prerequisite event for all the subsequent steps is the binding of the Ku70/Ku80 heterodimer to DNA ends [53]. In the most recent model, drawn from live cell imaging following nuclear laser micro-irradiation experiments, the other core components of the reaction are then independently recruited to Ku-bound DSBs [54].

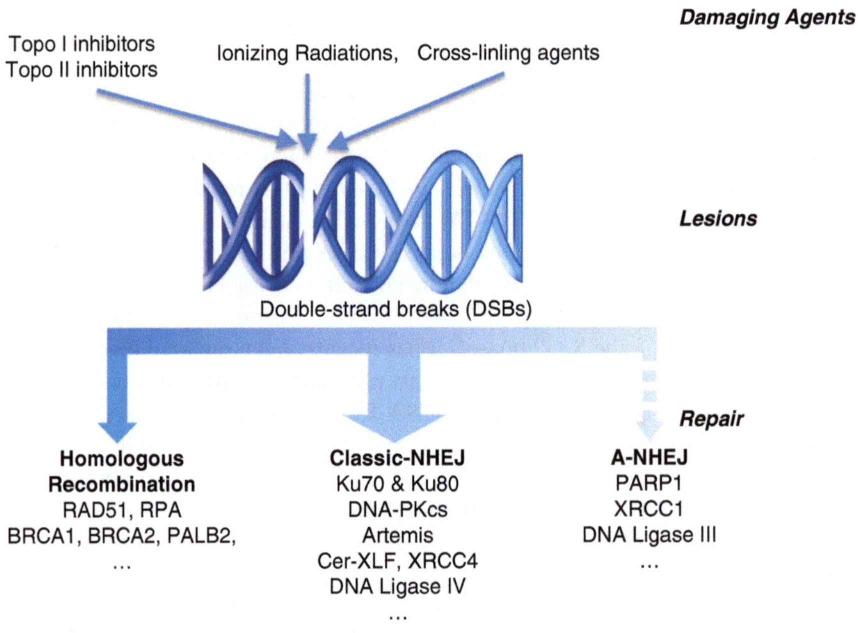

Fig. 1 DNA double-strand breaks damage and repair mechanisms. In normal mammalian cells, the classic NHEJ (C-NHEJ) pathway is the major repair pathway, as Homologous Recombination needs the sister chromatid and preferentially takes place in the S/G2 phases. At the DSBs, C-NHEJ proceeds through the recruitment of Ku70/80 and DNA-PK catalytic subunit (cs), XLF, XRCC4 and DNA Ligase IV. Although A-NHEJ is a minor DSB repair pathway, it may take over in specific situations (e.g. when Ku is absent) therefore leading to error-prone repair of DNA-damage

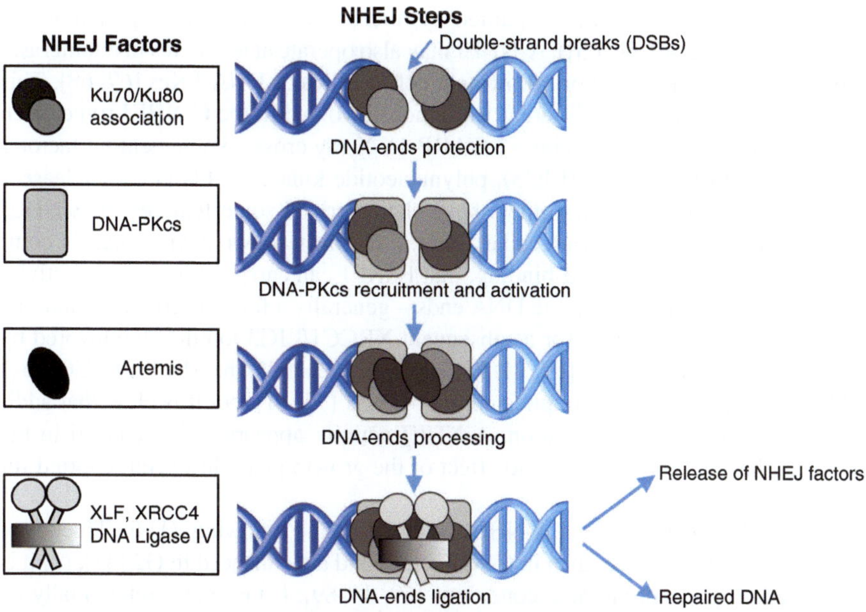

Fig. 2 NHEJ repair of DNA double-strand breaks (DSBs) DSBs can be produced by endogenous or exogenous damaging agents or during physiological processes as V(D)J recombination. Classic-NHEJ proceeds through the recruitment of Ku70/80 heterodimer at the ends of the DSBs, followed by the DNAPK catalytic subunit (cs) recruitment and activation. The DNA damage processing involves Artemis, through its DNA-PK interaction, or the Mre11 protein. Finally, a DNA polymerase (as Pol μ, λ) may synthesize new DNA ends before the ligation step, involving XLF, XRCC4 and DNA Ligase IV. The NHEJ factors and the repaired DNA are then released

These include the DNA-PKcs subunit, Cernunnos-XLF (Cer-XLF) and the XRCC4/DNA Ligase IV (LIG4) complex, which is preassembled by a tight association between the two partners [55]. Multiple interactions then take place among these factors resulting in a stable assembly of the NHEJ machinery. As a result, the NHEJ complex associates more tightly with damaged sites and becomes resistant to biochemical extraction from the damaged chromatin, at least during the time of the repair [56–58]. The DNA-PK holoenzyme (Ku/DNA-PKcs) recognizes, protects and bridges the DNA-ends in addition to having a serine/threonine protein kinase activity [59]. DNAPK conformational change mediated by autophosphorylation is necessary for activation of end-processing enzymes, such as the Artemis nuclease [60]. Ligation requires the concerted action of LIG4, XRCC4, and Cer-XLF, the latter promoting readenylation of LIG4 [61]. The ligation complex also has a role upstream of the ligation reaction, since it stimulates processing of DNA ends [62, 63]. At a later stage in the NHEJ process, this molecular machinery must be disassembled and released from the re-ligated DNA by a still unknown mechanism. A-NHEJ is not a robust or a particularly important DSB repair pathway because it has been detected in the absence of C-NHEJ. A-NHEJ mechanistically results in deletions that are often accompanied by microhomology at the repair junction (for reviews, see [49, 64, 65]). A-NHEJ may also operate at telomeres in telomerase-deficient mouse cells [66] or following a defect in Ku or DNA-PKcs [67, 68]. This pathway relies on factors different from those involved in the C-NHEJ route, such as poly (ADPribose) polymerase-1 (PARP-1), X-ray cross complementing factor 1 (XRCC1), DNA ligase III (LIG3), polynucleotide kinase, or Flap endonuclease 1 [48, 69–73]. Our group and others have characterized some features of A-NHEJ using biochemical assays with cell extracts. It has been shown in vitro that Ku competes with PARP1 DNA end-binding, that PARP1 can carry out a synapsis activity thanks to short homology at the DNA ends—generally a few nucleotides—and that PARP1 activity is required for a subsequent XRCC1/LIG3 joining step favored by regions of microhomology [69, 71, 73, 74]. More recently, the Mre11:Rad50:Nbs1 (MRN) complex has been implicated in A-NEHJ [75–81], but it is clear that additional factors await identification. A-NHEJ activity appears to be reduced in the plateau phase of growth, while no effect of the growth phase has been reported for C-NHEJ [82].

Established features of the A-NHEJ pathway are the following: (1) the kinetics of DSB repair appear slower than in C-NHEJ [69] and are enhanced in G2 [83]; (2) it is repressed by Ku under normal conditions [69, 84–89]; (3) it relies preferentially on resection of the DNA ends and end annealing driven by microhomology >4 bp for intrachromosomal substrates [88–91], V(D)J junctions [92] or CSR joins [86, 93], although this feature has been questioned in some reports [94]. This alternative pathway may be particularly relevant to genomic instability associated with tumor development. For example, frequent translocations lead to a high level of lymphomagenesis and other cancers in C-NHEJ deficient animal models [50, 95]. In addition, chromosomal translocations, like those at the origin of leukemia, are mediated by a rejoining pathway described as Ku- and XRCC4/LIG4-independent [84, 93, 96, 97].

5 Structure–Function of the DNA-PK Repair Complex

The DNA-PK holoenzyme (Ku/DNA-PKcs) recognizes, protects and bridges the DNA ends. The Ku70/Ku80 heterodimer (Ku), present in the cell as a preassembled heterodimer, recognizes and binds the DNA ends of the DSB [53, 98]. The recruitment of catalytic subunit DNA-PKcs occurs via the Ku80 C-terminal domain (Ku80-CTD). A truncated form of Ku70/Ku80 has been crystallized and shows a ringshaped form [98]. In addition, NMR studies of Ku80-CTD show a helical structure for the fragment comprising residues 592–709, although the extreme C-terminal portion of Ku80-CTD (residues 720–732) is disordered [99, 100]. A structural model of the functions of the C-terminal domains in the context of the full-length Ku70/Ku80 protein has also been reported [101].

When bound to DNA-ends, Ku recruits the DNA-PKcs, which by itself has a weak protein kinase activity, strongly stimulated through the Ku interaction [102, 103].

The phosphorylation occurs on an S/T-Q motif, although some serine/threonine in other sequences/targets could be phosphorylated [59, 104]. The phosphorylation of Artemis may help to activate its endonuclease activity [60, 105]. However, although DNA-PK also phosphorylates Ku, XRCC4, and Cer-XLF in the cell, mutational studies concluded that these phosphorylations are not functionally important, at least for NHEJ [59]. Thus, like ATM, DNA-PK may phosphorylate unknown substrates in vivo involved in processes other than DNA repair [106, 107]. More likely, DNA-PKcs is the relevant target of its own enzymatic activity. Indeed, DNA-PKcs is autophosphorylated after ionizing radiation treatment [108, 109]. Sixteen in vitro autophosphorylation sites in DNA-PKcs were identified and classified as two major clusters: the ABCDE cluster [110] and the PQR cluster [111, 112]. A further autophosphorylation site was identified at Thr 3950, within the kinase domain, involved in the regulation of the kinase activity of DNA-PKcs [113]. It has been suggested that the structural plasticity of DNA-PK is highly affected by autophosphorylation at those two clusters [110, 111]. Moreover, it has recently been reported that there are more than 30 autophosphorylation sites within DNA-PKcs; a model was proposed in which phosphorylation-induced conformational changes regulate the interaction of DNAPKcs with its partners Ku and DNA [59, 114]. This confirmed previous results showing that autophosphorylation of DNA-PKcs was a key event in the dissociation of DNA-PK from DNA [108, 109, 113–116]. On the other hand, biochemical studies on the mechanism of DNA-PK autophosphorylation indicate that it occurs in trans, both in vitro and in vivo [109].

Knowledge of the 3-D structure of DNA-PKcs contributes to a better understanding of its role in the NHEJ mechanism, illustrated for instance by the autophosphorylation reaction. However, structural studies of DNA-PKcs are challenging, due to its large size and poor recombinant protein production hence requiring complex purification from natural sources. Electron microscopy (EM) studies of the catalytic subunit DNAPKcs, a 469 kDa single-polypeptide chain, have produced a structure at 20 Å resolution, defining the general architecture of DNA-PKcs into three main regions, namely a head, a palm and a connecting arm [117, 118]. More recently, a 13 Å resolution cryo-electron microscopy (cryo-EM) structure of DNA-PKcs has revealed α helices throughout the molecule and a model has been proposed which localized the kinase domain in the head region [119, 120]. Studies have shown that up to eight repeats of the HEAT domain can fit into the cryoEM density model [120]. The HEAT domain (*H*untington, *E*longation factor 3, α regulatory subunit of protein phosphatase 2*A*, *T*OR1) consists of repeats of 37–47 residues forming a rod-like helical structure.

In addition, DNA-PKcs has recently been crystallized with Ku80-CTD at 6.6 Å, highlighting the overall topology and the formation of synaptic dimers [121, 122]. Negative staining electron microscopy, single particle or X-ray analysis indicates that DNA-PKcs autophosphorylation induced significant conformational changes that were postulated to function as a DNA release mechanism [114, 123, 124].

Finally, the phosphorylated form of DNA-PKcs is a substrate for serine/threonine phosphatases that play a role in the DDR. The catalytic subunits of PP2A (PP2Ac),

PP4 (PP4c), and PP6 (PP6c) belong to a subgroup referred to as the PP2A-like protein phosphatases (reviewed in [125]). Inhibition of PP2A-like protein phosphatases increases the phosphorylation status of DNA-PKcs and reduces its protein kinase activity [126]. In parallel, PP2A-like phosphatases (PP4 and PP2A) are involved in γ-H2AX dephosphorylation and have been shown to play a role in the DDR [127–130]. More recently, PP6 has been reported to be recruited by DNA-PKcs to DSBs, a step involved in the regulation of dephosphorylation of γ-H2AX, the dissolution of IR-induced foci and the release from the G2/M checkpoint [131]. Thus, DNA-PKcs is involved in the recruitment of multiple protein phosphatases to DSB sites and might interact through the series of HEAT repeats [131].

6 Inhibition of DNA-PK

Regardless of what the physiological substrates of DNA-PK are, the ability of small molecule inhibitors of DNA-PKcs to radiosensitize cells suggests that DNA-PK may be a good therapeutic target as a radiation sensitizer (reviewed in [132, 133]). In addition, DNA-PK has been implicated in the repair of chlorambucil-induced cross links, because increased DNA-PK activity in CLL cells correlates with clinical resistance to chlorambucil [134–137]. Furthermore, NHEJ and DNA-PK activity are increased or upregulated in radioresistant compared with radiosensitive CLL cells.

Thus, the inhibition of NHEJ through DNA-PK may rely on different strategies (Fig. 3): regulation of Ku or DNA-PKcs expression, inhibition of Ku/DNA-PKcs interaction, modulation of DNA-PKcs kinase activity, regulation of DNA-PKcs autophosphorylation, modulation of phosphatases activity. However, in the search for drugs useful in therapy, almost all the research activity is being devoted to the specific inhibition of the kinase activity.

6-1 Protein expression
RNA interference is being investigated as a therapeutic mechanism in the treatment of cancer, despite intrinsic problems like concentration, targeting to cancer cells and the metabolic stability of the miRNAs [138]. However, a decreased expression following RNAi treatment may not be sufficient to induce a strong phenotype, as reported for LIG3 and LIG4 functions [139]. A radiosensitizing effect has recently been reported, in vitro and in vivo, for miR-101 that targets DNA-PKcs and ATM via its binding to the 3′- UTR of DNA-PKcs or ATM mRNA [140]. In the case of Ku, in addition to the difficulty inherent in its very high level of expression and with pleiotropic localization and activities [141–148], targeting its expression does not appear to be of interest because it negatively controls the activity of the A-NHEJ mutagenic pathway [81].

6-2 Ku/DNA-PKcs interaction
Since the C-terminal portion of Ku80-CTD that recruits DNA-PKcs is structurally disordered [99, 100], the approach of drug design modelization is impractical. An indirect way has been developed by using short DNA

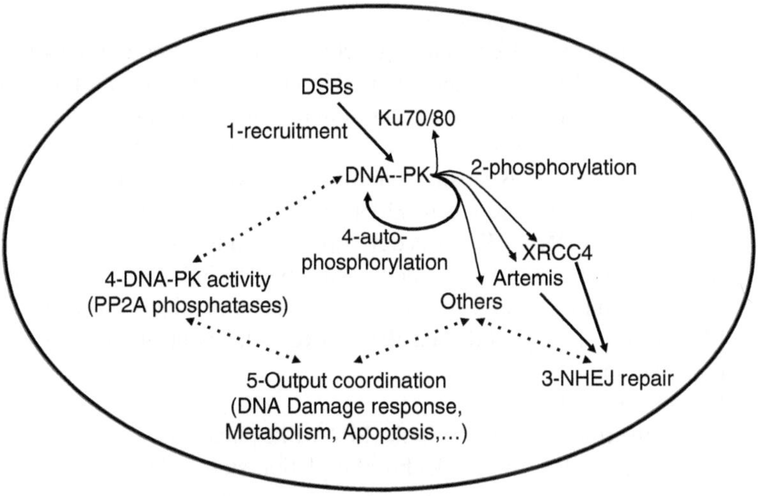

Fig. 3 DNA double-strand break repair and DNA-PK as pharmaceutical target. Different steps may be used as targets in the NHEJ pathway. Recruitment of Ku70/Ku80 is essential for DNA-PK interaction (1) and activation; autophosphorylation of DNA-PK and phosphorylations of substrates (2) are involved in the repair process. Regulation of the DNA-PK activity is key and could be achieved via phosphatases (4) or by inhibiting the previous steps. The output signals may coordinate different cell processes (5) such as the DNA damage response and some metabolic adaptations. (see text for details)

molecules (Dbait) that mimic DSB in order to down regulate the kinase activity by competing with DNA-PKc [149].

Dbait molecules sensitize xenografted tumors to radiotherapy, not by inhibiting the kinase activity of DNA-PK, but by acting through the induction of "false" DNA damage signaling [149–151].

6-3 DNA-PKcs kinase activity

Preliminary investigations of the inhibition of DNA-PK were undertaken by using wortmannin and LY294002, two nonselective PIKK inhibitors [152]. These drugs were shown to sensitize tumor cells to radiotherapy and chemotherapeutic agents and were used as a basis to develop more specific compounds. A flavone derivative, IC87361, led to tumor radiosensitization in both in vitro cell models and tumor xenograft in vivo models [133]. A more specific DNA-PK inhibitor, NU7026, has been reported to radiosensitize tumor cells [153]; similarly, NU7026 increased chlorambucil-sensitivity in CLL, correlated with DNA-PK inhibition and sensitization to chlorambucil [154]. Subsequently, a highly potent and selective DNA-PK inhibitor (NU7441) has been identified and showed an IC50 of 13 nM [155]. NU7441 induced sensitization of CLL when treated with fludarabine and chlorambucil [156] or mitoxanthorone [157]. In addition, DNA-PKcs inhibitors synergize with irinotecan to improve the killing of colon cancer cell lines in vitro [158]. A number of other agents are currently in preclinical trials [159].

6-4 DNA-PKcs autophosphorylation
A radiosensitizing effect of a single chain variable antibody fragment (scFv) against DNA-PKcs has been reported in vitro [160]. Very recently, epitopes in the autophosphorylation cluster domain have been expressed as antigens to screen a phage antibody library. The selected antibody increased sensitivity to IR, decreased DSB repair capability along with decreased kinase activity and autophosphorylation on S2056 induced by radiation [161]. Another way to inhibit DNA-PKcs was developed by using a subtractive combinatorial selection to identify peptide ligands able to bind DNA-PKcs. A peptide was selected that specifically bound and noncompetitively inactivated DNA-PKcs [162]. This peptide sensitizes BRCA-deficient tumor cells to genotoxic therapy.

6-5 Phosphatase activity
Due to the role of DNA-PK in the PP2A-like phosphatase recruitment at the break site involved in the dephosphorylation step of DNA-PKcs itself and g-H2AX turnover these proteins might be considered as pharmacological targets. However, various drawbacks could be raised such as the multiplicity of phosphatases, their lack of specificity and, as in the case of kinases, the difficulty in obtaining highly specific inhibitors.

7 Discussion

DNA-PKcs is required for C-NHEJ, V(D)J recombination and telomere length maintenance but it has recently been shown to contribute to other pathways: (1) it is involved in the G2 checkpoint in response to IR [163]; (2) it mediates metabolic gene activation in response to insulin [164]; (3) it may also function outside DNA repair through phosphorylation of other substrates [165, 166]. Also, and unexpectedly, the activation of cellular DDR pathways (ATM and DNA-PK) does not always require DNA damage but can be triggered by the stable association of single repair factors with chromatin [40]. Thus, hypoxia, by modifying higher-order chromatin structure and chromatin-remodeling complexes [167], triggers a DNA-PK-dependent DDR pathway.

A key regulator of the cellular response to oxygen deprivation is the transcription factor, hypoxia-inducible factor 1 (HIF-1), whose function results in the induction of a plethora of target genes that collectively confer cellular adaptation to hypoxia [168].

Indeed, DNA-PK protects HIF-1α from degradation, indicating that DNA-PK controls the amplitude of HIF-1α accumulation under hypoxia [44]. These novel findings expand the cellular importance of DNA-PK [169] but paradoxically, compromise the therapeutic interest of its inhibition that may therefore induce side effects in uncharacterized metabolic networks.

In some cases, DNA-PK either shows variation in expression or is mutated in tumor cells. Despite a high level of expression of Ku and DNA-PKcs, an up-regulation of

DNA-PKcs was reported in some tumors or IR-resistant cell lines, suggesting a role in tumor growth and survival [170–173]. Moreover, overexpression or increased activity of DNA-PKcs in various cancers is closely associated with metastasis, poor prognosis and radioresistance [156, 171, 174, 175]. Indeed, up-regulation of DNA-PK activity was shown to impair apoptosis in B-cell chronic lymphocytic leukemia [176]. Finally, in colorectal mismatch repair-deficient tumor cells (MSI), mutations in genes involved in DDR and DNA repair, including DNA-PKcs, have been reported [177].

Taken together, all these alterations in DNA-PK expression or activity suggest that the consequences of its inhibition should be useful against tumors. However, in tumor tissues, the expression of DNA-PK shows intratumor heterogeneity, suggesting difficulty in predicting the radio- or chemo-sensitivity of the tumor as well as when a DNA-PK inhibitor may be beneficial [174].

Strategies that block DNA repair will increase damage in the treated cells and result in increased cell death. Such approaches enhance sensitivity to treatment, although they do not provide selectivity against cancer cells as they increase the radiosensitivity or chemosensitivity of normal cells as well. Therefore the use of a DNA-PK inhibitor, in combination with genotoxic treatment, would allow the dose of irradiation or drug to be lowered without any gain in selectivity. It has recently been reported that monotherapy with DNA repair inhibitors could be successful with PARP inhibitors that can selectively kill BRCA1- and BRCA2-defective tumors [178, 179], with promising results in phase II/III clinical studies [180, 181]. The BRCA1 and BRCA2 genes encode large proteins that coordinate the HR DSB repair pathway [182]. Since BRCA1/2-deficient cells cannot repair DSBs by HR, PARP inhibitors will lead to the accumulation of DNA damage, genomic instability, and cell death. Interestingly, these effects may rely on DNA-PK-dependent NHEJ activity [183]. This is the first example of a successful monotherapy, where the strategy is reminiscent of the synthetic lethality process [184–186]. Synthetic lethality is obtained when the simultaneous loss of two nonessential mutations results in cell death, which does not occur if either gene product is present and functional. Treatment of solid tumors partially deficient for DNA repair pathways opens a therapeutic window of opportunity. In contrast, for patients without inherited defects in DNA repair pathways, the combination of DNA repair inhibitors with genotoxic chemotherapy remains logical [187]. However, many tumor cells have specific genetic lesions, which could then be exploited by targeting synthetic lethal partner genes [188].

In the case of NHEJ inhibition, XRCC4/XLF/LigIV may be a better pharmacological target than DNA-PK itself, since the inhibition of the ligation step will not allow the A-NHEJ pathway to proceed due to the remaining Ku binding to DNA. However, despite our knowledge of the structure of the ligation complex, inhibition of protein/protein interactions is a difficult task as is the specific inhibition of LigIV activity [55, 58].

Indeed, most drugs bind at the biological sites of action and this implies, for a compound to be biologically active on LigIV, it must be similar to its endogenous ligand [189], that is in this case, the DNA molecule.

In conclusion, in cancer chemotherapy, the target is usually thought to be the tumor cells. Because of the lack of selectivity against the tumor cells, a recent alternative research field now favors the use of drugs directed against the tumor microenvironment. Results combining these two approaches are beginning to appear, for instance, in radiotherapy. Nonetheless, in the search for DNA-PK inhibitors more needs to be learnt about the structure of the repair complex, currently only understood at a low level of resolution that cannot help in drug design and docking approaches. Moreover, we need new insights on postranslational modifications—other than phosphorylations—of partners or substrates, on their eventual roles in metabolic pathways relevant to cell survival and adaptability, and on the coordinated manner in which they repair DSBs. Indeed, this may help in understanding the alternative pathways the tumor cells may find and avoid unpredictable outputs. Particularly, in the case of DSB repair, the A-NHEJ pathway should be bypassed; different strategies may help to reach this goal, such as synthetic lethality approaches or targeting downstream C-NHEJ effectors. Altogether, all these different possibilities indicate that integrated programs will be the key in the future.

Acknowledgments The work from the team "Radiobiology and DNA repair" at CNRS IPBS has been supported by grants from Ligue nationale contre le cancer (équipe labelisée), Electricité de France (EDF), Institut National du Cancer (INCA), and Cancéropole Grand-Sud Ouest. Bernard Salles and Gladys Mirey are Professor and Associate Professor, respectively, at Faculty of Pharmacy, and Patrick Calsou is a scientist from INSERM.

References

1. Jeggo PA, Lobrich M (2007) DNA double-strand breaks: their cellular and clinical impact? Oncogene 26(56):7717–7719
2. Jackson SP, Bartek J (2009) The DNA-damage response in human biology and disease. Nature 461(7267):1071–1078
3. Kastan MB (2008) DNA damage responses: mechanisms and roles in human disease: 2007 G.H.A. Clowes Memorial Award Lecture. Mol Cancer Res 6(4):517–524
4. Lobrich M, Jeggo PA (2007) The impact of a negligent G2/M checkpoint on genomic instability and cancer induction. Nat Rev Cancer 7(11):861–869
5. Jeggo P, Lavin MF (2009) Cellular radiosensitivity: how much better do we understand it? Int J Radiat Biol 85(12):1061–1081
6. O'Driscoll M, Jeggo PA (2006) The role of double-strand break repair—insights from human genetics. Nat Rev Genet 7(1):45–54
7. Begg AC, Stewart FA, Vens C (2011) Strategies to improve radiotherapy with targeted drugs. Nat Rev Cancer 11(4):239–253
8. Helleday T, Petermann E, Lundin C et al (2008) DNA repair pathways as targets for cancer therapy. Nat Rev Cancer 8(3):193–204
9. Guzi TJ, Paruch K, Dwyer MP et al (2011) Targeting the replication checkpoint using SCH 900776, a potent and functionally selective CHK1 inhibitor identified via high content screening. Mol Cancer Ther 10(4):591–602
10. McNeely S, Conti C, Sheikh T et al (2010) Chk1 inhibition after replicative stress activates a double strand break response mediated by ATM and DNA-dependent protein kinase. Cell Cycle 9(5):995–1004
11. Morgan MA, Parsels LA, Zhao L et al (2010) Mechanism of radiosensitization by the Chk1/2 inhibitor AZD7762 involves abrogation of the G2 checkpoint and inhibition of homologous recombinational DNA repair. Cancer Res 70(12):4972–4981

12. Walton MI, Eve PD, Hayes A et al (2010) The preclinical pharmacology and therapeutic activity of the novel CHK1 inhibitor SAR-020106. Mol Cancer Ther 9(1):89–100
13. Ouyang H, Nussenzweig A, Kurimasa A et al (1997) Ku70 is required for DNA repair but not for T cell antigen receptor gene recombination in vivo. J Exp Med 186(6):921–929
14. Lees-Miller SP, Godbout R, Chan DW et al (1995) Absence of p350 subunit of DNA activated protein kinase from a radiosensitive human cell line. Science 267(5201):1183–1185
15. Burma S, Chen BP, Chen DJ (2006) Role of non-homologous end joining (NHEJ) in maintaining genomic integrity. DNA Repair (Amst) 5(9–10):1042–1048
16. Goodhead DT (1994) Initial events in the cellular effects of ionizing radiations: clustered damage in DNA. Int J Radiat Biol 65(1):7–17
17. Ward JF (1988) DNA damage produced by ionizing radiation in mammalian cells: identities, mechanisms of formation, and reparability. Prog Nucleic Acid Res Mol Biol 35:95–125
18. Blaisdell JO, Harrison L, Wallace SS (2001) Base excision repair processing of radiation induced clustered DNA lesions. Radiat Prot Dosimetry 97(1):25–31
19. Asaithamby A, Hu B, Chen DJ (2011) Unrepaired clustered DNA lesions induce chromosome breakage in human cells. Proc Natl Acad Sci U S A 108(20):8293–8298
20. Sedelnikova OA, Redon CE, Dickey JS et al (2010) Role of oxidatively induced DNA lesions in human pathogenesis. Mutat Res 704(1–3):152–159
21. Singleton BK, Griffin CS, Thacker J (2002) Clustered DNA damage leads to complex genetic changes in irradiated human cells. Cancer Res 62(21):6263–6269
22. Pommier Y, Leo E, Zhang H et al (2010) DNA topoisomerases and their poisoning by anticancer and antibacterial drugs. Chem Biol 17(5):421–433
23. Strumberg D, Pilon AA, Smith M et al (2000) Conversion of topoisomerase I cleavage complexes on the leading strand of ribosomal DNA into 5′-phosphorylated DNA double-strand breaks by replication runoff. Mol Cell Biol 20(11):3977–3987
24. Akkari YM, Bateman RL, Reifsteck CA et al (2000) DNA replication is required to elicit cellular responses to psoralen-induced DNA interstrand cross-links. Mol Cell Biol 20(21):8283–8289
25. De Silva IU, McHugh PJ, Clingen PH et al (2000) Defining the roles of nucleotide excision repair and recombination in the repair of DNA interstrand cross-links in mammalian cells. Mol Cell Biol 20(21):7980–7990
26. Raschle M, Knipscheer P, Enoiu M et al (2008) Mechanism of replication-coupled DNA interstrand crosslink repair. Cell 134(6):969–980
27. Kennedy DR, Gawron LS, Ju J et al (2007) Single chemical modifications of the C-1027 enediyne core, a radiomimetic antitumor drug, affect both drug potency and the role of ataxia-telangiectasia mutated in cellular responses to DNA double-strand breaks. Cancer Res 67(2):773–781
28. Povirk LF (1996) DNA damage and mutagenesis by radiomimetic DNA-cleaving agents: bleomycin, neocarzinostatin and other enediynes. Mutat Res 355(1–2):71–89
29. Sutherland BM, Bennett PV, Sidorkina O et al (2000) Clustered damages and total lesions induced in DNA by ionizing radiation: oxidized bases and strand breaks. Biochemistry 39(27):8026–8031
30. Van Lanen SG, Shen B (2008) Biosynthesis of enediyne antitumor antibiotics. Curr Top Med Chem 8(6):448–459
31. Xu YJ, Zhen YS, Goldberg IH (1994) C1027 chromophore, a potent new enediyne antitumor antibiotic, induces sequence-specific double-strand DNA cleavage. Biochemistry 33(19):5947–5954
32. Elmroth K, Nygren J, Martensson S et al (2003) Cleavage of cellular DNA by calicheamicin gamma1. DNA Repair (Amst) 2(4):363–374
33. Mao Z, Bozzella M, Seluanov A et al (2008) DNA repair by nonhomologous end joining and homologous recombination during cell cycle in human cells. Cell Cycle 7(18):2902–2906
34. Shrivastav M, De Haro LP, Nickoloff JA (2008) Regulation of DNA double-strand break repair pathway choice. Cell Res 18(1):134–147

35. Neal JA, Dang V, Douglas P et al (2011) Inhibition of homologous recombination by DNA-dependent protein kinase requires kinase activity, is titratable, and is modulated by autophosphorylation. Mol Cell Biol 31(8):1719–1733
36. Bonner WM, Redon CE, Dickey JS et al (2008) GammaH2AX and cancer. Nat Rev Cancer 8(12):957–967
37. Lord CJ, Ashworth A (2009) Bringing DNA repair in tumors into focus. Clin Cancer Res 15(10):3241–3243
38. Banuelos CA, Banath JP, Kim JY et al (2009) gammaH2AX expression in tumors exposed to cisplatin and fractionated irradiation. Clin Cancer Res 15(10):3344–3353
39. Rogakou EP, Pilch DR, Orr AH et al (1998) DNA double-stranded breaks induce histone H2AX phosphorylation on serine 139. J Biol Chem 273(10):5858–5868
40. Soutoglou E, Misteli T (2008) Activation of the cellular DNA damage response in the absence of DNA lesions. Science 320(5882):1507–1510
41. Misteli T, Soutoglou E (2009) The emerging role of nuclear architecture in DNA repair and genome maintenance. Nat Rev Mol Cell Biol 10(4):243–254
42. Pospelova TV, Demidenko ZN, Bukreeva EI et al (2009) Pseudo-DNA damage response in senescent cells. Cell Cycle 8(24):4112–4118
43. Toledo LI, Murga M, Gutierrez-Martinez P et al (2008) ATR signaling can drive cells into senescence in the absence of DNA breaks. Genes Dev 22(3):297–302
44. Bouquet F, Ousset M, Biard D et al (2011) A DNA-dependent stress response involving DNA-PK occurs in hypoxic cells and contributes to cellular adaptation to hypoxia. J Cell Sci 124(Pt 11):1943–1951
45. Wyman C, Kanaar R (2006) DNA double-strand break repair: all's well that ends well. Annu Rev Genet 40:363–383
46. Pardo B, Gomez-Gonzalez B, Aguilera A (2009) DNA repair in mammalian cells: DNA double-strand break repair: how to fix a broken relationship. Cell Mol Life Sci 66(6):1039–1056
47. Ma JL, Kim EM, Haber JE et al (2003) Yeast Mre11 and Rad1 proteins define a Ku-independent mechanism to repair double-strand breaks lacking overlapping end sequences. Mol Cell Biol 23(23):8820–8828
48. Wang H, Perrault AR, Takeda Y et al (2003) Biochemical evidence for Ku-independent backup pathways of NHEJ. Nucleic Acids Res 31(18):5377–5388
49. Nussenzweig A, Nussenzweig MC (2007) A backup DNA repair pathway moves to the forefront. Cell 131(2):223–225
50. Soulas-Sprauel P, Rivera-Munoz P, Malivert L et al (2007) V(D)J and immunoglobulin class switch recombinations: a paradigm to study the regulation of DNA end-joining. Oncogene 26(56):7780–7791
51. Weterings E, Chen DJ (2008) The endless tale of non-homologous end-joining. Cell Res 18(1):114–124
52. Lieber MR (2010) The mechanism of double-strand DNA break repair by the nonhomologous DNA end-joining pathway. Annu Rev Biochem 79:181–211
53. Downs JA, Jackson SP (2004) A means to a DNA end: the many roles of Ku. Nat Rev Mol Cell Biol 5(5):367–378
54. Yano K, Morotomi-Yano K, Wang SY et al (2008) Ku recruits XLF to DNA double-strand breaks. EMBO Rep 9(1):91–96
55. Wu PY, Frit P, Meesala S et al (2009) Structural and functional interaction between the human DNA repair proteins DNA ligase IV and XRCC4. Mol Cell Biol 29(11):3163–3172
56. Drouet J, Delteil C, Lefrancois J et al (2005) DNA-dependent protein kinase and XRCC4-DNA ligase IV mobilization in the cell in response to DNA double strand breaks. J Biol Chem 280(8):7060–7069
57. Drouet J, Frit P, Delteil C et al (2006) Interplay between Ku, Artemis, and the DNA dependent protein kinase catalytic subunit at DNA ends. J Biol Chem 281(38):27784–27793
58. Wu PY, Frit P, Malivert L et al (2007) Interplay between Cernunnos-XLF and nonhomologous end-joining proteins at DNA ends in the cell. J Biol Chem 282(44):31937–31943

59. Meek K, Dang V, Lees-Miller SP (2008) DNA-PK: the means to justify the ends? Adv Immunol 99:33–58
60. Goodarzi AA, Yu Y, Riballo E et al (2006) DNA-PK autophosphorylation facilitates Artemis endonuclease activity. EMBO J 25(16):3880–3889
61. Riballo E, Woodbine L, Stiff T et al (2009) XLF-Cernunnos promotes DNA ligase IV-XRCC4 re-adenylation following ligation. Nucleic Acids Res 37(2):482–492
62. Budman J, Kim SA, Chu G (2007) Processing of DNA for nonhomologous end-joining is controlled by kinase activity and XRCC4/ligase IV. J Biol Chem 282(16): 11950–11959
63. Akopiants K, Zhou RZ, Mohapatra S et al (2009) Requirement for XLF/Cernunnos in alignment-based gap filling by DNA polymerases lambda and mu for nonhomologous end joining in human whole-cell extracts. Nucleic Acids Res 37(12):4055–4062
64. Haber JE (2008) Alternative endings. Proc Natl Acad Sci U S A 105(2):405–406
65. McVey M, Lee SE (2008) MMEJ repair of double-strand breaks (director's cut): deleted sequences and alternative endings. Trends Genet 24(11):529–538
66. Maser RS, Wong KK, Sahin E et al (2007) DNA-dependent protein kinase catalytic subunit is not required for dysfunctional telomere fusion and checkpoint response in the telomerase-deficient mouse. Mol Cell Biol 27(6):2253–2265
67. Bombarde O, Boby C, Gomez D et al (2010) TRF2/RAP1 and DNA-PK mediate a double protection against joining at telomeric ends. EMBO J 29(9):1573–1584
68. Riha K, Heacock ML, Shippen DE (2006) The role of the nonhomologous end-joining DNA double-strand break repair pathway in telomere biology. Annu Rev Genet 40:237–277
69. Wang M, Wu W, Rosidi B et al (2006) PARP-1 and Ku compete for repair of DNA double strand breaks by distinct NHEJ pathways. Nucleic Acids Res 34(21):6170–6182
70. Liang L, Deng L, Chen Y et al (2005) Modulation of DNA end joining by nuclear proteins. J Biol Chem 280(36):31442–31449
71. Audebert M, Salles B, Calsou P (2004) Involvement of poly(ADP-ribose) polymerase-1 and XRCC1/DNA ligase III in an alternative route for DNA double-strand breaks rejoining. J Biol Chem 279(53):55117–55126
72. Audebert M, Salles B, Weinfeld M et al (2006) Involvement of polynucleotide kinase in a poly(ADP-ribose) polymerase-1-dependent DNA double-strand breaks rejoining pathway. J Mol Biol 356(2):257–265
73. Wang H, Rosidi B, Perrault R et al (2005) DNA ligase III as a candidate component of backup pathways of nonhomologous end joining. Cancer Res 65(10):4020–4030
74. Audebert M, Salles B, Calsou P (2008) Effect of double-strand break DNA sequence on the PARP-1 NHEJ pathway. Biochem Biophys Res Commun 369(3):982–988
75. Deng Y, Guo X, Ferguson DO et al (2009) Multiple roles for MRE11 at uncapped telomeres. Nature 460(7257):914–918
76. Deriano L, Stracker TH, Baker A et al (2009) Roles for NBS1 in alternative nonhomologous end-joining of V(D)J recombination intermediates. Mol Cell 34(1):13–25
77. Dinkelmann M, Spehalski E, Stoneham T et al (2009) Multiple functions of MRN in end-joining pathways during isotype class switching. Nat Struct Mol Biol 16(8):808–813
78. Rass E, Grabarz A, Plo I et al (2009) Role of Mre11 in chromosomal nonhomologous end joining in mammalian cells. Nat Struct Mol Biol 16(8):819–824
79. Xie A, Kwok A, Scully R (2009) Role of mammalian Mre11 in classical and alternative nonhomologous end joining. Nat Struct Mol Biol 16(8):814–818
80. Zhuang J, Jiang G, Willers H et al (2009) Exonuclease function of human Mre11 promotes deletional nonhomologous end joining. J Biol Chem 284(44):30565–30573
81. Cheng Q, Barboule N, Frit P et al (2011) Ku counteracts mobilization of PARP1 and MRN in chromatin damaged with DNA double-strand breaks. Nucleic Acids Res 39:9605–9619
82. Singh SK, Wu W, Zhang L et al (2011) Widespread dependence of backup NHEJ on growth state: ramifications for the use of DNA-PK inhibitors. Int J Radiat Oncol Biol Phys 79(2):540–548

83. Wu W, Wang M, Singh SK et al (2008) Repair of radiation induced DNA double strand breaks by backup NHEJ is enhanced in G2. DNA Repair (Amst) 7(2):329–338
84. Weinstock DM, Brunet EJasin M (2007) Formation of NHEJ-derived reciprocal chromosomal translocations does not require Ku70. Nat Cell Biol 9(8):978–981
85. Fattah F, Lee EH, Weisensel N et al (2010) Ku regulates the non-homologous end joining pathway choice of DNA double-strand break repair in human somatic cells. PLoS Genet 6(2):e1000855
86. Boboila C, Yan C, Wesemann DR et al (2010) Alternative end-joining catalyzes class switch recombination in the absence of both Ku70 and DNA ligase 4. J Exp Med 207(2):417–427
87. Mansour WY, Schumacher S, Rosskopf R et al (2008) Hierarchy of nonhomologous end-joining, single-strand annealing and gene conversion at site-directed DNA double-strand breaks. Nucleic Acids Res 36(12):4088–4098
88. Guirouilh-Barbat J, Rass E, Plo I et al (2007) Defects in XRCC4 and KU80 differentially affect the joining of distal nonhomologous ends. Proc Natl Acad Sci U S A 104(52):20902–20907
89. Schulte-Uentrop L, El-Awady RA, Schliecker L et al (2008) Distinct roles of XRCC4 and Ku80 in non-homologous end-joining of endonuclease- and ionizing radiation-induced DNA double-strand breaks. Nucleic Acids Res 36(8):2561–2569
90. Kuhfittig-Kulle S, Feldmann E, Odersky A et al (2007) The mutagenic potential of nonhomologous end joining in the absence of the NHEJ core factors Ku70/80, DNA-PKcs and XRCC4-LigIV. Mutagenesis 22(3):217–233
91. Guirouilh-Barbat J, Huck S, Bertrand P et al (2004) Impact of the KU80 pathway on NHEJ-induced genome rearrangements in mammalian cells. Mol Cell 14(5):611–623
92. Corneo B, Wendland RL, Deriano L et al (2007) Rag mutations reveal robust alternative end joining. Nature 449(7161):483–486
93. Yan CT, Boboila C, Souza EK et al (2007) IgH class switching and translocations use a robust non-classical end-joining pathway. Nature 449(7161):478–482
94. Mansour WY, Rhein T, Dahm-Daphi J (2010) The alternative end-joining pathway for repair of DNA double-strand breaks requires PARP1 but is not dependent upon microhomologies. Nucleic Acids Res 38(18):6065–6077
95. Rooney S, Chaudhuri J, Alt FW (2004) The role of the non-homologous end-joining pathway in lymphocyte development. Immunol Rev 200:115–131
96. Boboila C, Jankovic M, Yan CT et al (2010) Alternative end-joining catalyzes robust IgH locus deletions and translocations in the combined absence of ligase 4 and Ku70. Proc Natl Acad Sci U S A 107(7):3034–3039
97. Simsek D, Jasin M (2010) Alternative end-joining is suppressed by the canonical NHEJ component Xrcc4-ligase IV during chromosomal translocation formation. Nat Struct Mol Biol 17(4):410–416
98. Walker JR, Corpina RA, Goldberg J (2001) Structure of the Ku heterodimer bound to DNA and its implications for double-strand break repair. Nature 412(6847):607–614
99. Zhang Z, Hu W, Cano L et al (2004) Solution structure of the C-terminal domain of Ku80 suggests important sites for protein-protein interactions. Structure 12(3):495–502
100. Harris R, Esposito D, Sankar A et al (2004) The 3D solution structure of the C-terminal region of Ku86 (Ku86CTR). J Mol Biol 335(2):573–582
101. Rivera-Calzada A, Spagnolo L, Pearl LH et al (2007) Structural model of full-length human Ku70-Ku80 heterodimer and its recognition of DNA and DNA-PKcs. EMBO Rep 8(1):56–62
102. DeFazio LG, Stansel RM, Griffith JD et al (2002) Synapsis of DNA ends by DNA-dependent protein kinase. EMBO J 21(12):3192–3200
103. Hammarsten O, Chu G (1998) DNA-dependent protein kinase: DNA binding and activation in the absence of Ku. Proc Natl Acad Sci U S A 95(2):525–530
104. Lees-Miller SP, Sakaguchi K, Ullrich SJ et al (1992) Human DNA-activated protein kinase phosphorylates serines 15 and 37 in the amino-terminal transactivation domain of human p53. Mol Cell Biol 12(11):5041–5049

105. Niewolik D, Pannicke U, Lu H et al (2006) DNA-PKcs dependence of Artemis endonucleolytic activity, differences between hairpins and 5′ or 3′ overhangs. J Biol Chem 281(45):33900–33909
106. Shiloh Y (2006) The ATM-mediated DNA-damage response: taking shape. Trends Biochem Sci 31(7):402–410
107. Lavin MF, Kozlov S (2007) ATM activation and DNA damage response. Cell Cycle 6(8):931–942
108. Chan DW, Lees-Miller SP (1996) The DNA-dependent protein kinase is inactivated by autophosphorylation of the catalytic subunit. J Biol Chem 271(15):8936–8941
109. Meek K, Douglas P, Cui X et al (2007) trans Autophosphorylation at DNA-dependent protein kinase's two major autophosphorylation site clusters facilitates end processing but not end joining. Mol Cell Biol 27(10):3881–3890
110. Douglas P, Sapkota GP, Morrice N et al (2002) Identification of in vitro and in vivo phosphorylation sites in the catalytic subunit of the DNA-dependent protein kinase. Biochem J 368(Pt 1):243–251
111. Cui X, Yu Y, Gupta S et al (2005) Autophosphorylation of DNA-dependent protein kinase regulates DNA end processing and may also alter double-strand break repair pathway choice. Mol Cell Biol 25(24):10842–10852
112. Chen BP, Chan DW, Kobayashi J et al (2005) Cell cycle dependence of DNA-dependent protein kinase phosphorylation in response to DNA double strand breaks. J Biol Chem 280(15):14709–14715
113. Douglas P, Cui X, Block WD et al (2007) The DNA-dependent protein kinase catalytic subunit is phosphorylated in vivo on threonine 3950, a highly conserved amino acid in the protein kinase domain. Mol Cell Biol 27(5):1581–1591
114. Hammel M, Yu Y, Mahaney BL et al (2010) Ku and DNA-dependent protein kinase dynamic conformations and assembly regulate DNA binding and the initial non-homologous end joining complex. J Biol Chem 285(2):1414–1423
115. Merkle D, Douglas P, Moorhead GB et al (2002) The DNA-dependent protein kinase interacts with DNA to form a protein-DNA complex that is disrupted by phosphorylation. Biochemistry 41(42):12706–12714
116. Uematsu N, Weterings E, Yano K et al (2007) Autophosphorylation of DNA-PKCS regulates its dynamics at DNA double-strand breaks. J Cell Biol 177(2):219–229
117. Chiu CY, Cary RB, Chen DJ et al (1998) Cryo-EM imaging of the catalytic subunit of the DNA-dependent protein kinase. J Mol Biol 284(4):1075–1081
118. Leuther KK, Hammarsten O, Kornberg RD et al (1999) Structure of DNA-dependent protein kinase: implications for its regulation by DNA. EMBO J 18(5):1114–1123
119. Rivera-Calzada A, Maman JD, Spagnolo L et al (2005) Three-dimensional structure and regulation of the DNA-dependent protein kinase catalytic subunit (DNA-PKcs). Structure 13(2):243–255
120. Williams DR, Lee KJ, Shi J et al (2008) Cryo-EM structure of the DNA-dependent protein kinase catalytic subunit at subnanometer resolution reveals alpha helices and insight into DNA binding. Structure 16(3):468–477
121. Sibanda BL, Chirgadze DY, Blundell TL (2010) Crystal structure of DNA-PKcs reveals a large open-ring cradle comprised of HEAT repeats. Nature 463(7277):118–121
122. Ochi T, Sibanda BL, Wu Q et al (2010) Structural biology of DNA repair: spatial organisation of the multicomponent complexes of nonhomologous end joining. J Nucleic Acids pii:621695
123. Dobbs TA, Tainer JA, Lees-Miller SP (2010) A structural model for regulation of NHEJ by DNA-PKcs autophosphorylation. DNA Repair (Amst) 9(12):1307–1314
124. Morris EP, Rivera-Calzada A, da Fonseca PC et al (2011) Evidence for a remodelling of DNAPK upon autophosphorylation from electron microscopy studies. Nucleic Acids Res 39(13):5757–5767
125. Moorhead GB, Trinkle-Mulcahy L, Ulke-Lemee A (2007) Emerging roles of nuclear protein phosphatases. Nat Rev Mol Cell Biol 8(3):234–244

126. Douglas P, Moorhead GBG, Ye R et al (2001) Protein phosphatases regulate DNA-dependent protein kinase activity. J Biol Chem 276(22):18992–18998
127. Nakada S, Chen GI, Gingras AC et al (2008) PP4 is a gamma H2AX phosphatase required for recovery from the DNA damage checkpoint. EMBO Rep 9(10):1019–1026
128. Wang B, Zhao A, Sun L et al (2008) Protein phosphatase PP4 is overexpressed in human breast and lung tumors. Cell Res 18(9):974–977
129. Chowdhury D, Xu X, Zhong X et al (2008) A PP4-phosphatase complex dephosphorylates gamma-H2AX generated during DNA replication. Mol Cell 31(1):33–46
130. Keogh MC, Kim JA, Downey M et al (2006) A phosphatase complex that dephosphorylates gammaH2AX regulates DNA damage checkpoint recovery. Nature 439(7075):497–501
131. Douglas P, Zhong J, Ye R et al (2010) Protein phosphatase 6 interacts with the DNA-dependent protein kinase catalytic subunit and dephosphorylates gamma-H2AX. Mol Cell Biol 30(6):1368–1381
132. O'Connor MJ, Martin NM, Smith GC (2007) Targeted cancer therapies based on the inhibition of DNA strand break repair. Oncogene 26(56):7816–7824
133. Shinohara ET, Geng L, Tan J et al (2005) DNA-dependent protein kinase is a molecular target for the development of noncytotoxic radiation-sensitizing drugs. Cancer Res 65(12):4987–4992
134. Muller C, Dusseau C, Calsou P et al (1998) Human normal peripheral blood B-lymphocytes are deficient in DNA-dependent protein kinase activity due to the expression of a variant form of the Ku86 protein. Oncogene 16(12):1553–1560
135. Muller C, Christodoulopoulos G, Salles B et al (1998) DNA-Dependent protein kinase activity correlates with clinical and in vitro sensitivity of chronic lymphocytic leukemia lymphocytes to nitrogen mustards. Blood 92(7):2213–2219
136. Muller C, Calsou P, Salles B (2000) The activity of the DNA-dependent protein kinase (DNAPK) complex is determinant in the cellular response to nitrogen mustards. Biochimie 82(1):25–28
137. Christodoulopoulos G, Muller C, Salles B et al (1998) Potentiation of chlorambucil cytotoxicity in B-cell chronic lymphocytic leukemia by inhibition of DNA-dependent protein kinase activity using wortmannin. Cancer Res 58(9):1789–1792
138. Weidhaas JB, Babar I, Nallur SM et al (2007) MicroRNAs as potential agents to alter resistance to cytotoxic anticancer therapy. Cancer Res 67(23):11111–11116
139. Windhofer F, Wu W, Iliakis G (2007) Low levels of DNA ligases III and IV sufficient for effective NHEJ. J Cell Physiol 213(2):475–483
140. Yan D, Ng WL, Zhang X et al (2010) Targeting DNA-PKcs and ATM with miR-101 sensitizes tumors to radiation. PLoS One 5(7):e11397
141. Takeda AA, de Barros AC, Chang CW et al (2011) Structural basis of importin-alpha-mediated nuclear transport for Ku70 and Ku80. J Mol Biol 412(2):226–234
142. Ohno M, Kunimoto M, Nishizuka M et al (2009) Ku proteins function as corepressors to regulate farnesoid X receptor-mediated gene expression. Biochem Biophys Res Commun 390(3):738–742
143. Kim E, Li K, Lieu C et al (2008) Expression of apolipoprotein C-IV is regulated by Ku antigen/peroxisome proliferator-activated receptor gamma complex and correlates with liver steatosis. J Hepatol 49(5):787–798
144. Bednarek R, Boncela J, Smolarczyk K et al (2008) Ku80 as a novel receptor for thymosin beta4 that mediates its intracellular activity different from G-actin sequestering. J Biol Chem 283(3):1534–1544
145. Shi L, Qiu D, Zhao G et al (2007) Dynamic binding of Ku80, Ku70 and NF90 to the IL-2 promoter in vivo in activated T-cells. Nucleic Acids Res 35(7):2302–2310
146. Munakata Y, Saito-Ito T, Kumura-Ishii K et al (2005) Ku80 autoantigen as a cellular coreceptor for human parvovirus B19 infection. Blood 106(10):3449–3456
147. Muller C, Paupert J, Monferran S et al (2005) The double life of the Ku protein: facing the DNA breaks and the extracellular environment. Cell Cycle 4(3):438–441
148. Monferran S, Paupert J, Dauvillier S et al (2004) The membrane form of the DNA repair protein Ku interacts at the cell surface with metalloproteinase 9. EMBO J 23(19):3758–3768

149. Quanz M, Berthault N, Roulin C et al (2009) Small-molecule drugs mimicking DNA damage: a new strategy for sensitizing tumors to radiotherapy. Clin Cancer Res 15(4):1308–1316
150. Quanz M, Chassoux D, Berthault N et al (2009) Hyperactivation of DNA-PK by doublestrand break mimicking molecules disorganizes DNA damage response. PLoS One 4(7):e6298
151. Berthault N, Maury B, Agrario C et al (2011) Comparison of distribution and activity of nanoparticles with short interfering DNA (Dbait) in various living systems. Cancer Gene Ther 18:695–706
152. Rosenzweig KE, Youmell MB, Palayoor ST et al (1997) Radiosensitization of human tumor cells by the phosphatidylinositol3-kinase inhibitors wortmannin and LY294002 correlates with inhibition of DNA-dependent protein kinase and prolonged G2-M delay. Clin Cancer Res 3(7):1149–1156
153. Veuger SJ, Curtin NJ, Richardson CJ et al (2003) Radiosensitization and DNA repair inhibition by the combined use of novel inhibitors of DNA-dependent protein kinase and poly(ADP-ribose) polymerase-1. Cancer Res 63(18):6008–6015
154. Amrein L, Loignon M, Goulet AC et al (2007) Chlorambucil cytotoxicity in malignant B lymphocytes is synergistically increased by 2-(morpholin-4-yl)-benzo[h]chomen-4-one (NU7026)-mediated inhibition of DNA double-strand break repair via inhibition of DNA-dependent protein kinase. J Pharmacol Exp Ther 321(3):848–855
155. Zhao Y, Thomas HD, Batey MA et al (2006) Preclinical evaluation of a potent novel DNA-dependent protein kinase inhibitor NU7441. Cancer Res 66(10):5354–5362
156. Willmore E, Elliott SL, Mainou-Fowler T et al (2008) DNA-dependent protein kinase is a therapeutic target and an indicator of poor prognosis in B-cell chronic lymphocytic leukemia. Clin Cancer Res 14(12):3984–3992
157. Elliott SL, Crawford C, Mulligan E et al (2011) Mitoxantrone in combination with an inhibitor of DNA-dependent protein kinase: a potential therapy for high risk B-cell chronic lymphocytic leukaemia. Br J Haematol 152(1):61–71
158. Davidson D, Coulombe Y, Martinez-Marignac VL et al (2011) Irinotecan and DNA-PKcs inhibitors synergize in killing of colon cancer cells. Invest New Drugs 30:1248–1256
159. Martin SA, Lord CJ, Ashworth A (2008) DNA repair deficiency as a therapeutic target in cancer. Curr Opin Genet Dev 18(1):80–86
160. Li S, Takeda Y, Wragg S et al (2003) Modification of the ionizing radiation response in living cells by an scFv against the DNA-dependent protein kinase. Nucleic Acids Res 31(20):5848–5857
161. Du L, Zhou LJ, Pan XJ et al (2010) Radiosensitization and growth inhibition of cancer cells mediated by an scFv antibody gene against DNA-PKcs in vitro and in vivo. Radiat Oncol 5:70
162. Moeller BJ, Sidman RL, Pasqualini R et al (2011) Discovery of DNA repair inhibitors by combinatorial library profiling. Cancer Res 71(5):1816–1824
163. Shang ZF, Huang B, Xu QZ et al (2010) Inactivation of DNA-dependent protein kinase leads to spindle disruption and mitotic catastrophe with attenuated checkpoint protein 2 phosphorylation in response to DNA damage. Cancer Res 70(9):3657–3666
164. Wong RH, Chang I, Hudak CS et al (2009) A role of DNA-PK for the metabolic gene regulation in response to insulin. Cell 136(6):1056–1072
165. Britton S, Froment C, Frit P et al (2009) Cell nonhomologous end joining capacity controls SAF-A phosphorylation by DNA-PK in response to DNA double-strand breaks inducers. Cell Cycle 8(22):3717–3722
166. Shi M, Vivian CJ, Lee KJ et al (2009) DNA-PKcs-PIDDosome: a nuclear caspase-2-activating complex with role in G2/M checkpoint maintenance. Cell 136(3):508–520
167. Johnson AB, Denko NBarton MC (2008) Hypoxia induces a novel signature of chromatin modifications and global repression of transcription. Mutat Res 640(1–2):174–179
168. Semenza GL (2003) Targeting HIF-1 for cancer therapy. Nat Rev Cancer 3(10):721–732
169. Kong X, Shen Y, Jiang N et al (2011) Emerging roles of DNA-PK besides DNA repair. Cell Signal 23:1273–1280

170. Soderlund Leifler K, Queseth S, Fornander T et al (2010) Low expression of Ku70/80, but high expression of DNA-PKcs, predict good response to radiotherapy in early breast cancer. Int J Oncol 37(6):1547–1554
171. Beskow C, Skikuniene J, Holgersson A et al (2009) Radioresistant cervical cancer shows upregulation of the NHEJ proteins DNA-PKcs, Ku70 and Ku86. Br J Cancer 101(5):816–821
172. Hosoi Y, Watanabe T, Nakagawa K et al (2004) Up-regulation of DNA-dependent protein kinase activity and Sp1 in colorectal cancer. Int J Oncol 25(2):461–468
173. Ader I, Muller C, Bonnet J et al (2002) The radioprotective effect of the 24 kDa FGF-2 isoform in HeLa cells is related to an increased expression and activity of the DNA dependent protein kinase (DNA-PK) catalytic subunit. Oncogene 21(42):6471–6479
174. Tonotsuka N, Hosoi Y, Miyazaki S et al (2006) Heterogeneous expression of DNA-dependent protein kinase in esophageal cancer and normal epithelium. Int J Mol Med 18(3):441–447
175. Shintani S, Mihara M, Li C et al (2003) Up-regulation of DNA-dependent protein kinase correlates with radiation resistance in oral squamous cell carcinoma. Cancer Sci 94(10):894–900
176. Deriano L, Guipaud O, Merle-Beral H et al (2005) Human chronic lymphocytic leukemia B cells can escape DNA damage-induced apoptosis through the nonhomologous end-joining DNA repair pathway. Blood 105(12):4776–4783
177. Miquel C, Jacob S, Grandjouan S et al (2007) Frequent alteration of DNA damage signalling and repair pathways in human colorectal cancers with microsatellite instability. Oncogene 26(40):5919–5926
178. Bryant HE, Schultz N, Thomas HD et al (2005) Specific killing of BRCA2-deficient tumours with inhibitors of poly(ADP-ribose) polymerase. Nature 434(7035):913–917
179. Farmer H, McCabe N, Lord CJ et al (2005) Targeting the DNA repair defect in BRCA mutant cells as a therapeutic strategy. Nature 434(7035):917–921
180. Fong PC, Boss DS, Yap TA et al (2009) Inhibition of poly(ADP-ribose) polymerase in tumors from BRCA mutation carriers. N Engl J Med 361(2):123–134
181. Fong PC, Yap TA, Boss DS et al (2010) Poly(ADP)-ribose polymerase inhibition: frequent durable responses in BRCA carrier ovarian cancer correlating with platinum-free interval. J Clin Oncol 28(15):2512–2519
182. Carden CP, Yap TA, Kaye SB (2010) PARP inhibition: targeting the Achilles' heel of DNA repair to treat germline and sporadic ovarian cancers. Curr Opin Oncol 22(5):473–480
183. Patel AG, Sarkaria JN, Kaufmann SH (2011) Nonhomologous end joining drives poly(ADPribose) polymerase (PARP) inhibitor lethality in homologous recombination-deficient cells. Proc Natl Acad Sci U S A 108(8):3406–3411
184. Helleday T (2010) Homologous recombination in cancer development, treatment and development of drug resistance. Carcinogenesis 31(6):955–960
185. Chan N, Pires IM, Bencokova Z et al (2010) Contextual synthetic lethality of cancer cell kill based on the tumor microenvironment. Cancer Res 70(20):8045–8054
186. Helleday T (2011) The underlying mechanism for the PARP and BRCA synthetic lethality: clearing up the misunderstandings. Mol Oncol 5:387–393
187. Bolderson E, Richard DJ, Zhou BB et al (2009) Recent advances in cancer therapy targeting proteins involved in DNA double-strand break repair. Clin Cancer Res 15(20):6314–6320
188. Yap TA, Sandhu SK, Carden CP et al (2011) Poly(ADP-ribose) polymerase (PARP) inhibitors: exploiting a synthetic lethal strategy in the clinic. CA Cancer J Clin 61(1):31–49
189. Fauman EB, Rai BK, Huang ES (2011) Structure-based druggability assessment-identifying suitable targets for small molecule therapeutics. Curr Opin Chem Biol 15(4):463–468

Growth Factor Receptor Signaling, DNA Damage Response, and Cancer Cell Susceptibility to Chemotherapy and Relapses

Moulay Alaoui-Jamali, Amine Saad, and Gerald Batist

Abbreviations

Ask	Apoptosis signal regulated kinase
BER	Base excision repair
ErbB/Her	Tyrosine kinase receptor of the EGFR family
ERCC1	Excision repair cross-complementing gene 1
Erk	Extracellular signal regulated kinase
Gadd45	Growth arrest and DNA damage 45
JNK	c-jun N-terminal kinase
MAPK	Mitogen-activated protein kinase
NER	Nucleotide excision repair
Sapk	Stress-activated protein kinase

1 Introduction

Tumor relapse following primary chemotherapy treatment is an omnious event in most cancer patients undergoing treatment and represents a major clinical challenge in part because relapsed tumors often express aggressive behavior and develop cross-resistance to a wide range of structurally and functionally unrelated agents, which limits the benefit of alternative regimens. In the case of chemotherapy drugs targeting DNA, e.g., alkylating agents and platinums, drug resistant cells develop an impressive arsenal of constitutive and inducible DNA-damage response mechanisms with a broad impact on cell cycle checkpoint and DNA repair mechanisms to

M. Alaoui-Jamali (✉) • A. Saad • G. Batist
Departments of Medicine and Oncology, Lady Davis Institute for Medical Research and Segal Cancer Centre, McGill University, Montreal, QC H3T 1E2, Canada
e-mail: moulay.alaoui-jamali@mcgill.ca

L. Panasci et al. (eds.), *Advances in DNA Repair in Cancer Therapy*, Cancer Drug Discovery and Development 72, DOI 10.1007/978-1-4614-4741-2_3,
© Springer Science+Business Media, LLC 2013

escape chemotherapy-induced cell death. In many instances, DNA damage response signals originate at the plasma membrane as a result of growth factor receptor activation and the signals are then propagated via signal transduction cascades involving a high level of cross talks and feedback loops among distinct signaling pathways. It is therefore predictable that the overexpression of growth factor receptors frequently seen in cancer greatly impacts on chemotherapy response and relapses.

Mechanisms by which growth factor receptor-coupled signaling promote chemotherapy resistance are multifactorial and deregulation of DNA repair pathways represents a major mechanism for certain DNA-interacting drugs, in particular alkylating agents and platinums (cisplatin and carboplatin). The great progress in the characterization of the cell's major DNA repair processes, namely, base excision repair, nucleotide excision repair, double-strand break repair, and recombination repair revealed that most of the DNA repair pathways have protein kinase components directly modulating their activity and are regulated by upstream growth factor receptors, as well as by epigenetic mechanisms [1]. This chapter focuses on DNA damage-activated signaling cascades coupled to growth factor receptors and their connection to chemotherapy-induced DNA damage response and drug resistance. A particular emphasis is given to the family of mitogen activated kinases (Mapk) known to regulate DNA repair mechanisms. The potential impact of Mapk signaling inhibitors on the modulation of DNA damage response and DNA repair in the context of overcoming drug resistance is discussed.

2 Growth Factor Receptor-Coupled Signal Transduction Pathways that Are Points of Convergence for DNA Damage Response and DNA Repair

DNA damage response comprises a network of integrated signaling pathways that regulate a multifaceted response, and its components can be broadly divided as sensors, transducers, and effectors (Fig. 1). Sensors are believed to sense aberrant DNA structures and initiate the global DNA damage response. Unlike yeast, the identity of DNA damage sensors in mammalian cells remains partially understood although Atm, Brca1, the Nbs1-Mre11-Rad50 complex, and some mismatch proteins have been implicated (reviewed in 2). The transducers and effectors involved in regulating the cellular response to DNA damage stress include a variety of kinases and substrates implicated in the regulation of DNA repair, transcription, chromatin remodeling, and cell cycle checkpoints; together they constitute the core of the DNA damage response network. In this context, deregulation of cell transduction pathways secondary to deregulation of upstream growth factor receptors greatly impact on the DNA damage response to promote chemotherapy resistance.

Among various growth factor receptors commonly deregulated in cancer, aberrant expression of EGFR/ErbB receptors (most commonly overexpression/amplification) and/or their ligands has been widely investigated in relation to relapses and progres-

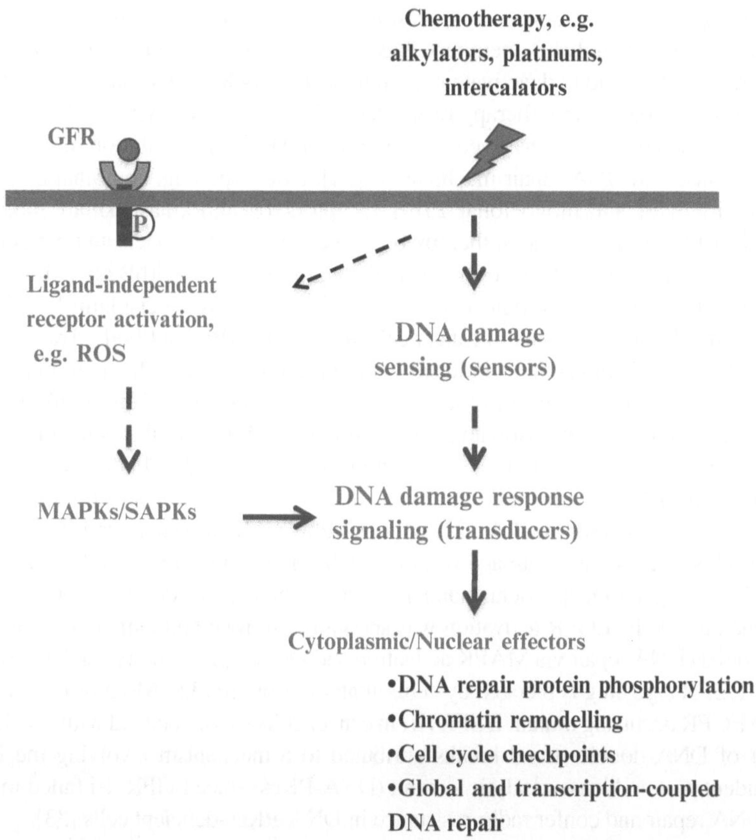

Fig. 1 Tumor cell response to DNA-interacting chemotherapy drugs. The scheme shows major mechanisms by which chemotherapy initiate the global DNA damage response, including DNA lesion sensing by sensors, regulation of DNA damage-associated cell signaling (transducers) and effectors; the later include chromatin remodeling, cell cycle checkpoints, and global and transcription-coupled DNA repair. Growth factor receprtor signaling can contribute to DNA damage response via modulation of DNA damage signaling). Moreover, chemotherapy can active growth factor receptors in a ligand independent manner, e.g., via ROS, or induces growth factor shedding , e.g., pro-EGF. Chemotherapy can also promote receptor nuclear translocation, e.g., phospho-EGFR, leading to regulation of DNA repair enzymes such as DNA-PK. In this context, deregulation of cell transduction pathways secondary to deregulation of upstream growth factor receptors can greatly impact on the DNA damage response including DNA repair to promote either chemotherapy sensitivity or resistance

sion. The EGFR receptor family includes EGFR (ErbB1), ErbB2, ErbB3, and ErbB4 receptors. The existence of a multitude of ErbB receptor ligands (e.g., EGF, HB-EGF, heregulins, BTC, and EPR) and the propensity of these receptors to homo- and heterodimerize lead to the activation of a broad and diverse signaling network [3, 4], which extends beyond members of the ErbB family to include cross talks with receptors such as G protein-coupled receptors [5–7], IGFR [8–10], PDGFR [11, 12]; interleukin receptors [13], and the urokinase-type plasminogen activator (uPAR) [14, 15].

Among ErbB receptors, overexpression of ErbB-2, the preferential dimerization partner for the other ErbB members, has been shown to deregulate the kinetics of MAPK activation and to deactivate in particular the Ras-Raf-ERK and JNK pathways [16], and promotes chemotherapy resistance [17, 23]. The Raf-MEK-ERK, p38, and JNK pathways have been involved in various contexts in the regulation of DNA damage response and DNA repair mechanisms [24]. Not surprising, combination of chemotherapy with anti-monoclonal ErbB receptors or anti-kinase small molecules sensitizes tumor cells to chemotherapy both in experimental models and patients [21, 22, 25–27]. We and others have shown an association between ErbB2 hyperactivation and upregulation of cell repair activity following exposure to cisplatin [18–20, 28, 30–32]. In addition, exposure of cancer cells to the anti-ErbB2 antibody, Trastuzumab, delayed the repair of cisplatin-induced interstrand cross links, which are believed to be critical for cisplatin anticancer activity in contrast to intrastrand cross links [29]. As well, the combination of trastuzumab and cisplatin in clinical trials results in response rates higher than that reported for either single agent alone [27]. This synergistic activity involving DNA repair modulation was also demonstrated using the high-energy α-particle emitting radionuclide (212)Bi (212)Pb-TCMC-trastuzumab [30].

Of relevance to DNA damage response, UV light irradiation of cells can activate EGFR in a ligand-independent manner via a mechanism involving reactive oxygen intermediates [31]. EGFR activation was shown to promote nonhomologous end-joining (NHEJ) DNA repair via MAPK activation and DNA repair activity can be prevented when EGFR signaling is blocked by cetuximab or erlotinib [32]. Moreover, expression of the EGFR occurring mutant EGFRvIII in cancer cells was associated with accelerated repair of DNA double-strand breaks attributed to a mechanism involving the DNA-dependent protein kinase catalytic subunit (DNA-PKcs) since EGFRvIII failed to regulate DNA repair and confer radio-resistance in DNA-PKcs-deficient cells [33].

An alternative mechanism by which ErbB receptors can regulate DNA repair mechanisms is via cell cycle checkpoints. For instance, stimulation with EGF produces proliferative signals in large part due to the activation of the transcription factor AP-1 [34, 35]. The cylin dependent kinase inhibitor $p21^{waf1}$,which is positively regulated by EGFR, is involved in the resistance to bulky adducts induced by cisplatin and its disruption preferentially sensitizes some cell types to cisplatin and nitrogen mustard [36]. However, the relationship between EGFR expression and resistance to cisplatin or -radiation seems to depend on cellular contexts and it has been proposed that a critical level of EGFR signaling, including MAPK activation, is necessary for the regulation of the switch between repair of cisplatin adducts and apoptosis in tumor cells [37–40].

3 Signal Transduction Pathways that Regulate Effectors of the DNA-Damage Response

The protein kinases that are coupled to growth factor receptors to phosphorylate DNA repair and effectors of cell cycle arrest checkpoints can be divided in two major groups. The first group includes kinases activated by damaged DNA and associated with inherent

DNA repair and cell cycle arrest disorders. As such, deficiencies in the gene products of *a*taxia *t*elangiectasia *m*utated (Atm), its homologue Atr, or the *DNA*-dependent *p*rotein *k*inase (Dna-pk) predispose to cancer and correlate with high radiosensitivity and abnormal cell cycle arrest. The role of these kinases in the DNA damage-induced checkpoints has been extensively reviewed [2], but as described in some of the accompanying chapters there is a growing body of information expanding our understanding of the roles of these kinases in the regulation of DNA repair processes. A second very broad group includes the kinases of MAPK and stress-activated signal transduction pathways that are activated by several stress stimuli, including chemotherapy. These kinases include Erks (Fig. 2a, b) as well as the p38 and Sapk serine/threonine kinases (Fig. 3a–d). Depending on the stress, activation of the Sapks of the Mapk superfamily can be the result of growth factor receptor activation, cytoskeletal alterations, or of the signals emanating from the damage-activated kinases, e.g., via the Atm ⇒ Abl ⇒ Map3k pathway. It is therefore evident that the cellular response to stress depends on a multitude of factors, including the unique characteristics of the stress itself, as well as the expression patterns of a vast number of proteins with highly integrated yet often opposing functions. The pathways described herein reflect the complexity and the diversity of phosphorylation-dependent mechanisms that mammalian cells use to deal with chemotherapy-induced DNA damage response.

3.1 Sapk/Mapk Transduction Pathways and the Stress Response

The stress-activated protein kinase (Sapk) has been generally referred to as the c-*j*un *N*-terminal *k*inases (Jnks) and p38. The Jnk (henceforth referred to as Sapk) and p38 kinases are members of the Mapk superfamily. The hierarchical nature of signaling through the superfamily is illustrated by the *r*eceptor *t*yrosine *k*inase (RTK) ⇒ *E*xtracellular signal *r*egulated *k*inase (Erk) pathway (Fig. 2a). In the prototypical cascade, ligand binding stimulates receptor activation (e.g., through dimerization) which results in auto- and trans-phosphorylation of multiple tyrosine residues. These residues are bound by adapter proteins such as the SH2-containing *g*rowth factor *r*eceptor *b*ound protein-*2* (Grb2). Grb2 is bound to *s*on *o*f *s*evenless (Sos) via SH3 domains, and the latter acts as a *g*uanine nucleotide *e*xchange *f*actor (GEF), stimulating the exchange of GDP for GTP bound by Ras, resulting in activation of this kinase. Ras is a farnesylated protein and therefore membrane-bound; upon its activation, it becomes an adaptor itself, recruiting Raf to the membrane. The precise mechanism of Raf-1 activation by Ras is not yet clear, but localization to the membrane as well as other signals emanating from activated Ras are necessary. Raf-1 represents the first component of the Mapk core pathway, a signaling module that is reiterated in several parallel forms responding to a variety of stimuli. The highest-level component of this module is variously labeled Mapkkk, Map3k, or Mekk, and, in the case of Raf-1, its activation results in the amplification of the extracellular signal through Mek1/2 and Erk1/2. As noted above, aberrant growth factor receptors can result in signal amplification and provides upstream control points to modulate both the duration and specificity signaling.

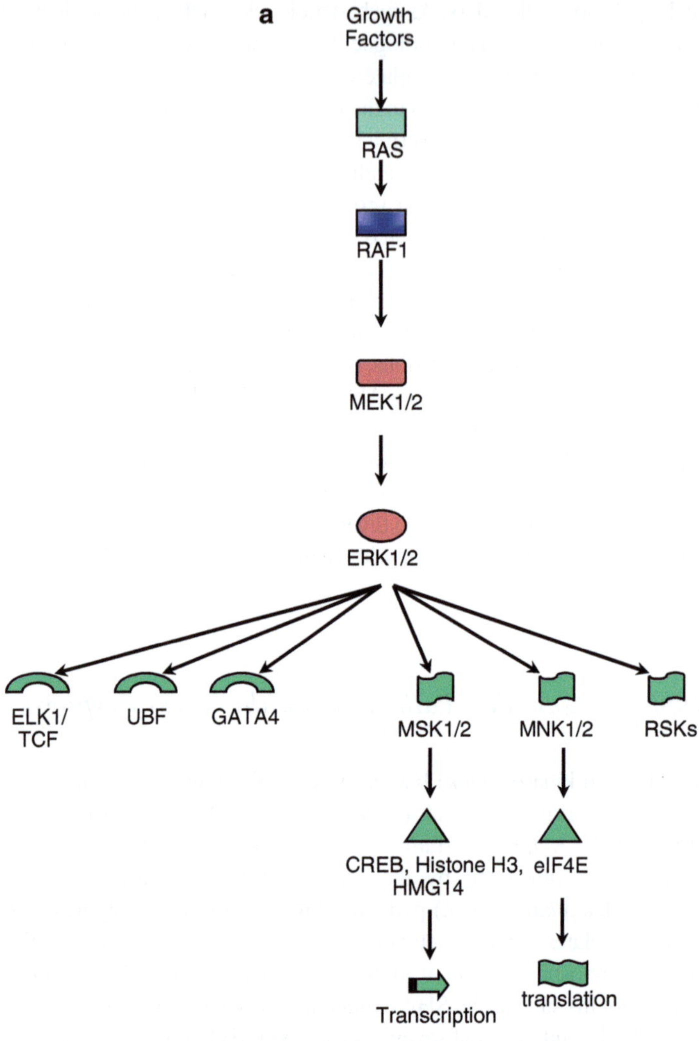

Fig. 2 (**a**) A pared-down representation of the MAP3K>MEK>MAPK cassette. (**b**) Inhibitors of the ERK pathway include a number of MAPK phosphatases, the commonly used MEK1 inhibitor PD98059, and several effector kinases activated by the p38 pathways, such as PRAK and MAPKAP-K2. Note also that PP1 and PP2A have been reported as targets of p38 (see Fig. 3c)

The Sapks (Fig. 3a, b) and p38 (Fig. 3c, d) are strongly activated by cellular stresses, including DNA damage by chemotherapy drugs, oxidative stresses, hyper- and hypoosmolarity, heat shock, anisomycin, heavy metals, and other insults. Indeed, the c-jun N-terminal kinase, Jnk1, was cloned and identified as a kinase phosphorylating c-jun on Ser-63 and Ser-73 [41] following UV irradiation. A simplified impression is that Erk activation results from growth factor stimulation and promotes survival/proliferation, whereas cytotoxic agents activate the Sapks,

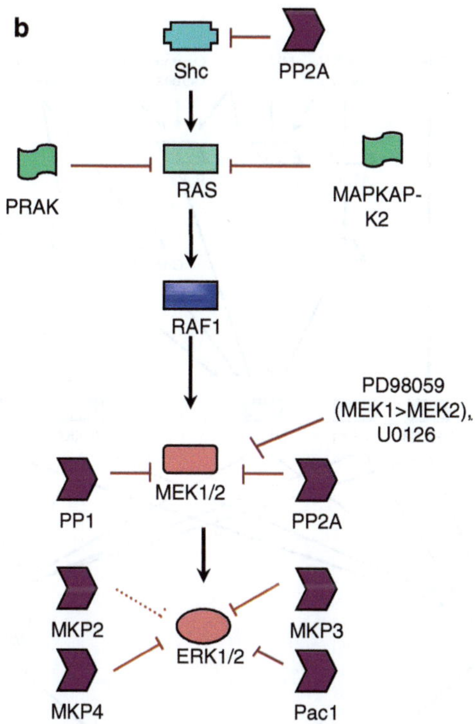

Fig. 2 (continued)

leading to "damage control" or apoptotic responses. Unfortunately, there is a great deal of reiteration between the various pathways, each making distinct (though not yet fully resolved) contributions to survival in response to various stress conditions. Furthermore, a broad "stress," such as chemotherapy, is often multifactorial. These stimuli, for instance, all have concomitant elements of oxidative stress, which also modulate SAPK activation.

Common laboratory models provide a second example of the complexity of DNA damage response . For instance, UV can potentially activate Sapks through at least three mechanisms: direct DNA/protein damage; as a consequence of RTK oligomerization; or via inactivation of phosphatases or other effects of oxidative stress. These phenomena were examined in Rat1 fibroblasts [42, 43] treated with UVB (λ 280–320 nm), with the goal of discerning the contribution of oxidative stress to Sapk activation. Interestingly, the induction of Sapkγ activity was strong and rapid in response to UVB, and the kinetics of this response were similar to those seen upon treatment with the ribotoxin anisomycin, but differed from the slower and more gradual induction seen upon arsenite treatment (an inducer of oxidative stress). These kinetics were mimicked by those of Sek1activation (Sapk activator). Further, the activation of Sapkγ in response to UVB, anisomycin, and IL-1α was not blunted by 30 mM n-acetyl cysteine (NAC) pretreatment, though this was sufficient to completely ablate the

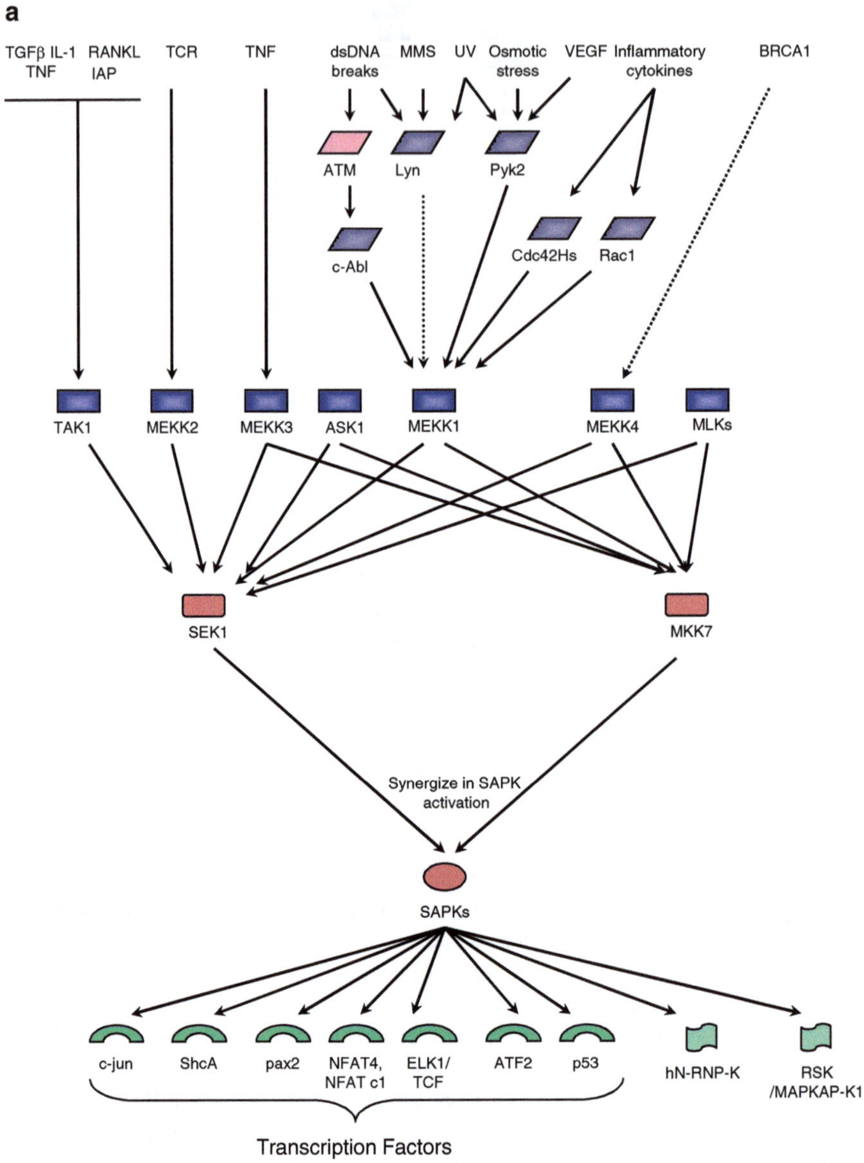

Fig. 3 (**a**) Major activators of SAPK pathways. Activators stimulate MAP3Ks either directly or via kinases upstream of the prototypical MAP3k>MEK>MAPK cassette. A given genotoxic/chemothereapeutic agent may activate one or several MAP3Ks. MEKK1 represents a major point of convergence for signals arising from genotoxic agents. Signals are then transmitted to SEK1 and MKK7 which synergize in the activation of the SAPKs, resulting in effects on transcriptional activity via SAPK-target transcription factors. (**b**) Inhibitors of the SAPK pathway and upstream kinases include physiological inhibitors such as phosphatases and kinases, as well as pharmacological inhibitors. (**c**) Many of the MAP3Ks that stimulate SAPK activation appear to be shared with the p38 pathway, e.g., ASK1, MEKK3, MEKK4, TAK1. Each of these activates MKK3 or MKK6 (or both). MKK3 appears to activate p38α and -β, while MKK6 activates all four p38 isoforms.

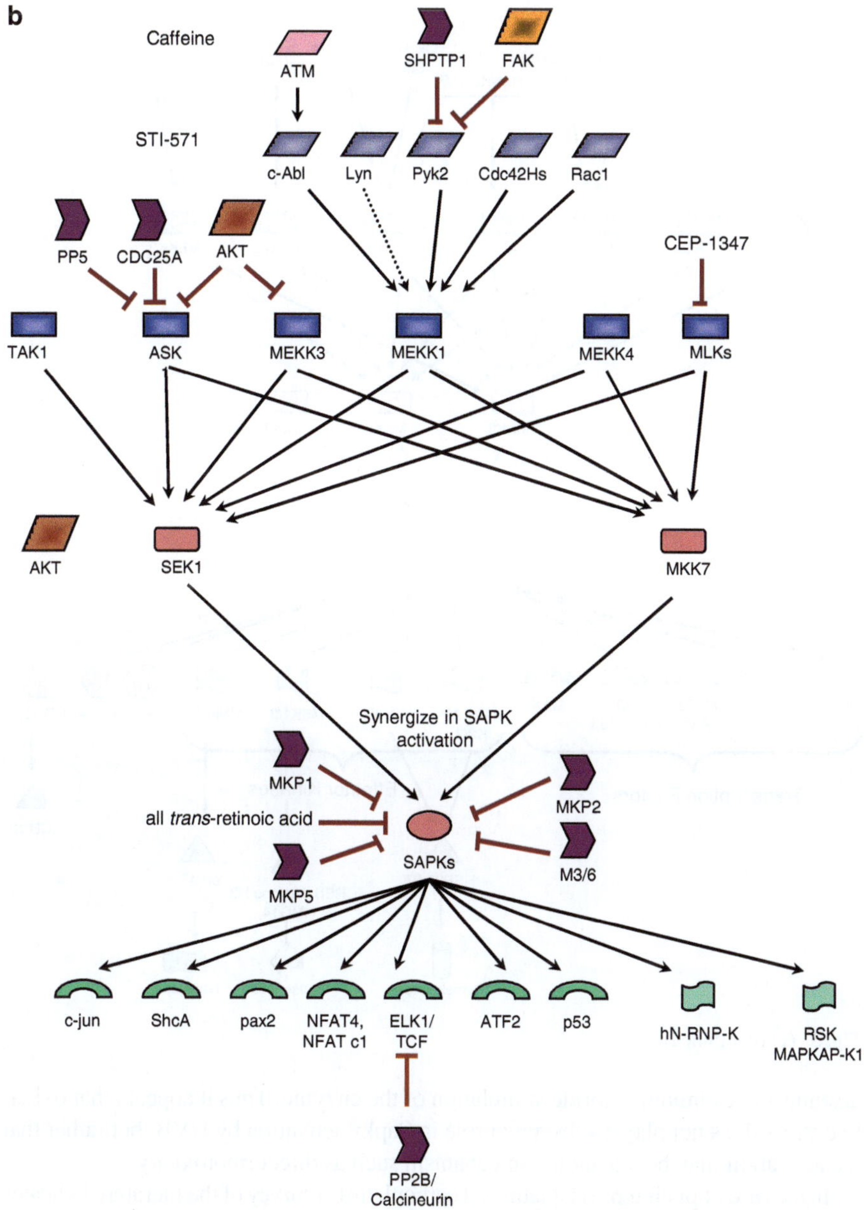

SEK1 has also been reported to activate p38s; its upstream activators are shown in Fig. 3a. p38 kinases exert their impact via several mechanisms, including activation of transcription factors, effector kinases, and phosphatases. In the case of Cdc25, phosphorylation results in inactivation and degradation. (**d**) Inhibitors of the p38 pathway include a wide array of MAPK phosphatases, kinases, and pharmacological inhibitors such as the CSAIDs. A notable characteristic of the CSAIDs is that they inhibit p38α and -β, but have little activity towards p38γ or -δ (i.e., they inhibit the same subset of isoforms activated by MKK3)

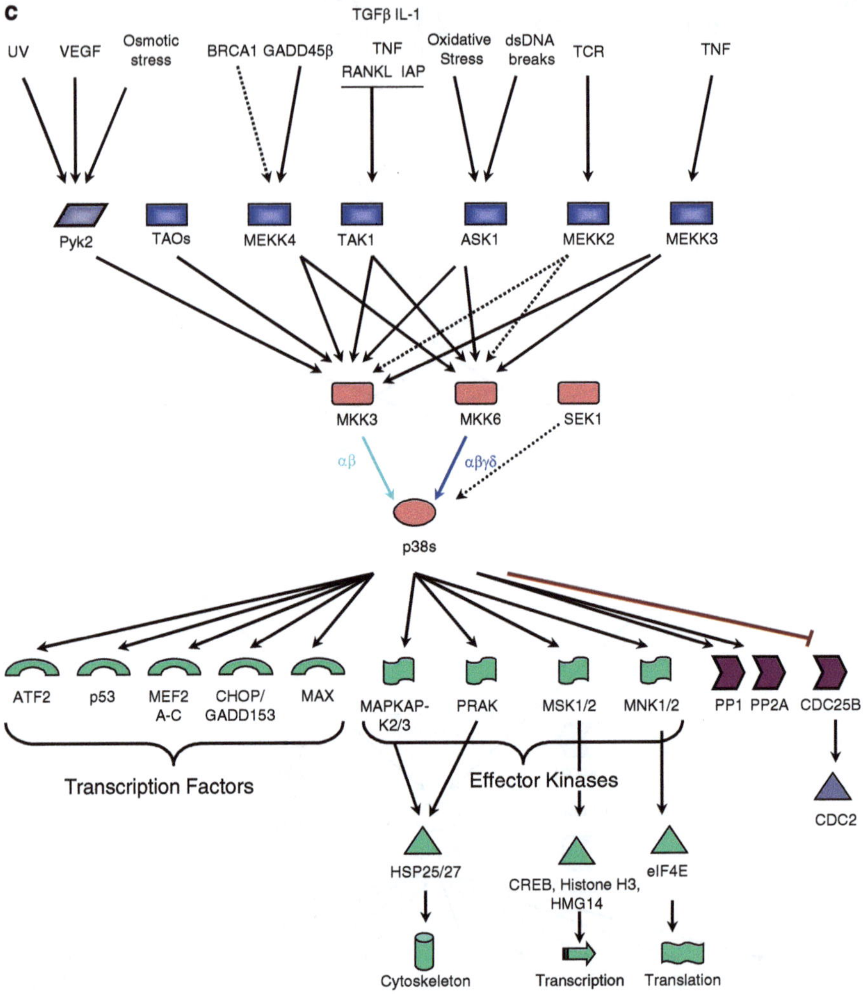

Fig. 3 (**c**) (continued)

arsenite and cadmium chloride stimulation of the enzyme. Thus it appears that oxidative stress does not play a substantial role in Sapkγ activation by UVB, but rather that the activation may be via another mechanism such as direct ribotoxicity.

In the case of platinums (cisplatin and carboplatin), a survey of the literature indicates a broad and varied activation of Mapk/Sapks in response to different chemotherapeutics and DNA-damaging agents. Mitogen-activated protein kinase (MAPK) pathway and DUSP6, a phosphatase involved in dephosphorylation of extracellular signal-regulated kinase (ERK) were reported to regulate the induction of the NER gene, ERCC1,following exposure to cisplatin in melanoma resistant cells [44]. Furthermore, prolonged activation of c-jun N-terminal kinase activity was reported in cells treated with cisplatin but not with transplatin (a therapeutically inactive isomer of cisplatin) [45]. While transplatin

d

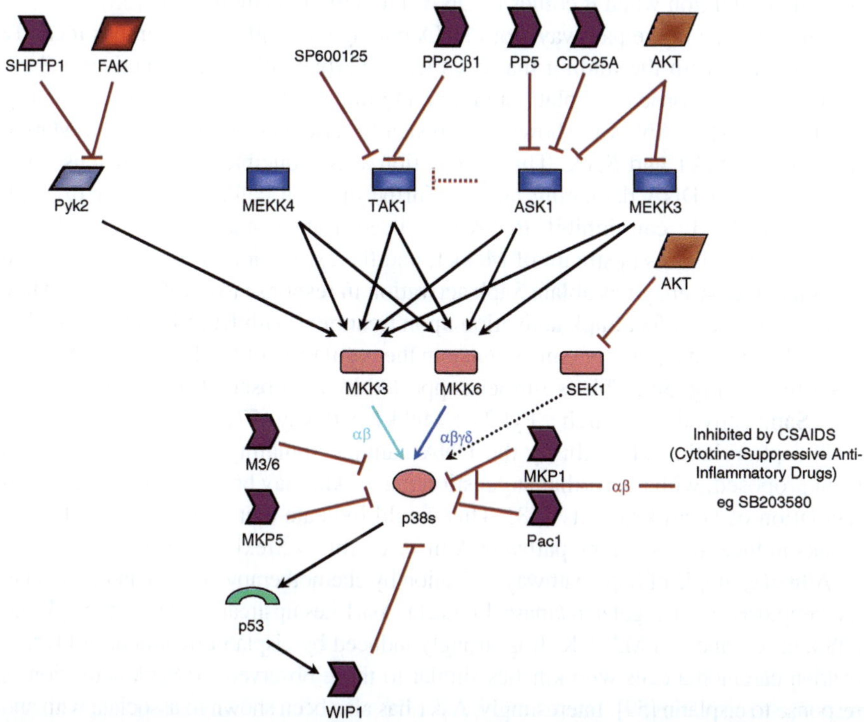

Fig. 3 (**d**) (continued)

produced a rapid and transient increase in c-jun phosphorylation, cisplatin stimulated a more prolonged increase. Furthermore, transplatin was a very effective inducer of MKP-1, (a dual-specificity phosphatase which inactivates p38 and Sapk), while cisplatin induced only a marginal increase in MKP-1 protein levels.

The amplitude of Sapk activation by cisplatin may vary depending on cell lines used, but most reports indicate a similarly prolonged Sapk activation in response to this drug [45–47]. In contrast to the c-jun kinases, there are discrepancies with respect to the reported effects of cisplatin on p38 activity. While some studies reported no effect [48], others observed a strong induction of p38 [46], notably p38γ [49]. This induction is significant with respect to analyses of Sapk pathways because, although it is often considered a minor isoform, p38γ has been suggested to be more efficient in phosphorylating ATF2 than p38α [50, 51]. Equally important, inhibition of the p38 mitogen-activated protein kinase signal, or knockdown of p38 expression was reported to significantly decrease etoposide-induced ERCC1 protein levels and DNA repair capacity in lung cancer cells [52]. Noticeable, the *c*ytokine-*s*uppressive *a*nti-*i*nflammatory *d*rugs (CSAIDs) such as SB203580, which are commonly used to inhibit the p38 kinases, do not inhibit p38γ or -δ, but rather exert their effect exclusively through the predominant isoforms, p38α and -β [51]. Furthermore, the emerging role for p38 as a principal mediator of UV-induced G2/M arrest deserves particular attention when it comes to DNA interacting chemotherapy [53].

Another step in the pathway from DNA damage to Sapk activation has recently been filled in with the finding that cells from c-Abl$^{-/-}$ mice are defective in Sapk activation in response to cisplatin and ionizing radiation (IR), but not inflammatory cytokines [54]. c-Abl was shown to physically associate with Mekk1, a Map3k upstream of Sek1 and Sapk. This association was inducible in the nucleus upon treatment with DNA damaging agents. Furthermore, a cellular inhibitor of Sapk activation, Jip-1, can inhibit Bcr/Abl-induced transformation [55]. Rac1 and Cdc42Hs are kinases upstream of Mekk1, and the expression of dominant negative mutants of these enzymes ablate Sapk activation in response to cytokines [56]. They do not, however, affect Sapk activation upon treatment with IR [54], indicating that the Mekk1 sits at a point of convergence in the regulation of Sapk responses to various stresses (Fig. 3a). This is further supported by the observation that UV stimulates Sapk activation through a Pyk2 ⇒ Mekk1 pathway [57].

Phosphorylation of c-Abl by the DNA damage-signaling kinase Atm has also been proposed, with the further suggestion that c-Abl may be involved in the down-regulation of Dna-pk activity [58]. This would indicate that double-stranded DNA breaks induce Sapks via the pathway Atm ⇒ c-Abl ⇒ Mekk1 ⇒ Sek1 ⇒ Sapk.

A final example of Sapk pathway activation by chemotherapy agents can be found in the *a*poptosis *s*ignal regulated *k*inase-1 (Ask1). Ask1 lies upstream of both the Sapk and p38 kinases, and is a MAP3K. It is strongly induced by cisplatin treatment of Ovcar3 ovarian carcinoma cells with kinetics similar to those observed for Sapk activation in response to cisplatin [59]. Interestingly, Ask1 has also been shown to associate with and phosphorylate Cdc25A [60] (Fig. 2b), a proto-oncogene which is overexpressed in several cancers. In this case, however, the interaction between Cdc25A and Ask1 appears to be independent of the former enzyme's phosphatase activity. Overexpression of Cdc25A or phosphatase-deficient Cdc25A (C430S) resulted in decreased activation of Ask1 in response to the oxidant H_2O_2. This also led to suppression of Sapk and p38 activation in response to this stress. This apparent inhibition of Ask1 by Cdc25A may be a negative feedback mechanism for p38 or Sapks. Though Cdc25A is predominantly nuclear, it has previously been shown to associate with cytoplasmic Raf1 [61], and all three Cdc25 isoforms conditionally associating with 14-3-3 proteins, with the phosphorylation and nuclear export of Cdc25A being a mechanism of Chk1 and Chk2 regulation of its activity in response to UV and γ-radiation, respectively [62, 63]. Furthermore, p38 phosphorylates Cdc25B in the cytoplasm at the G2/M checkpoint in response to UV [53]; thus all three Cdc25 isoforms could potentially participate in cytoplasmic interactions to regulate Sapk signaling.

In summary, the activation of multiple Sapk/Mapks in response to chemotherapy stress can trigger multiple signals whose specificity is often cell type-dependent. The precise role of each kinase in the DNA damage response is, however, somewhat more difficult to discern. This is further complicated by the broad range of DNA damage types induced by anticancer drugs, the high degree of cross talk between the mitogen- and stress-activated protein kinase pathways as well as by the cell heterogeneity observed in cancer tissue.

3.2 *Biological Significance of SAPKs Activation to Chemotherapy Response*

As seen above, DNA damaging agents, including chemotherapy drugs, produce strong activation of Sapk/Mapks through a number of different mechanisms. The activation of these kinases modulate DNA damage response to contribute to either cell death or survival, depending on the context. Expression of a dominant negative (dn) (non-phosphorylatable) c-jun construct was shown to sensitize a cisplatin-resistant cell line proposed to be the result of a repair defect in the dn-c-jun-expressing cells; this association was not observed with the therapeuticaly inactive analogue transplatin [64]. The lack of induction of c-jun kinase activity by transplatin conflicts with the result of Sanchez-Perez et al. [45], though this is in agreement with Hayakawa et al. [47], suggesting that differences in cell lines and assay conditions may be responsible for some of the discrepancy. Of particular relevance, activation of SAPK/JNK was reported to be induced by non-repaired cisplatin adducts in transcribed genes and this led to activation of DNA repair factors including Ataxia telangiectasia mutated- and Rad3-related kinase, and replication protein A [65]. In contrast to the suggested protective role of c-jun in response to cisplatin, a paper by Sanchez-Perez et al. [66] indicates a pro-apoptotic role for c-jun in response to cisplatin. Using a knockout mouse embryonic fibroblast cell model, the authors show that c-jun$^{-/-}$ cells are resistant to cisplatin, but can be sensitized by restoration of c-jun by transfection.

Clearly, some of the effects of Sapk/Mapk activation in response to DNA damaging agents remain to be established, with particular attention given to choice of the cell line, method measuring kinase activation and mechanism of pathway inhibition. The importance of the latter issue is emphasized by the disparate results of studies using dominant negative mutants [47, 64] or knockout cell lines [66] to study the function of jun in response to cisplatin. Moreover, the effect of Sapk can be complicated by the impact on other DNA damage responsive genes such as the *G*rowth *a*rrest and *D*NA *d*amage 45 (Gadd45), a stress-inducible protein regulated by Sapk/Mapks and implicated in G2/M checkpoints, and possibly in DNA repair by modifying chromatin structure [67, 68].

3.3 *Modulation of Sapk/Mapk Activation by DNA Damage, the Case of p53*

p53 regulation (via phosphorylation) by Sapks has broad implications for the regulation for DNA damage response, including DNA repair. The multifunctional tumor suppressor p53 is involved in both DNA repair and cell cycle arrest [69]. Transcriptional control of gene expression by p53 [70] is essential for the cellular response after DNA damage and phosphorylation is limiting to this regulation. In DNA-damaged cells, p53 is phosphorylated on many Serine/Threonine residues resulting in modulation of its affinity for different transcriptional targets. For example, phosphorylation of Ser15 is increased following UV-induced DNA damage and correlates with nuclear shuttling of p53 [71]. Phosphorylation on Ser residues enhances transcription of the Cdk inhibitor p21^{waf1}, which contributes to cell cycle arrest [72]. This implies that this is one mechanism by which deficiencies in kinases upstream of p53, such as Atm and Chk2, result in impairment of DNA damaged-induced cell cycle arrest [73]. Also, loss of p53 function can compromise induction of apoptosis and DNA damage repair resulting in drug resistance, increased mutation, and neoplastic progression.

During genotoxic stress p53 is subject to multiple phosphorylations. Sapk phosphorylation of p53 on Thr-81 is important for p53 stabilization and for its transcriptional activities in response to stress [74]. Both Erk1/2 and p38 have been implicated in the regulation of p53 function in response to NO [75]. However, the phosphorylation of p53 by Pka, Sapks, and CKII is conformation-dependent [76]. The mutations affecting the p53 tumor suppressor genes in Li-Fraumeni syndrome and more than 50% of all sporadic cancers are clustered in the DNA binding domain and affect the transcriptional activity and conformation which in turn is likely to affect its phosphorylation, resulting in inactive forms of p53 [76]. Furthermore, viral oncoproteins functionally inactivate p53 in a large proportion of tumors with genetically intact p53 locus [77–79]. Notably, p53 was shown to enhance sensitivity to EGFR inhibitors via induction of cell-cycle arrest, apoptosis, and DNA damage repair [80]. As such, p53-dependant pathways are attractive targets to manipulate cancer cell response to chemotherapy drugs.

The association of p38 and Erks with p53 in untreated UVB- and UVC treated cell lysates has been reported [81, 82], while the dissociation of p38 from p53 following UV or cisplatin [83] has also been observed. As is often the case, some of these differences may be due to the particular cell lines studied or to the types of UV (UVB versus UVC) used. The most significant difference, however, is that one report suggests p38 and Erk phosphorylation of p53 Ser15 in response to UV and cisplatin [81], while another states that the phosphorylation is primarily on Ser33, not Ser15 [82]. While the former study shows that p38 and Erk can co-precipitate p53, and that their inhibition blocks phosphorylation of p53 on Ser15, the latter shows a similar co-precipitation, and an absence of kinase activity towards an artificial p53 substrate consisting of the first 25 aa of this protein. The solution to the apparent conflict seems to lie in the phosphorylation of Ser33 by p38, which appears to be required for phosphorylation at surrounding sites. In a similar vein, another study examined the effect of Erk inhibition on p53 Ser15 phosphorylation in response to cisplatin [84]. This report suggests that the MAPK/ERK inhibitor PD98059 is more effective than wortmannin (DNA-PK, Atm inhibitor), caffeine (Atr inhibitor) or the p38 inhibitor SB202190 at inhibiting phosphorylation of p53 at Ser15. Further, PD98059 completely ablates both p21waf1 and Mdm2 induction following cisplatin treatment, suggesting a strong effect on p53 transactivation. Again, however, these extensive inhibitor studies involve mostly whole-cell treatments with inhibitors, supporting the conclusion of Bulavin et al. [82], namely, that Mapk phosphorylation of p53 Ser33 coordinates further N-terminal phosphorylations.

The involvement of other Sapks in the response to cisplatin was further investigated using dn-Sek1 to inhibit the Sapk pathway, and SB202190 or SB203580 to inhibit p38α/β [46]. Neither of these treatments had an observable effect on apoptosis, indicating that Erk, but not Sapk or p38, plays a role in inducing apoptosis in response to cisplatin. As is often the case, however, these results do not apply to all cell lines, as PC3 cells show no effect of Erk inhibition on apoptosis, and the in fact these results conflict with the above studies using dn-c-jun and c-jun knockouts [47, 64, 66]. Interestingly, the fact that PC3 cells are p53 mutant may suggest that the mechanism of Erk-dependent apoptosis is via p53, as suggested for p38, and as would be expected given the results of Persons et al., [84] as noted above. This too, must be appraised cautiously, however, given the contrasting findings that PD980159 *sensitizes* Caov-3 (p53 mutant) and A2780 (p53 wt) ovarian carcinoma cells to cisplatin [47], as well as C8161 melanoma cells (p53 wt) [85].

3.4 p38 As a Checkpoint Kinase: Regulation of Two Steps

Consequent to DNA damage, the dual-specificity phosphatase Cdc25A is rapidly degraded, resulting in maintenance of inhibitory phosphorylation on Cdks and delayed transition from G1 to S phase. In the absence of functional p53, the cell cycle resumes concomitant with restoration of Cdc25A expression while arrest can be completely avoided by Cdc25A overexpression [62]. This arrest is therefore reinforced by p53-dependent $p21^{waf1}$ expression, which similarly targets the Cdks, resulting in Rb hypophosphorylation. Phosphorylation of Cdc25 phosphatases creates 14-3-3 binding sites leading to their sequestration in the cytosol following various stresses [86]. Chk1, Chk2, and p38 have been shown to phosphorylate the various Cdc25 isoforms in response to several cellular stresses (reviewed in [87]). In response to ionizing radiation, Cdc25A is phosphorylated by Chk2, while Chk1 phosphorylates Cdc25C [88]. In contrast, upon UV irradiation, Cdc25A is phosphorylated in a Chk1-dependent fashion, representing the first wave of a bipartite G1/S checkpoint [62]. Cdc25B is phosphorylated by p38 following UV treatment, initiating the G2/M checkpoint [53]. As mentioned above, UV-induced checkpoints are reinforced by p38 phosphorylation of p53, coordinating subsequent phosphorylations around the N-terminus of p53. Similarly, p38, Chk1, and Chk2 play a dual role in phosphorylating both p53 and Cdc25s.

In a negative feedback loop, p53 downregulates Chk1 transcription [89], while p38 is inactivated by the p53-inducible Wip1 [90]. Repression of Chk1 by p53 requires p21waf1, since p21waf1 alone is sufficient for this to occur and cells lacking p21waf1 cannot downregulate Chk1[89]. Interestingly, pRb is also required for Chk1 downregulation. p53 and Chk1 play interdependent and complementary roles in regulating both the arrest and resumption of G2 after DNA damage [89]. p53/p21waf1/pRb are also required for maintenance of G2 arrest [91, 92]. Another transcriptional target of p53, the 14-3-3 phospho-binding proteins, is involved in the initiation and maintenance of the G2 arrest by sequestering Cdc25C in the cytoplasm [93]. Although 14-3-3 proteins are not kinases, their cell cycle arrest function relies mostly on kinase activities since they bind phosphorylated proteins with much greater affinity. Thus, the Mapk/Sapks are emerging as regulators of this pathway at several points. Erk and p38 can phosphorylate p53 [81–84], resulting in induction of p21, as well as Gadd45, which is proposed to regulate the G2/M checkpoint by disruption of the Cdc2–cyclinB1 complex [94]. Gadd45 induction following UV is also proposed to be directly mediated by Sapks and Erks, though not p38, in a p53-independent manner [67]. We therefore see each of the major Mapk/Sapk family members playing an important role in checkpoint regulation: Sapk and Erks through Gadd45 induction, p38 and Erks through p53 phosphorylation, and p38 through Cdc25B phosphorylation. Additional effects of these kinases on apoptosis (terminal cell cycle exit), are also apparent, but are beyond the scope of this chapter.

4 Kinases Involved in Phosphorylation of DNA Repair Proteins

In general, mammalian DNA repair proteins are not thought to be transcriptionally inducible, though some show minor induction in specific circumstances. This may be because basal levels of genomic insult are sufficient to require a constantly functioning repair system. It is also logical that DNA repair proteins are primarily regulated posttranscriptionally, since DNA lesions would impede their expression. ERCC1, for example, is induced following cisplatin treatment of A2780 ovarian carcinoma cells, reportedly by a combination of increased transcription and mRNA stabilization [95]. Additionally, some of the enzymes providing the basic building blocks required for repair may be induced following UV irradiation [96]. However, it is likely safe to assume that the major part of repair activity modulation derives from posttranslational modification or association with proteins (e.g., p53), which are stabilized in response to genotoxic insult. Indeed, posttranslational modification of cell cycle checkpoint and DNA repair proteins can stimulate arrest and repair via several mechanisms [97, 98]. The phosphorylation status of these proteins can modulate their stability [99], complex formation, subcellular localization [100, 101], catalytic activity [102], DNA binding affinity and transcriptional activity [103], as well as structural remodeling affecting both the protein and chromatin structure [104, 105]. As such, the kinases of the signal transduction pathways activated by genotoxic stress will directly or indirectly modulate DNA repair and cell cycle.

Nucleotide excision repair (NER) is modulated by phosphorylation and kinase inhibitors.

Activation of p53 by phosphorylation is important for efficient DNA repair. To date, however, little is know regarding the modulation of DNA repair activities due to phosphorylation of DNA repair proteins per se. It has been shown that nucleotide excision repair is inhibited by phosphorylation (via CAK phosphorylation of repair components) and that the inhibition of CAK by the cyclic nucleotide protein kinase inhibitor, H-8, restores the NER activity to original levels [106], suggesting that the activity of the NER can be downregulated by phosphorylation. This is an important finding to understand the controversial role of the p53-regulatory pathway and specifically its downstream effector, p21waf1, in the regulation of NER. The role of CAK in NER provides a link between p21waf1 and NER since high levels of p21waf1 can inhibit CAK in vivo [92], which in turn should increase NER activity. This model would support studies describing the contribution of p21waf1 in NER [107, 108]. This is in apparent contrast with other studies showing that p21waf1 has little effect [108–110] or an inhibitory role in NER [111]. In addition, a recent study showed by LM-PCR that basal levels of p21waf1 inhibited NER in a p53-deficient background [112]. This deficiency in NER may be interpreted on the basis of the CAK/NER complex interaction.

The inhibition of NER has been used as an approach to improve cisplatin-based chemotherapy, particularely in cisplatin resistant tumors. Unfortunately, treatment with cisplatin and other bulky adduct inducing drugs, e.g., alkylating agents, is inconsistently successful despite frequent low NER capacity in tumor cells due to p53 deficiencies (50% of all cancers) which impairs both global genomic NER [113] and transcription–coupled NER [114]. Other repair pathways such as base excision repair (BER) can act on damage preferentially repaired by NER, perhaps representing a mechanism by which to overcome NER-deficiency. It should be noted, however, that p53 plays a direct role in BER by stabilizing the interaction between DNApolβ and abasic DNA [115, 116]. Therefore, p53 deficiencies would compromise BER as well, and this repair mechanism is unlikely to compensate for NER lost in a p53 deficient background.

4.1 p53-Dependent DNA Repair

Gadd45 and p21waf1 are two DNA-damage inducible genes that can be induced via both p53-dependent and –independent pathways. Many studies implicate those two stress-inducible proteins in NER and apoptosis [117]. In vivo p21waf1 can be phosphorylated by protein kinase B (Akt/Pkb) [118], an anti-apoptotic kinase. Both Gadd45 and p21waf1 interact with PCNA [119, 120], which is known to affect cell cycle progression by supporting DNA repair and, indirectly, survival. An additional function of Gadd45 is to bind to UV-damaged chromatin, which affect lesion accessibility [117]. A direct role for p21waf1 phosphorylation in NER has not been addressed; however, phosphorylation by mitogen-activated protein (MAP) kinases is involved in the induction of the Gadd45 promoter after DNA damage [67]. Similarly, inhibition of Sapkγ and Erk kinase activities either by expression of a dominant negative mutant Sapkγ or by treatment with a selective chemical inhibitor of Erk (PD098059) substantially abrogates the UV induction of the Gadd45 promoter [67]. P53-independent induction of Gadd45 [121] and p21waf1 [122] has been described following DNA damage, including treatment with cisplatin [123]. Notably, colon carcinoma is characterized by frequent p53 and mismatch repair deficiencies. The p53-dependent upregulation of human mismatch repair gene MSH2 in UV-irradiated colon carcinoma cells depends on a functional interaction with c-jun [124],(Although UV is not a therapeutic agent, some of its properties may reflect those of more relevant chemotherapeutic agents). As described above, the c-jun kinases (Sapks) are activated by many cellular stresses, including cisplatin.

4.2 Replication Protein A

Among the many proteins involved in NER, replication protein A (RPA) is one factor known to be phosphorylated after DNA damage, though the kinase(s) responsible have not yet been determined. The single stranded DNA binding protein RPA is a multifunctional hetero-trimer involved in NER [106, 125] replication, and repair of strand breaks [126, 127]. RPA is modified by phosphorylation during replication [128] and the DNA damage response [129]. In particular, the 32-kDa subunit is phosphorylated following UVC [130]. Hyperphosphorylation of RPA has been observed in cells from patients with either GGR or transcription-coupled repair (TCR) deficiency (A, C, and G complementation groups of Xeroderma pigmentosum and A and B groups of Cockayne syndrome, respectively). This excludes both intermediates in the NER pathway and signals from stalled transcription as essential signals for RPA hyperphosphorylation. However, UV-sensitive cells deficient in NER and TCR require lower doses of UV irradiation to induce RPA32 hyperphosphorylation than normal cells, suggesting that persistent unrepaired lesions contribute to RPA phosphorylation. UVC irradiation experiments on nonreplicating cells and S-phase-synchronized cells emphasize a role for DNA replication arrest in the presence of UV-induced lesions in RPA UV-induced hyperphosphorylation in mammalian cells [130]. One might therefore speculate that inhibition of RPA phosphorylation could improve treatments inducing NER-substrate lesions.

4.3 O^6-alkylguanine-DNA Alkyltransferase

The expression of O^6-alkylguanine-DNA alkyltransferase (AGT) (26) a DNA repair protein that confers tumor resistance to many anticancer alkylating agents is upregulated in the absence of p53 [131] and frequently overexpressed in oral cancer cells genetically and functionally deficient for p53 [132]. p53 thus acts as a repressor of AGT expression, whereas the activators of Pkc, phorbol-12-myristate-13-acetate (PMA), and 1,2-diacyl-sn-glycerol (DAG), as well as the protein phosphatase inhibitor, okadaic acid (OA), increase the transcriptional level of AGT(27). The activity of AGT is inhibited by phosphorylation that can be catalyzed by Pka, Pkc, and/or CKII [133]. Thus the activation of these kinases may impair the elimination of akylated DNA lesions.

4.4 Blm Helicase

Bloom's syndrome (BS), a rare genetic disease, arises through mutations in both alleles of the Blm gene which encodes a 3′-5′ DNA helicase. BS patients exhibit a high predisposition to development of all types of cancer affecting the general population and Blm-deficient cells display a strong genetic instability. Blm participates in the cellular response to ionizing radiation. The Blm defect is associated with a partial escape of cells from the γ-irradiation-induced G2/M cell cycle checkpoint. In response to ionizing radiation, Blm protein is phosphorylated and accumulates through an Atm-dependent pathway [134]. Caffeine, by inhibiting Atm and its homologue Atr [135, 136] enhances the radiosensitivity of cells in part by altering the phosphorylation of Blm, in addition to its effects on Chk1 and Chk2, as outlined above.

5 Histone Modifiers Implicated in the Regulation of DNA Damage Response and DNA Repair

In response to DNA damage, detection of lesions and repair of DNA must occur in the context of chromatin ultrastructure. Folding into chromatin alters the accessibility of the DNA to proteins involved in DNA transactions. Likewise, several mechanisms have evolved to regulate the chromatin-packaged state of DNA under stress response. These include: covalent histone modifications, ATP-dependent chromatin remodeling and histone variant incorporation. In this context, genetic studies have revealed that mutants of histone modifying proteins and chromatin remodellers often show sensitivity to genotoxic agents. Covalent histone modification, including histone phosphorylation, methylation, acertylation, or ubiquitination are associated with DNA damage response, including checkpoint mechanisms. As reviewed by Costelloea et al. [137], several enzymes associated with chromatin remodeling are involved in the DNA damage response, in particular, the DNA double strand break (DSB). These include: Mec1 (Stable retention of DDR checkpoint proteins at DSB, DSB repair), CK2 (DNA damage regulated kinase that phosphorylates H4S1, linked to histone deacetylation), Dot 1 (Required for 53Bp1, Rad9 recruitment to DSB, checkpoint activation in S. cerevisiae, marks active chromatin), Set1 (Involved in checkpoint activation in *S. cerevisiae*, H3K4me2, H3K4me3, mark 5′ region of active genes, H3K4me1 localized to silenced chromatin), Esa1, Gcn5, and Hat1. For instance, extensive phosphorylation of H2A(X) is amongst the early events following DSB. γH2A(X) is necessary for the damage-induced focal accumulation of proteins involved in checkpoint signaling, DNA repair, as well as chromatin remodeling. Importantly, this modification is not needed for the initial recruitment to DSBs of key DDR proteins believed to be involved in DNA damage sensing, such as Nbs1 or 53Bp1. Some proteins, for example Mdc1, bind directly to γH2A(X) via an interaction between the Ser139 phosphate and the BRCT domains of Mdc1. However, direct interaction with γH2A(X) has not been demonstrated for all proteins recruited to the site of DNA damage. The recruitment of many of these proteins may be facilitated through other histone modifications.

6 Circumventing Chemotherapy Resistance by Targeting Signaling Molecules That Modulate DNA Damage Response and DNA Repair

The concerted role of plasma membrane growth factor receptors in the regulation of multiple proliferative and survival pathways, including response to chemotherapy-induced DNA damage and DNA repair, make growth factor receptor and their coupled signaling components attractive targets to modulate chemotherapy response in refractory/drug resistant cancer. Several anti-receptors are currently approved for clinical use, either small molecules or antibodies. Also an increasing number of novel inhibitors targeting components of downstream signaling, including Mapk/Sapk, have been identified, some of which have entered or completed clinical trials (reviewed in 138, 139). This exciting progress provides opportunities to exploit this knowledge in formulating alternative combinatorial regimens to sensitize resistant cancer cells to chemotherapy. As noted above, combination of chemotherapy with anti-monoclonal ErbB receptors or small molecule kinase inhibitors sensitize tumor cells to specific DNA-interacting chemotherapy drugs in particular platinums [21, 22, 25–27]. The anti-ErbB2 antibody,Trastuzumab, inhibited or delayed the repair of cisplatin-induced DNA damage, as well as γ-radiation, and enhanced cytotoxicity of cisplatin in preclinical models and in patients [18, 19, 27–30]. A similar result was reported for the EGFR inhibitor, Gefitinib, and cisplatin in ovarian cancer cells [140]. Modulation of DNA repair by ErbB receptors was found to be mediated, at least in part, via the MAK pathway [19, 32], suggesting that targeting MAPKs should achieve therapeutic benefit as well. Interestingly, a variety of MAPK inhibitors are being developed, including noncompetitive inhibitors of MEK1 and MEK2 such as PD98059, PD184352, and U0126; competitive inhibitors of MEK1 and MEK2 such as Ro092210 and LLZ16402; AZD6244 (Selumetinib; locks MEK1/2 into an inactive conformation to prevent ERK phosphorylation), RDEA-119 (an allosteric inhibitor of MEK1/2 with activity when administered by oral route), SP600125 which inhibits Jun-N-terminal kinase 2 (JNK2); CEP1347 (KT7515) which inhibit multiple MAPK kinases (MLK1, 2 and 3), and others targeting p38 (e.g., several p38 inhibitors are being evaluated in clinical trials mostly for rheumatoid arthritis or psoriasis, including Vertex 745 (VX745), RPR200765A, SB235699, and SCIO469. The availability of these targeted molecules certainly opens-up exciting directions to investigate their relevance to the DNA damage response and DNA repair in the context of improving therapeutic response in chemotherapy refractory cancers. In order to selectively sensitize tumors and not normal tissue to chemotherapy, the overexpression of these receptor targets for "chemosensitization" would have to be determined in the tumor. This represents an emerging theme of tumor profiling that ensures that the correct patient sub-group is treated.

The use of modulators that target directly DNA repair mechanisms has also provided encouraging results to modulate chemotherapy response. For instance, the triple-negative breast cancer (TNBC), which accounts for up to 20% of all breast cancers, is an aggressive subtype of breast cancer where targeted therapies used for hormone receptor-positive and HER2-overexpressing breast cancers are ineffective and with a high incidence of relapse to conventional chemotherapy. DNA reacting drugs such as cisplatin have been shown to be effective in the neoadjuvant setting for TNBC but again relapses are common. The poly (ADP-ribose) polymerase-1 (PARP1), an enzyme involved in DNA repair, is significantly increased in TNBC and other cancer types [141]. Encouraging results have been achieved when chemotherapeutic agents such as platinums are combined with PARP1 inhibitors in TNBC [142]. Of note, potential benefits of a combination of EGFR antibody, Cetuximab, and PARP1 inhibitors was reported in head and neck cancer; this synergistic effects was shown to occur via modulation of nonhomologous end-joining (NHEJ)- and homologous recombination (HR)-mediated DNA double strand break (DSB) repair. Once again, profiling the tumor for the overexpression of the target will help ensure that patient selection is optimal for clinical benefit. [143].

Finally, targeting cell cycle checkpoints required for DNA repair via inhibition of checkpoint mechanisms addressed above is another potential strategy to interfere with DNA repair, e.g., p38, CDC25B, and protein kinase C to abrogate G2 arrest through a Cdc2-dependent pathway. The Atm and Atr kinases and their downstream effectors Chk1 and Chk2 all are also appealing targets to enhance chemotherapy response.

7 Concluding Remarks and Perspectives

The progress in the characterization of DNA damage response signaling and resolution of crystal structures of DNA repair proteins provides exciting avenues toward discovery of selective molecules targeting specific levels of the DNA damage response and with the potential to overcome resistance to DNA-interacting chemotherapy such as platinums, which represent a major class of anticancer agents. An alternative strategy is based on the evidence that receptor signaling-mediated chemo-resistance involves, at least in part, impaired cell cycle checkpoints, increased DNA repair, and/or downregulation of the apoptotic threshold. As noted above, growth factor receptor upregulation is a frequent cause of innate resistance in many types of cancers. Inhibition of the upstream receptors can not only interfere with the proliferative signals but also render cells more susceptible to drug-induced apoptosis. Finally, inhibitors of cell cycle checkpoints are certainly an important Achille's heel of tumor resistance to genotoxic chemotherapy drugs and therefore represent a promising avenue for future therapies. Finally, the potential of incorporating DNA damage response modulators in chemotherapy regimens for refractory or relapsed cancers is not without dilemma. Clinical experience with all anticancer drugs, including targheted agents and therapeutic antibodies revealed to cancer cells are masters of developping alteranative mechanisms to escape cell death. Resistance to inhibitors of DNA-damage associated signaling inhibitors has been documented to occurs via mutations in the target genes, feedback regulatory or compensatory mechanisms [144–148]. Also, heterogeneity in the level and activity of signaling molecules between tumor cell subpopulations can result in differential effect of the inhibitors. Therefore, resistance to signaling molecules can underscore the potential of a combined therapeutic approach and add to the endless cycle of drug resistance. Certainly, defining genetic alterations of individual tumors is becoming prerequisite to maximizing therapeutic efficacy in the era of individualized medicine, and targeting DNA damage response signaling is no exception. Here too, the emerging tenets of personalized medicine require some means to assess tumors for the overexpression of the targets of chemosensitization, wherther by biopsy, in blood, or by a yet-to-be-defined noninvasive imaging approach, so that there is selective effects on tumor and not normal tissue.

Acknowledgments Work cited from this laboratory on the regulation of chemotherapy stress response was supported by grants from the Canadian Institutes for Health Research (CIHR), the Canadian Cancer Research Society, and the Quebec Breast Cancer Foundation.

References

1. Wilting RH, Dannenberg JH (2012) Epigenetic mechanisms in tumorigenesis, tumor cell heterogeneity and drug resistance. Drug Resist Updat 15(1–2):21–38
2. Shiloh Y (2001) ATM and ATR: networking cellular responses to DNA damage. Curr Opin Genet Dev 11(1):71–77
3. Yarden Y, Sliwkowski MX (2001) Untangling the ErbB signalling network. Nat Rev Mol Cell Biol 2(2):127–137
4. Riese DJ 2nd, Stern DF (1998) Specificity within the EGF family/ErbB receptor family signaling network. Bioessays 20(1):41–48
5. Gschwind A, Zwick E, Prenzel N, Leserer M, Ullrich A (2001) Cell communication networks: epidermal growth factor receptor transactivation as the paradigm for interreceptor signal transmission. Oncogene 20(13):1594–1600
6. Luttrell LM, Daaka Y, Lefkowitz RJ (1999) Regulation of tyrosine kinase cascades by G-protein-coupled receptors. Curr Opin Cell Biol 11(2):177–183
7. Carpenter G (1999) Employment of the epidermal growth factor receptor in growth factor-independent signaling pathways. J Cell Biol 146(4):697–702
8. Balañá ME, Labriola L, Salatino M, Movsichoff F, Peters G, Charreau EH, Elizalde PV (2001) Activation of ErbB-2 via a hierarchical interaction between ErbB-2 and type I insulin-like growth factor receptor in mammary tumor cells. Oncogene 20(1):34–47
9. Albanell J, Baselga J (2001) Unraveling resistance to trastuzumab (Herceptin): insulin-like growth factor-I receptor, a new suspect. J Natl Cancer Inst 93(24):1830–1832
10. Yu H, Rohan T (2000) Role of the insulin-like growth factor family in cancer development and progression. J Natl Cancer Inst 92(18):1472–1489
11. Saito Y, Haendeler J, Hojo Y, Yamamoto K, Berk BC (2001) Receptor heterodimerization: essential mechanism for platelet-derived growth factor-induced epidermal growth factor receptor transactivation. Mol Cell Biol 21(19):6387–6394
12. Li J, Kim YN, Bertics PJ (2000) Platelet-derived growth factor-stimulated migration of murine fibroblasts is associated with epidermal growth factor receptor expression and tyrosine phosphorylation. J Biol Chem 275(4):2951–2958
13. Qiu Y, Ravi L, Kung HJ (1998) Requirement of ErbB2 for signalling by interleukin-6 in prostate carcinoma cells. Nature 393(6680):83–85
14. Chandrasekar N, Mohanam S, Gujrati M, Olivero WC, Dinh DH, Rao JS (2003) Downregulation of uPA inhibits migration and PI3k/Akt signaling in glioblastoma cells. Oncogene 22(3):392–400

15. Liu D, Aguirre Ghiso J, Estrada Y, Ossowski L (2002) EGFR is a transducer of the urokinase receptor initiated signal that is required for in vivo growth of a human carcinoma. Cancer Cell 1(5):445–457
16. Olayioye MA, Neve RM, Lane HA, Hynes NE (2000) The ErbB signaling network: receptor heterodimerization in development and cancer. EMBO J 19(13):3159–3167
17. Arteaga CL, Winnier AR, Poirier MC, Lopez-Larraza DM, Shawver LK, Hurd SD, Stewart SJ (1994) p185c-erbB-2 signal enhances cisplatin-induced cytotoxicity in human breast carcinoma cells: association between an oncogenic receptor tyrosine kinase and drug-induced DNA repair. Cancer Res 54(14):3758–3765
18. Yen L, Nie ZR, You XL, Richard S, Langton-Webster BC, Alaoui-Jamali MA (1997) Regulation of cellular response to cisplatin-induced DNA damage and DNA repair in cells overexpressing p185(erbB-2) is dependent on the ras signaling pathway. Oncogene 14(15):1827–1835
19. You XL, Yen L, Zeng-Rong N, Al Moustafa AE, Alaoui-Jamali MA (1998) Dual effect of erbB-2 depletion on the regulation of DNA repair and cell cycle mechanisms in non-small cell lung cancer cells. Oncogene 17(24):3177–3186
20. Pietras RJ, Fendly BM, Chazin VR, Pegram MD, Howell SB, Slamon DJ (1994) Antibody to HER-2/neu receptor blocks DNA repair after cisplatin in human breast and ovarian cancer cells. Oncogene 9(7):1829–1838
21. Pietras RJ, Pegram MD, Finn RS, Maneval DA, Slamon DJ (1998) Remission of human breast cancer xenografts on therapy with humanized monoclonal antibody to HER-2 receptor and DNA-reactive drugs. Oncogene 17(17):2235–2249
22. Pietras RJ, Poen JC, Gallardo D, Wongvipat PN, Lee HJ, Slamon DJ (1999) Monoclonal antibody to HER-2/neureceptor modulates repair of radiation- induced DNA damage and enhances radiosensitivity of human breast cancer cells overexpressing this oncogene. Cancer Res 59(6):1347–1355
23. Tsai CM, Chang KT, Li L, Perng RP, Yang LY (2000) Interrelationships between cellular nucleotide excision repair, cisplatin cytotoxicity, HER-2/neu gene expression, and epidermal growth factor receptor level in non-small cell lung cancer cells. Jpn J Cancer Res 91(2):213–222
24. Wei B, Zhou Y, Xu Z, Xi B, Cheng H, Ruan J, Zhu M, Hu Q, Wang Q, Wang Z, Yan Z, Jin K, Zhou D, Xuan F, Huang X, Shao J, Lu P (2011) The effect of hOGG1 Ser326Cys polymorphism on cancer risk: evidence from a meta-analysis. PLoS One 6(11):e27545
25. Christmann M, Pick M, Lage H, Schadendorf D, Kaina B (2001) Acquired resistance of melanoma cells to the antineoplastic agent fotemustine is caused by reactivation of the DNA repair gene MGMT. Int J Cancer 92(1):123–129
26. Boldogh I, Ramana CV, Chen Z, Biswas T, Hazra TK, Grösch S, Grombacher T, Mitra S, Kaina B (1998) Regulation of expression of the DNA repair gene O6-methylguanine-DNA methyltransferase via protein kinase C-mediated signaling. Cancer Res 58(17):3950–3956
27. Pegram MD, Slamon DJ (1999) Combination therapy with trastuzumab (Herceptin) and cisplatin for chemoresistant metastatic breast cancer: evidence for receptor-enhanced chemosensitivity. Semin Oncol 26(4 Suppl 12):89–95
28. Zeng-Rong N, Paterson J, Alpert L, Tsao MS, Viallet J, Alaoui-Jamali MA (1995) Elevated DNA repair capacity is associated with intrinsic resistance of lung cancer to chemotherapy. Cancer Res 55(21):4760–4764
29. Boone JJ, Bhosle J, Tilby MJ, Hartley JA, Hochhauser D (2009) Involvement of the HER2 pathway in repair of DNA damage produced by chemotherapeutic agents. Mol Cancer Ther 8(11):3015–3023
30. Yong KJ, Milenic DE, Baidoo KE, Brechbiel MW (2012) 212Pb-Radioimmunotherapy Induces G2 Cell-Cycle Arrest and Delays DNA Damage Repair in Tumor Xenografts in a Model for Disseminated Intraperitoneal Disease. Mol Cancer Ther 11(3):639–648
31. Huang RP, Wu JX, Fan Y, Adamson ED (1996) UV activates growth factor receptors via reactive oxygen intermediates. J Cell Biol 133(1):211–220

32. Kriegs M, Kasten-Pisula U, Rieckmann T, Holst K, Saker J, Dahm-Daphi J, Dikomey E (2010) The epidermal growth factor receptor modulates DNA double-strand break repair by regulating non-homologous end-joining. DNA Repair (Amst) 9(8):889–897
33. Mukherjee B, McEllin B, Camacho CV, Tomimatsu N, Sirasanagandala S, Nannepaga S, Hatanpaa KJ, Mickey B, Madden C, Maher E, Boothman DA, Furnari F, Cavenee WK, Bachoo RM, Burma S (2009) EGFRvIII and DNA double-strand break repair: a molecular mechanism for radioresistance in glioblastoma. Cancer Res 69(10):4252–4259
34. Chin YE, Kitagawa M, Su WC, You ZH, Iwamoto Y, Fu XY (1996) Cell growth arrest and induction of cyclin-dependent kinase inhibitor p21 WAF1/CIP1 mediated by STAT1. Science 272(5262):719–722
35. Bromberg JF, Fan Z, Brown C, Mendelsohn J, Darnell JE Jr (1998) Epidermal growth factor-induced growth inhibition requires Stat1 activation. Cell Growth Differ 9(7):505–512
36. Fan S, Chang JK, Smith ML, Duba D, Fornace AJ Jr, O'Connor PM (1997) Cells lacking CIP1/WAF1 genes exhibit preferential sensitivity to cisplatin and nitrogen mustard. Oncogene 14(18):2127–2136
37. Dixit M, Yang JL, Poirier MC, Price JO, Andrews PA, Arteaga CL (1997) Abrogation of cisplatin-induced programmed cell death in human breast cancer cells by epidermal growth factor antisense RNA. J Natl Cancer Inst 89(5):365–373
38. Lammering G, Hewit TH, Hawkins WT, Contessa JN, Reardon DB, Lin PS, Valerie K, Dent P, Mikkelsen RB, Schmidt-Ullrich RK (2001) Epidermal growth factor receptor as a genetic therapy target for carcinoma cell radiosensitization. J Natl Cancer Inst 93(12):921–929
39. Reardon DB, Contessa JN, Mikkelsen RB, Valerie K, Amir C, Dent P, Schmidt-Ullrich RK (1999) Dominant negative EGFR-CD533 and inhibition of MAPK modify JNK1 activation and enhance radiation toxicity of human mammary carcinoma cells. Oncogene 18(33):4756–4766
40. Harari PM, Huang SM (2001) Head and neck cancer as a clinical model for molecular targeting of therapy: combining EGFR blockade with radiation. Int J Radiat Oncol Biol Phys 49(2):427–433
41. Derijard B, Hibi M, Wu IH, Barrett T, Su B, Deng T, Karin M, Davis RJ (1994) JNK1: a protein kinase stimulated by UV light and Ha-Ras that binds and phosphorylates the c-Jun activation domain. Cell 76(6):1025–1037
42. Butterfield L, Storey B, Maas L, Heasley LE (1997) c-Jun NH2-terminal kinase regulation of the apoptotic response of small cell lung cancer cells to ultraviolet radiation. J Biol Chem 272(15):10110–10116
43. Iordanov MS, Magun BE (1999) Different mechanisms of c-Jun NH(2)-terminal kinase-1 (JNK1) activation by ultraviolet-B radiation and by oxidative stressors. J Biol Chem 274(36):25801–25806
44. Li W, Melton DW (2011) Cisplatin regulates the MAPK kinase pathway to induce increased expression of DNA repair gene ERCC1 and increase melanoma chemoresistance. Oncogene 17(24):3177–3186. doi:10.1038/onc.2011.426
45. Sanchez-Perez I, Murguia JR, Perona R (1998) Cisplatin induces a persistent activation of JNK that is related to cell death. Oncogene 16(4):533–540
46. Wang X, Martindale JL, Holbrook NJ (2000) Requirement for ERK activation in cisplatin-induced apoptosis. J Biol Chem 275(50):39435–39443
47. Hayakawa J, Ohmichi M, Kurachi H, Ikegami H, Kimura A, Matsuoka T, Jikihara H, Mercola D, Murata Y (1999) Inhibition of extracellular signal-regulated protein kinase or c-Jun N-terminal protein kinase cascade, differentially activated by cisplatin, sensitizes human ovarian cancer cell line. J Biol Chem 274(44):31648–31654
48. Persons DL, Yazlovitskaya EM, Cui W, Pelling JC (1999) Cisplatin-induced activation of mitogen-activated protein kinases in ovarian carcinoma cells: inhibition of extracellular signal-regulated kinase activity increases sensitivity to cisplatin. Clin Cancer Res 5(5):1007–1014
49. Pillaire MJ, Nebreda AR, Darbon JM (2000) Cisplatin and UV radiation induce activation of the stress-activated protein kinase p38gamma in human melanoma cells. Biochem Biophys Res Commun 278(3):724–728

50. Kumar S, McDonnell PC, Gum RJ, Hand AT, Lee JC, Young PR (1997) Novel homologues of CSBP/p38 MAP kinase: activation, substrate specificity and sensitivity to inhibition by pyridinyl imidazoles. Biochem Biophys Res Commun 235(3):533–538
51. Goedert M, Cuenda A, Craxton M, Jakes R, Cohen P (1997) Activation of the novel stress-activated protein kinase SAPK4 by cytokines and cellular stresses is mediated by SKK3 (MKK6); comparison of its substrate specificity with that of other SAP kinases. EMBO J 16(12):3563–3571
52. Tsai MS, Weng SH, Chen HJ, Chiu YF, Huang YC, Tseng SC, Kuo YH, Lin YW (2012) Inhibition of p38 MAPK-Dependent Excision Repair Cross-Complementing 1 Expression Decreases the DNA Repair Capacity to Sensitize Lung Cancer Cells to Etoposide. Mol Cancer Ther 11(3):561–571
53. Bulavin DV, Higashimoto Y, Popoff IJ, Gaarde WA, Basrur V, Potapova O, Appella E, Fornace AJ Jr (2001) Initiation of a G2/M checkpoint after ultraviolet radiation requires p38 kinase. Nature 411(6833):102–107
54. Kharbanda S, Pandey P, Yamauchi T, Kumar S, Kaneki M, Kumar V, Bharti A, Yuan ZM, Ghanem L, Rana A, Weichselbaum R, Johnson G, Kufe D (2000) Activation of MEK kinase 1 by the c-Abl protein tyrosine kinase in response to DNA damage. Mol Cell Biol 20(14):4979–4989
55. Dickens M, Rogers JS, Cavanagh J, Raitano A, Xia Z, Halpern JR, Greenberg ME, Sawyers CL, Davis RJ (1997) A cytoplasmic inhibitor of the JNK signal transduction pathway. Science 277(5326):693–696
56. Coso OA, Chiariello M, Yu JC, Teramoto H, Crespo P, Xu N, Miki T, Gutkind JS (1995) The small GTP-binding proteins Rac1 and Cdc42 regulate the activity of the JNK/SAPK signaling pathway. Cell 81(7):1137–1146
57. Tokiwa G, Dikic I, Lev S, Schlessinger J (1996) Activation of Pyk2 by stress signals and coupling with JNK signaling pathway. Science 273(5276):792–794
58. Shangary S, Brown KD, Adamson AW, Edmonson S, Ng B, Pandita TK, Yalowich J, Taccioli GE, Baskaran R (2000) Regulation of DNA-dependent protein kinase activity by ionizing radiation-activated abl kinase is an ATM-dependent process. J Biol Chem 275(39):30163–30168
59. Chen Z, Seimiya H, Naito M, Mashima T, Kizaki A, Dan S, Imaizumi M, Ichijo H, Miyazono K, Tsuruo T (1999) ASK1 mediates apoptotic cell death induced by genotoxic stress. Oncogene 18(1):173–180
60. Zou X, Tsutsui T, Ray D, Blomquist JF, Ichijo H, Ucker DS, Kiyokawa H (2001) The cell cycle-regulatory CDC25A phosphatase inhibits apoptosis signal- regulating kinase 1. Mol Cell Biol 21(14):4818–4828
61. Galaktionov K, Jessus C, Beach D (1995) Raf1 interaction with Cdc25 phosphatase ties mitogenic signal transduction to cell cycle activation. Genes Dev 9(9):1046–1058
62. Mailand N, Falck J, Lukas C, Syljuasen RG, Welcker M, Bartek J, Lukas J (2000) Rapid destruction of human Cdc25A in response to DNA damage. Science 288(5470):1425–1429
63. Falck J, Mailand N, Syljuasen RG, Bartek J, Lukas J (2001) The ATM-Chk2-Cdc25A checkpoint pathway guards against radioresistant DNA synthesis. Nature 410(6830):842–847
64. Potapova O, Haghighi A, Bost F, Liu C, Birrer MJ, Gjerset R, Mercola D (1997) The Jun kinase/stress-activated protein kinase pathway functions to regulate DNA repair and inhibition of the pathway sensitizes tumor cells to cisplatin. J Biol Chem 272(22):14041–14044
65. Helbig L, Damrot J, Hülsenbeck J, Köberle B, Brozovic A, Osmak M, Fiket Z, Kaina B, Fritz G (2011) Late activation of stress-activated protein kinases/c-Jun N-terminal kinases triggered by cisplatin-induced DNA damage in repair-defective cells. J Biol Chem 286(15):12991–13001
66. Sanchez-Perez I, Perona R (1999) Lack of c-Jun activity increases survival to cisplatin. FEBS Lett 453(1–2):151–158
67. Tong T, Fan W, Zhao H, Jin S, Fan F, Blanck P, Alomo I, Rajasekaran B, Liu Y, Holbrook NJ, Zhan Q (2001) Involvement of the map kinase pathways in induction of gadd45 following uv radiation. Exp Cell Res 269(1):64–72

68. Takekawa M, Saito H (1998) A family of stress-inducible GADD45-like proteins mediate activation of the stress-responsive MTK1/MEKK4 MAPKKK. Cell 95(4):521–530
69. Sanchez Y, Elledge SJ (1995) Stopped for repairs. Bioessays 17(6):545–548
70. Chao C, Saito S, Kang J, Anderson CW, Appella E, Xu Y (2000) p53 transcriptional activity is essential for p53-dependent apoptosis following DNA damage. EMBO J 19(18):4967–4975
71. Ljungman M, O'Hagan HM, Paulsen MT (2001) Induction of ser15 and lys382 modifications of p53 by blockage of transcription elongation. Oncogene 20(42):5964–5971
72. Cuddihy AR, Li S, Tam NW, Wong AH, Taya Y, Abraham N, Bell JC, Koromilas AE (1999) Double-stranded-RNA-activated protein kinase PKR enhances transcriptional activation by tumor suppressor p53. Mol Cell Biol 19(4):2475–2484
73. Waterman MJ, Stavridi ES, Waterman JL, Halazonetis TD (1998) ATM-dependent activation of p53 involves dephosphorylation and association with 14-3-3 proteins. Nat Genet 19(2):175–178
74. Buschmann T, Potapova O, Bar-Shira A, Ivanov VN, Fuchs SY, Henderson S, Fried VA, Minamoto T, Alarcon-Vargas D, Pincus MR, Gaarde WA, Holbrook NJ, Shiloh Y, Ronai Z (2001) Jun NH2-terminal kinase phosphorylation of p53 on Thr-81 is important for p53 stabilization and transcriptional activities in response to stress. Mol Cell Biol 21(8):2743–2754
75. Kim SJ, Ju JW, Oh CD, Yoon YM, Song WK, Kim JH, Yoo YJ, Bang OS, Kang SS, Chun JS (2002) ERK-1/2 and p38 kinase oppositely regulate nitric oxide-induced apoptosis of chondrocytes in association with p53, caspase-3, and differentiation status. J Biol Chem 277(2):1332–1339
76. Adler V, Pincus MR, Minamoto T, Fuchs SY, Bluth MJ, Brandt-Rauf PW, Friedman FK, Robinson RC, Chen JM, Wang XW, Harris CC, Ronai Z (1997) Conformation-dependent phosphorylation of p53. Proc Natl Acad Sci USA 94(5):1686–1691
77. Barbosa MS (1996) The oncogenic role of human papillomavirus proteins. Crit Rev Oncog 7(1–2):1–18
78. Malkin D, Chilton-MacNeill S, Meister LA, Sexsmith E, Diller L, Garcea RL (2001) Tissue-specific expression of SV40 in tumors associated with the Li- Fraumeni syndrome. Oncogene 20(33):4441–4449
79. Zur HH (1996) Papillomavirus infections–a major cause of human cancers. Biochim Biophys Acta 1288(2):F55–F78
80. Huang S, Benavente S, Armstrong EA, Li C, Wheeler DL, Harari PM (2011) p53 modulates acquired resistance to EGFR inhibitors and radiation. Cancer Res 71(22):7071–7079
81. She QB, Chen N, Dong Z (2000) ERKs and p38 kinase phosphorylate p53 protein at serine 15 in response to UV radiation. J Biol Chem 275(27):20444–20449
82. Bulavin DV, Saito S, Hollander MC, Sakaguchi K, Anderson CW, Appella E, Fornace AJ Jr (1999) Phosphorylation of human p53 by p38 kinase coordinates N-terminal phosphorylation and apoptosis in response to UV radiation. EMBO J 18(23):6845–6854
83. Sanchez-Prieto R, Rojas JM, Taya Y, Gutkind JS (2000) A role for the p38 mitogen-acitvated protein kinase pathway in the transcriptional activation of p53 on genotoxic stress by chemotherapeutic agents. Cancer Res 60(9):2464–2472
84. Persons DL, Yazlovitskaya EM, Pelling JC (2000) Effect of extracellular signal-regulated kinase on p53 accumulation in response to cisplatin. J Biol Chem 275(46):35778–35785
85. Mandic A, Viktorsson K, Heiden T, Hansson J, Shoshan MC (2001) The MEK1 inhibitor PD98059 sensitizes C8161 melanoma cells to cisplatin- induced apoptosis. Melanoma Res 11(1):11–19
86. Fu H, Subramanian RR, Masters SC (2000) 14-3-3 proteins: structure, function, and regulation. Annu Rev Pharmacol Toxicol 40:617–647
87. Walworth NC (2001) DNA damage: Chk1 and Cdc25, more than meets the eye. Curr Opin Genet Dev 11(1):78–82
88. Hirose Y, Berger MS, Pieper RO (2001) Abrogation of the Chk1-mediated G(2) checkpoint pathway potentiates temozolomide-induced toxicity in a p53-independent manner in human glioblastoma cells. Cancer Res 61(15):5843–5849

89. Gottifredi V, Karni-Schmidt O, Shieh SS, Prives C (2001) p53 down-regulates CHK1 through p21 and the retinoblastoma protein. Mol Cell Biol 21(4):1066–1076
90. Takekawa M, Adachi M, Nakahata A, Nakayama I, Itoh F, Tsukuda H, Taya Y, Imai K (2000) p53-inducible wip1 phosphatase mediates a negative feedback regulation of p38 MAPK-p53 signaling in response to UV radiation. EMBO J 19(23):6517–6526
91. Bunz F, Dutriaux A, Lengauer C, Waldman T, Zhou S, Brown JP, Sedivy JM, Kinzler KW, Vogelstein B (1998) Requirement for p53 and p21 to sustain G2 arrest after DNA damage. Science 282(5393):1497–1501
92. Smits VA, Klompmaker R, Vallenius T, Rijksen G, Makela TP, Medema RH (2000) p21 inhibits Thr161 phosphorylation of Cdc2 to enforce the G2 DNA damage checkpoint. J Biol Chem 275(39):30638–30643
93. Chan TA, Hwang PM, Hermeking H, Kinzler KW, Vogelstein B (2000) Cooperative effects of genes controlling the G(2)/M checkpoint. Genes Dev 14(13):1584–1588
94. Jin S, Antinore MJ, Lung FD, Dong X, Zhao H, Fan F, Colchagie AB, Blanck P, Roller PP, Fornace AJ Jr, Zhan Q (2000) The GADD45 inhibition of Cdc2 kinase correlates with GADD45-mediated growth suppression. J Biol Chem 275(22):16602–16608
95. Ferry KV, Hamilton TC, Johnson SW (2000) Increased nucleotide excision repair in cisplatin-resistant ovarian cancer cells: role of ERCC1-XPF. Biochem Pharmacol 60(9):1305–1313
96. Li D, Turi TG, Schuck A, Freedberg IM, Khitrov G, Blumenberg M (2001) Rays and arrays: the transcriptional program in the response of human epidermal keratinocytes to UVB illumination. FASEB J 15(13):2533–2535
97. Aguda BD (1999) Instabilities in phosphorylation-dephosphorylation cascades and cell cycle checkpoints. Oncogene 18(18):2846–2851
98. Steegenga WT, van der Eb AJ, Jochemsen AG (1996) How phosphorylation regulates the activity of p53. J Mol Biol 263(2):103–113
99. Fuchs SY, Fried VA, Ronai Z (1998) Stress-activated kinases regulate protein stability. Oncogene 17(11 Reviews):1483–1490
100. Huang HC, Nguyen T, Pickett CB (2000) Regulation of the antioxidant response element by protein kinase C- mediated phosphorylation of NF-E2-related factor 2. Proc Natl Acad Sci USA 97(23):12475–12480
101. Ben Levy R, Hooper S, Wilson R, Paterson HF, Marshall CJ (1998) Nuclear export of the stress-activated protein kinase p38 mediated by its substrate MAPKAP kinase-2. Curr Biol 8(19):1049–1057
102. Lawler S, Fleming Y, Goedert M, Cohen P (1998) Synergistic activation of SAPK1/JNK1 by two MAP kinase kinases in vitro. Curr Biol 8(25):1387–1390
103. Wang XZ, Ron D (1996) Stress-induced phosphorylation and activation of the transcription factor CHOP (GADD153) by p38 MAP Kinase. Science 272(5266):1347–1349
104. Modesti M, Kanaar R (2001) DNA repair: spot(light)s on chromatin. Curr Biol 11(6):R229–R232
105. Zhong S, Jansen C, She QB, Goto H, Inagaki M, Bode AM, Ma WY, Dong Z (2001) Ultraviolet B-induced phosphorylation of histone H3 at serine 28 is mediated by MSK1. J Biol Chem 276(35):33213–33219
106. Araujo SJ, Tirode F, Coin F, Pospiech H, Syvaoja JE, Stucki M, Hübscher U, Egly JM, Wood RD (2000) Nucleotide excision repair of DNA with recombinant human proteins: definition of the minimal set of factors, active forms of TFIIH, and modulation by CAK. Genes Dev 14(3):349–359
107. Stivala LA, Riva F, Cazzalini O, Savio M, Prosperi E (2001) p21(waf1/cip1)-null human fibroblasts are deficient in nucleotide excision repair downstream the recruitment of PCNA to DNA repair sites. Oncogene 20(5):563–570
108. Sheikh MS, Chen YQ, Smith ML, Fornace AJ Jr (1997) Role of p21Waf1/Cip1/Sdi1 in cell death and DNA repair as studied using a tetracycline-inducible system in p53-deficient cells. Oncogene 14(15):1875–1882

109. Adimoolam S, Lin CX, Ford JM (2001) The p53-regulated cyclin-dependent kinase inhibitor, p21 (cip1, waf1, sdi1), is not required for global genomic and transcription-coupled nucleotide excision repair of UV-induced DNA photoproducts. J Biol Chem 276(28):25813–25822
110. Shivji MK, Grey SJ, Strausfeld UP, Wood RD, Blow JJ (1994) Cip1 inhibits DNA replication but not PCNA-dependent nucleotide excision-repair. Curr Biol 4(12):1062–1068
111. Cooper MP, Balajee AS, Bohr VA (1999) The C-terminal domain of p21 inhibits nucleotide excision repair In vitro and In vivo. Mol Biol Cell 10(7):2119–2129
112. Therrien JP, Loignon M, Drouin R, Drobetsky EA (2001) Ablation of p21waf1cip1 expression enhances the capacity of p53- deficient human tumor cells to repair UVB-induced DNA damage. Cancer Res 61(9):3781–3786
113. Ford JM, Hanawalt PC (1997) Expression of wild-type p53 is required for efficient global genomic nucleotide excision repair in UV-irradiated human fibroblasts. J Biol Chem 272(44):28073–28080
114. Therrien JP, Drouin R, Baril C, Drobetsky EA (1999) Human cells compromised for p53 function exhibit defective global and transcription-coupled nucleotide excision repair, whereas cells compromised for pRb function are defective only in global repair. Proc Natl Acad Sci USA 96(26):15038–15043
115. Zhou J, Ahn J, Wilson SH, Prives C (2001) A role for p53 in base excision repair. EMBO J 20(4):914–923
116. Offer H, Milyavsky M, Erez N, Matas D, Zurer I, Harris CC, Rotter V (2001) Structural and functional involvement of p53 in BER in vitro and in vivo. Oncogene 20(5):581–589
117. Smith ML, Ford JM, Hollander MC, Bortnick RA, Amundson SA, Seo YR, Deng CX, Hanawalt PC, Fornace AJ Jr (2000) p53-mediated DNA repair responses to UV radiation: studies of mouse cells lacking p53, p21, and/or gadd45 genes. Mol Cell Biol 20(10):3705–3714
118. Rossig L, Jadidi AS, Urbich C, Badorff C, Zeiher AM, Dimmeler S (2001) Akt-dependent phosphorylation of p21(Cip1) regulates PCNA binding and proliferation of endothelial cells. Mol Cell Biol 21(16):5644–5657
119. Smith ML, Chen IT, Zhan Q, Bae I, Chen CY, Gilmer TM, Kastan MB, O'Connor PM, Fornace AJ Jr (1994) Interaction of the p53-regulated protein Gadd45 with proliferating cell nuclear antigen. Science 266(5189):1376–1380
120. Waga S, Hannon GJ, Beach D, Stillman B (1994) The p21 inhibitor of cyclin-dependent kinases controls DNA replication by interaction with PCNA. Nature 369(6481):574–578
121. O'Reilly MA, Staversky RJ, Watkins RH, Maniscalco WM, Keng PC (2000) p53-independent induction of GADD45 and GADD153 in mouse lungs exposed to hyperoxia. Am J Physiol Lung Cell Mol Physiol 278(3):L552–L559
122. Russo T, Zambrano N, Esposito F, Ammendola R, Cimino F, Fiscella M, Jackman J, O'Connor PM, Anderson CW, Appella E (1995) A p53-independent pathway for activation of WAF1/CIP1 expression following oxidative stress. J Biol Chem 270(49):29386–29391
123. Zamble DB, Jacks T, Lippard SJ (1998) p53-Dependent and -independent responses to cisplatin in mouse testicular teratocarcinoma cells. Proc Natl Acad Sci USA 95(11):6163–6168
124. Scherer SJ, Maier SM, Seifert M, Hanselmann RG, Zang KD, Muller-Hermelink HK, Angel P, Welter C, Schartl M (2000) p53 and c-Jun functionally synergize in the regulation of the DNA repair gene hMSH2 in response to UV. J Biol Chem 275(48):37469–37473
125. Aboussekhra A, Biggerstaff M, Shivji MK, Vilpo JA, Moncollin V, Podust VN, Proti M, Hübscher U, Egly JM, Wood RD (1995) Mammalian DNA nucleotide excision repair reconstituted with purified protein components. Cell 80(6):859–868
126. Wold MS (1997) Replication protein A: a heterotrimeric, single-stranded DNA-binding protein required for eukaryotic DNA metabolism. Annu Rev Biochem 66:61–92
127. Iftode C, Daniely Y, Borowiec JA (1999) Replication protein A (RPA): the eukaryotic SSB. Crit Rev Biochem Mol Biol 34(3):141–180

128. Dutta A, Stillman B (1992) cdc2 family kinases phosphorylate a human cell DNA replication factor, RPA, and activate DNA replication. EMBO J 11(6):2189–2199
129. Oakley GG, Loberg LI, Yao J, Risinger MA, Yunker RL, Zernik-Kobak M, Khanna KK, Lavin MF, Carty MP, Dixon K (2001) UV-induced hyperphosphorylation of replication protein a depends on DNA replication and expression of ATM protein. Mol Biol Cell 12(5):1199–1213
130. Rodrigo G, Roumagnac S, Wold MS, Salles B, Calsou P (2000) DNA replication but not nucleotide excision repair is required for UVC- induced replication protein A phosphorylation in mammalian cells. Mol Cell Biol 20(8):2696–2705
131. Srivenugopal KS, Shou J, Mullapudi SR, Lang FF Jr, Rao JS, Ali-Osman F (2001) Enforced expression of wild-type p53 curtails the transcription of the O(6)-methylguanine-DNA methyltransferase gene in human tumor cells and enhances their sensitivity to alkylating agents. Clin Cancer Res 7(5):1398–1409
132. Guo W, Liu X, Lee S, Park NH (1999) High O6-methylguanine methyl transferase activity is frequently found in human oral cancer cells with p53 inactivation. Int J Oncol 15(4):817–821
133. Srivenugopal KS, Mullapudi SR, Shou J, Hazra TK, Ali-Osman F (2000) Protein phosphorylation is a regulatory mechanism for O6-alkylguanine- DNA alkyltransferase in human brain tumor cells. Cancer Res 60(2):282–287
134. Ababou M, Dutertre S, Lecluse Y, Onclercq R, Chatton B, Amor-Gueret M (2000) ATM-dependent phosphorylation and accumulation of endogenous BLM protein in response to ionizing radiation. Oncogene 19(52):5955–5963
135. Blasina A, Price BD, Turenne GA, McGowan CH (1999) Caffeine inhibits the checkpoint kinase ATM. Curr Biol 9(19):1135–1138
136. Sarkaria JN, Busby EC, Tibbetts RS, Roos P, Taya Y, Karnitz LM, Abraham RT (1999) Inhibition of ATM and ATR kinase activities by the radiosensitizing agent, caffeine. Cancer Res 59(17):4375–4382
137. Costelloe T, Fitzgerald J, Murphy NJ, Flaus A, Lowndes NF (2006) Chromatin modulation and the DNA damage response. Exp Cell Res 312(14):2677–2686
138. English JM, Cobb MH (2002) Pharmacological inhibitors of MAPK pathways. Trends Pharmacol Sci 23(1):40–45
139. Grace D (2010) MEK/MAPK Inhibitors. J Thorac Oncol 5(12 Suppl 6):S474–S475
140. Ohta T, Ohmichi M, Shibuya T, Takahashi T, Tsutsumi S, Takahashi K, Kurachi H (2012) Gefitinib (ZD1839) increases the efficacy of cisplatin in ovarian cancer cells. Cancer Biol Ther 13(6):408–416
141. Ossovskaya V, Koo IC, Kaldjian EP, Alvares C, Sherman BM (2010) Upregulation of Poly (ADP-Ribose) Polymerase-1 (PARP1) in Triple-Negative Breast Cancer and Other Primary Human Tumor Types. Genes Cancer 1(8):812–821
142. Reeder-Hayes KE, Carey LA, Sikov WM (2010) Clinical trials in triple negative breast cancer. Breast Dis 32(1–2):123–136
143. Nowsheen S, Bonner JA, Lobuglio AF, Trummell H, Whitley AC, Dobelbower MC, Yang ES (2011) Cetuximab augments cytotoxicity with poly (adp-ribose) polymerase inhibition in head and neck cancer. PLoS One 6(8):e24148
144. Yang JY, Chang CJ, Xia W, Wang Y, Wong KK, Engelman JA, Du Y, Andreeff M, Hortobagyi GN, Hung MC (2010) Activation of FOXO3a is sufficient to reverse mitogen-activated protein/extracellular signal-regulated kinase kinase inhibitor chemoresistance in human cancer. Cancer Res 70:4709–4718
145. Meng J, Peng H, Dai B, Guo W, Wang L, Ji L, Minna JD, Chresta CM, Smith PD, Fang B, Roth JA (2009) High level of AKT activity is associated with resistance to MEK inhibitor AZD6244(ARY-142886). Cancer Biol Ther 8:2073–2080
146. Carracedo A, Ma L, Teruya-Feldstein J, Rojo F, Salmena L, Alimonti A, Egia A, Sasaki AT, Thomas G, Kozma SC, Papa A, Nardella C, Cantley LC, Baselga J, Pandolfi PP (2008) Inhibition of mTORC1 leads to MAPK pathway activation through a PI3K-dependent feedback loopin human cancer. J Clin Invest 118(9):3065–3074

147. Solit DB, Garraway LA, Pratilas CA, Sawai A, Getz G, Basso A, Ye Q, Lobo JM, She Y, Osman I, Golub TR, Sebolt-Leopold J, Sellers WR, Rosen N (2006) BRAF mutation predicts sensitivity to MEK inhibition. Nature 439(7074):358–362
148. Hatzivassiliou G, Song K, Yen I, Brandhuber BJ, Anderson DJ, Alvarado R, Ludlam MJ, Stokoe D, Gloor SL, Vigers G, Morales T, Aliagas I, Liu B, Sideris S, Hoeflich KP, Jaiswal BS, Seshagiri S, Koeppen H, Belvin M, Friedman LS, Malek S (2010) RAF inhibitors prime wild-type RAF to activate the MAPK pathway and enhance growth. Nature 464(7287):431–435

The Relationship Between DNA-Repair Genes, Cellular Radiosensitivity, and the Response of Tumors and Normal Tissues to Radiotherapy

David Murray and Matthew Parliament

1 The Use of Predictive Assays in Radiation Therapy for Cancer

Radiation therapy (XRT) continues to be an important component in the management of many cancer patients. The dose delivered to a tumor is calculated on the basis of the anticipated tolerance of the normal tissues within the irradiated field, which is determined empirically on the basis of population-averaged clinical data [1]. These calculations, which typically accept a severe late complication rate of ≤5%, do not account for the different susceptibilities of XRT patients to late normal tissue effects [2], which can be severe and sometimes life threatening. Such interindividual variations are substantial even though major advances have been made with respect to the conformality of XRT delivery. Similarly, the clinical responsiveness of tumors to XRT is often quite different even among tumors of the same pathology. One approach to further improving the therapeutic outcome of XRT is to develop individualized treatment plans that utilize pretreatment biomarkers that would predict both normal tissue tolerance levels and/or tumor responsiveness to therapy on a patient-by-patient basis. For such biomarkers to become used routinely in the clinic, they will have to be capable of rapidly and reliably identifying unusually or even

D. Murray (✉)
Division of Experimental Oncology, Department of Oncology,
Cross Cancer Institute, University of Alberta,
11560 University Ave, Edmonton, AB, Canada T6G 1Z2
e-mail: davem@cancerboard.ab.ca

M. Parliament
Division of Radiation Oncology, Department of Oncology,
Cross Cancer Institute, University of Alberta,
Edmonton, AB, Canada T6G 1Z2

L. Panasci et al. (eds.), *Advances in DNA Repair in Cancer Therapy*,
Cancer Drug Discovery and Development 72, DOI 10.1007/978-1-4614-4741-2_4,

© Springer Science+Business Media, LLC 2013

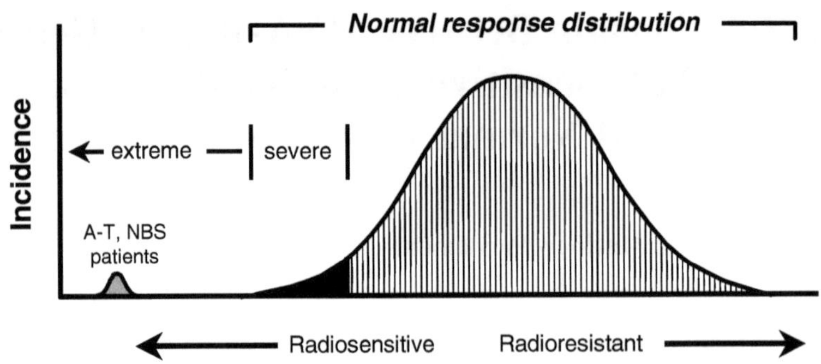

Fig. 1 Hypothetical cellular-molecular basis for inter-patient differences in susceptibility to normal-tissue complications following radiation therapy for cancer. The model is a simplified version of that presented by Burnet and colleagues [4]. A-T, ataxia telangiectasia; NBS, Nijmegen breakage syndrome

moderately radiosensitive patients or radioresistant tumors prior to treatment. This need has, for many years, driven intensive research into the development of assays and biomarkers for predicting individual radiosensitivity [3].

In the post-genomics era, interest in identifying biomarkers for interindividual variations in normal tissue and tumor response to XRT has intensified because the latest analytical platforms permit high-throughput testing of huge numbers of clinical samples. The generation and validation of a biomarker signature for normal-tissue sensitivity would allow clinicians to identify patients who might not tolerate dose escalation but who might instead benefit from alternative therapeutic options. It would also provide a stratification tool for clinical trials of novel XRT regimens in which normal tissue complications are an end point. Furthermore, a better understanding of the molecular basis of variability in radiosensitivity would facilitate the rational design of drugs that would enhance normal tissue recovery in patients who might otherwise be at risk of late tissue injury. These objectives are the primary focus of this chapter. First, however, we take a quick look back at the history of this field.

2 Prediction of XRT Outcome Using Cell-Based Assays

Figure 1, which has been adapted from Burnet and colleagues [4], illustrates the premise of this discussion from the perspective of cellular and genetic screening approaches to normal tissue hypersensitivity. In this simplified model (Fig. 1) we assume that radiosensitive individuals exhibit either (1) "extreme" overreactions to XRT, which probably have a monogenic origin or (2) "severe" overreactions, mostly occupying the sensitive tail of a normal distribution, and which probably have a polygenic origin. Similar and additional considerations such as microenvironment and genetic instability are relevant for tumor control.

2.1 *Clonogenic Survival Curves*

The in vitro clonogenic cell survival assay has been used to evaluate the radiation sensitivity of both normal tissues (using surrogate normal cell types such as skin fibroblasts and lymphocytes) and tumors [3, 5–7]. Although such assays have shown promise for predicting late (but not acute) reactions in some settings (e.g., [8, 9]), they are unlikely to be integrated into cancer management decisions, in part because: (1) they are insufficiently precise and reproducible for routine clinical use; (2) they require several replicate determinations, so it can take weeks to generate useable data; and (3) patient radiosensitivity is not always associated with cellular radiosensitivity (reviewed in [2]).

2.2 *Phenotypic Studies of DSB Rejoining*

Exposure of mammalian cells to ionizing radiation (IR) causes a number of types of damage to their genome, including base and sugar damage, single strand breaks (SSBs) and DNA cross links; however, DNA double-strand breaks (DSBs), which involve local scission of both strands of the DNA helix, are considered to be the "signature" lesions of IR exposure, and it is the unrepaired or misrepaired DSBs that are believed to be the primary cause of cell death [10]. Indeed, the ability of cells to repair DSBs is a critical determinant of their radiosensitivity. For this reason, there has been much interest in measuring DSB repair in normal and cancerous cells in the context of predictive assays in XRT. Such assays should have the advantages of being quicker and more reproducible than clonogenic cell survival assays. Their anticipated down side is that, unlike the clonogenic cell survival assay, they only discern one of many factors that could contribute to heterogeneity in cellular radiosensitivity, even though it may be the dominant one.

2.2.1 DSB Rejoining in Normal Cells

A relationship between patient sensitivity to normal tissue complications in the clinic and inefficient DSB rejoining in surrogate normal cell types, typically fibroblasts or lymphocytes, derived from these patients and irradiated *ex vivo* has been suggested in some studies (e.g., [11–13] and references therein) but not in others (e.g., [14, 15]). Studies using pulsed-field gel electrophoresis (PFGE) to measure DSB rejoining *ex vivo* in fibroblasts obtained from breast cancer patients who received standardized XRT illustrate the limitations of such assays for predicting normal tissue sensitivity. An initial study of 39 patients indicated a correlation between the level of unrejoined DSBs in their fibroblasts at 24 h after irradiation and the clinical severity of late fibrosis [16]. However, when a validation cohort of 50 patients was included, the relationship between residual DSBs in fibroblasts and late

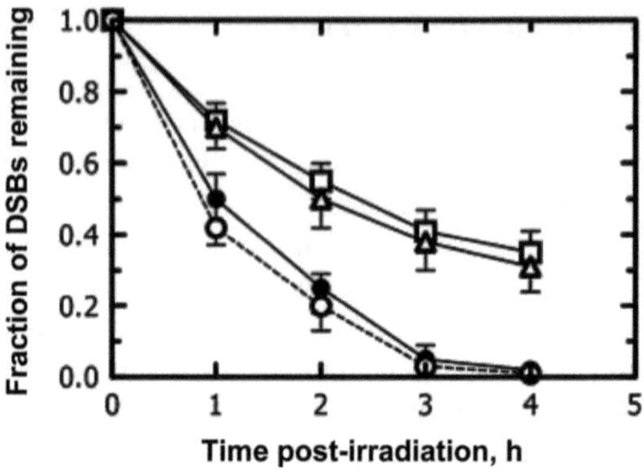

Fig. 2 Time course of DSB rejoining in GM38 (*circle*) and GM10 (*filled circle*) normal fibroblasts and in AT5BI (*square*) and AT2BE (*triangle*) Ataxia Telangiectasia cells, assessed using the neutral comet assay. The data represent the fraction of comet tail fluorescence (a measure of DSBs) remaining at up to 4 h after exposure to 8 Gy of γ-rays, compared to that measured immediately after irradiation. From Mirzayans et al. [20]

fibrosis for the combined cohort disappeared [17]. In these studies, DSBs were measured after a dose of 150 Gy. After such an overwhelming level of genomic insult, it is unlikely that measurements of DSB repair would reflect any subtle differences in pro-survival responses occurring after doses representative of those used in the clinic. Similarly, El-Awady and colleagues [18] did not see a significant correlation between acute clinical reaction scores (RTOG scale) in breast cancer patients receiving postsurgical XRT and un-rejoined DSB levels measured in their fibroblasts using constant-field gel electrophoresis at 24 h after a dose of 100 Gy.

On the positive side, the last decade has seen the emergence of more sensitive assays for studying responses to DNA damage after lower doses, i.e., where most of the cells remain functional during the period when their repair activity is being monitored. For example, with suitable optimization, the neutral comet assay can discriminate DSB rejoining after doses of 10 Gy or less [19, 20] while the IR-induced focus (IRIF) formation assay, usually involving γ-H2AX foci (see Sect. 2.3), offers the opportunity to interrogate cellular responses after even lower doses (e.g., [21]). The value of these approaches is illustrated in a study of DSB rejoining in fibroblasts derived from individuals with the genetic hyper-radiosensitivity syndrome ataxia telangiectasia (A-T) caused by mutation of the *ATM* gene. Using both the neutral comet and γ-H2AX assays, A-T cells were found to be markedly deficient in DSB rejoining compared to normal fibroblasts after exposure to ≤8 Gy of γ-rays [20]. This is illustrated for the neutral comet assay in Fig. 2. A similar deficiency in DSB rejoining based on the resolution of γ-H2AX foci was seen in mouse A-T cells after a 1-Gy exposure [22], in blood lymphocytes and normal tissues from $ATM^{-/-}$ homozygous mice after a 2 Gy in vivo exposure [23], and in cells from $ATM^{-/-}$ homozygous A-T patients [24]. Thus, whereas A-T cells do not typically exhibit an overt DSB repair defect after high-dose exposures [25], these cells are clearly repair deficient after doses in the survival curve range when the repair response is not overwhelmed. We return to the predictive potential of such assays in Sect. 2.3.

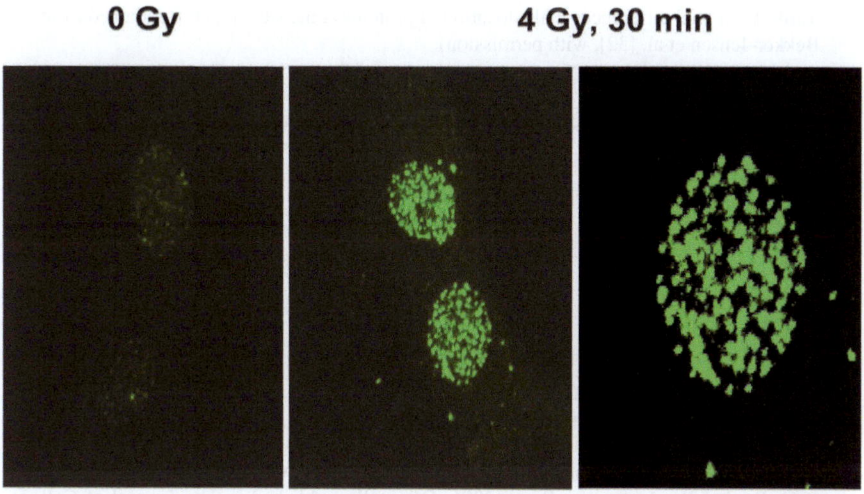

Fig. 3 Typical fluorescence image of nuclear γ-H2AX foci before and 30 min after exposure of GM38 normal human fibroblasts to 4 Gy of γ-rays. The slides were immunostained with an anti-γ-H2AX (serine-139) antibody from Cedarlane Laboratories (Hornby, ON, Canada). The *right-hand panel* is an enlargement of the image of the lower cell in the *middle panel*

2.2.2 DSB Rejoining in Tumor Cells

A relationship between cellular radiosensitivity in the clonogenic survival assay and ability to rejoin IR-induced DSBs assayed by "direct" assays in panels of human tumor cell lines using a variety of assays and IR doses has been reported in some studies, but again this is not a universal finding (e.g., [2, 26–28] and references therein). As noted above for normal tissues, the advent of more sensitive DSB assays has renewed interest in this area, as discussed below.

2.3 *Radiation-Induced DSB-Repair Foci*

Exposure of mammalian cells to IR causes the rapid relocalization of DSB repair-related proteins to the site of a DSB. The resulting "ionizing radiation-induced nuclear foci" or "IRIFs" are believed to be essential for the rejoining of DSBs. An early event in this process is the phosphorylation of the rare variant histone, H2AX, at the site of the DSB within minutes of irradiation [29]. The key effector kinase here is activated ATM, although DNA-PK appears to have an overlapping role [30]. Typically an average of ~2,000 H2AX molecules are phosphorylated at each DSB site such that, when the cells are stained with an antibody that recognizes phosphorylated H2AX (γ-H2AX), each DSB can be visualized as a discrete "repair focus" or IRIF [31]. Indeed, there appears to be an equivalency between initial levels of IR-induced γ-H2AX foci and physical DSBs [30]. The assay is illustrated in Fig. 3,

Table 1 Spatial differences in the location of proteins at the site of a DSB (adapted from Bekker-Jensen et al. [32], with permission)

DSB-flanking chromatin[a]	Single-stranded DNA microcompartments[b]	No retention at DSBs[c]
ATM	ATR	DNA-PK
NBS1	ATRIP	KU70
MRE11	RPA	SMC1
RAD50	RAD17	SMC3
MDC1	RAD9	CHK1
53BP1	RAD51/RAD52	CHK2
BRCA1	FANC-D2	TP53
	BRCA1/BRCA2	CDC25A
	NBS1, MRE11, RAD50	

[a]Interactions that operate throughout interphase
[b]Interactions restricted to S/G2 phase
[c]Based on local protein accumulation at physiologically relevant DSBs
©The Rockefeller University Press, 2006. Originally published in The Journal of Cell Biology. 173: 195–206

which shows typical γ-H2AX fluorescence images of normal fibroblasts before and 30 min after exposure to 4 Gy of γ-rays.

These γ-H2AX-flagged sites serve to recruit mediators and subsequently DSB-repair proteins to the lesion. The resulting IRIFs are dynamic entities that contain thousands of copies of each involved protein. Proteins that relocalize to IRIFs include MRE11-RAD50-NBS1 (i.e., the "MRN" complex), MDC1, SMC1, the BRCA1-BARD1 complex, BRCA2, 53BP1, RPA, the ubiquitin ligases RNF8 and RNF168, and the RAD51 family of proteins (e.g., [29, 30]). Not only are there temporal differences in the association of these various proteins, but also there are spatial differences in the location of proteins at the site of a DSB (see Table 1). Perhaps most intriguing in this context is the finding that the three phosphoinositide-3-kinase-related protein kinases (PIKKs) involved in DNA damage signaling—ATM, ATR, and DNA-PK—occupy distinct compartments at such foci (at DSB-flanking chromatin, single-stranded DNA and unprocessed DSB ends, respectively) [32]. MDC1 is the major factor that recognizes and binds to γ-H2AX, with 53BP1 also having the ability to detect early chromatin changes associated with a DSB. As discussed in Sect. 3.4, the cell has two major options for repairing a DSB—nonhomologous end joining (NHEJ) and homologous recombination repair (HRR)—that result in the formation of distinct protein complexes. For example, foci involving the key HRR protein RAD51 represent nuclear domains for HRR [33, 34].

2.3.1 Relationship Between γ-H2AX and Radiosensitivity in Model Systems

In the context of DSB rejoining, a number of investigators have evaluated whether the rate of γ-H2AX IRIF resolution or residual foci levels at later times (typically 18–24 h) after irradiation can provide a reliable measure of radiosensitivity. A relationship between cell survival and the resolution of γ-H2AX foci has indeed been reported using a variety of model systems, notably various normal cell lines, panels of unrelated tumor cell lines or lines derived from a single type of tumor, and tumor xenograft models [35–42]. In these studies, the DSB rejoining rate based on γ-H2AX foci resolution or residual levels was typically faster in radioresistant versus radiosensitive systems. There are, however, some exceptions to this behavior. Notably, Mahrhofer and colleagues [43] did not see a correlation between γ-H2AX foci resolution at up to 18 h post-irradiation and cellular radiosensitivity (surviving fraction at 2 Gy) in 10 human tumor and normal cell lines. Although Yoshikawa et al. [44] did see a correlation between residual IR-induced γ-H2AX foci and loss of clonogenic potential in two normal diploid cell lines and the HeLa tumor cell line, no such relationship was apparent in three other tumor cell lines that exhibited high background levels of foci.

Understandably, there has been much interest in the quantification of IRIFs from the perspective of predicting clinical response to XRT. The fact that this assay has unprecedented sensitivity [21, 30] makes it attractive for quantifying DSBs in clinical material, and it can be readily automated for high throughput purposes using flow cytometry (e.g., [39, 45–47]), tissue microarrays [48] and various imaging platforms, some of which can provide 3-D information and can be used to measure DSB repair in tumor biopsy samples [35, 48]. However, some caution must be exercised when using this assay because: (1) IRIFs represent an indirect measure of DSBs that actually reflect the cells' response to the DSB; (2) the assay may respond to damage other than DSBs, including changes in chromatin structure; (3) phosphorylation of H2AX is also observed during apoptosis [31] as well as senescence [49]; and (4) the assay sensitivity is limited by endogenous or "cryptogenic" foci levels, which can be problematic for tumor cells that exhibit genetic instability [30, 50]. Although it does not seem to be an issue with respect to predictive assays, residual γ-H2AX foci persist longer than unrejoined physical DSBs, suggesting that chromatin remodeling is still active at these post-repair sites [36, 51, 52]. Another consideration is the known dependence of γ-H2AX foci on the level of condensation of the chromatin [52]. An important strength of the IRIF assay is that, being an "intermediate" end point between clinical response/toxicity and individual molecular biomarkers, it interrogates the integrity of the entire functional response pathway to DSBs, the most important class of IR-induced DNA lesions.

2.3.2 Clinical Applications of γ-H2AX Assays

Early studies of γ-H2AX IRIFs in tissues following in vivo IR exposures have been reviewed elsewhere [48]. In one such study, γ-H2AX IRIFs were readily apparent in skin fibroblasts obtained by biopsy from prostate cancer patients following XRT [53]. γ-H2AX IRIFs were also quantifiable and linearly dependent on the dose in blood lymphocytes obtained from patients with benign or malignant tumors after a low dose diagnostic computed tomography (CT) exposure to the chest and/or abdomen

[54, 55]. Intriguingly, the resolution of these foci was abnormally slow in lymphocytes from a patient who had earlier exhibited severe late complications following XRT [54]. Simonsson et al. [56] also observed γ-H2AX IRIFs in normal skin biopsies from prostate cancer patients undergoing XRT, where the absorbed doses to skin were between 0.05 and 1.1 Gy; interestingly, there was no difference between γ-H2AX IRIF levels measured at 30 min vs. 2 h after an individual XRT fraction, indicating that no foci resolution occurred over this time frame. More recently, γ-H2AX foci have also been demonstrated in lymphocytes obtained from patients after angiography, where doses as low as 2 mGy could be detected [57].

With respect to DSB rejoining and adverse reactions, Olive and colleagues [58] reported that the rate of resolution of γ-H2AX foci in peripheral blood mononuclear cells from prostate cancer patients receiving brachytherapy was similar in patients who developed late normal tissue complications compared to those who did not. In that study the cells were exposed in vitro to 4 fractions of 0.7 Gy X-rays given at 3 h intervals, and residual IRIFs were measured at 18 h after the final fraction. Werbrouck and coworkers [59] also investigated the ability of the γ-H2AX IRIF assay to predict normal tissue complications using in vitro-irradiated peripheral blood T-lymphocytes from a group of 29 cervix and endometrial cancer patients being treated with XRT. Again, the kinetics of the resolution of γ-H2AX foci were similar in both patient groups (i.e., in those exhibiting none/mild versus moderate/severe complications) up to 24 h after exposure to either 0.5 Gy IR at high dose rate or to 2.2 Gy at low dose rate.

Vasireddy and colleagues [60] examined the resolution of γ-H2AX foci in lymphoblast cell lines derived from 18 severely radiosensitive patients (RTOG grade 3 or 4 acute or late effects) and non-radiosensitive controls (RTOG grade 0–1). After a 2-Gy exposure, there was no significant difference between radiosensitive samples and 11 control samples with respect to the average rate of γ-H2AX resolution up to 24 h post-irradiation. However, one cell line, RS1, established from a patient who exhibited severe acute erythema, did exhibit a slower resolution of γ-H2AX foci depletion up to 24 h that was confirmed using the PFGE assay.

Figure 4 shows some preliminary data from a case–control study in which we are evaluating various biomarkers of normal tissue complications. The case group comprises 40 patients who exhibited either a severe acute or late side effect (RTOG grade ≥3) following XRT or a secondary malignancy in the irradiated site within 10 years of treatment. Fibroblast strains from 5 of the 7 hypersensitive patients evaluated to date using the γ-H2AX assay showed elevated levels of residual foci at 24 h after in vitro exposure to 4 or 8 Gy of γ-rays when compared to normal fibroblasts. Only 1 of the 7 cell lines, RS008ED, displayed overt radiosensitivity in the clonogenic survival assay. Of interest in this regard are the above-mentioned data from Lobrich and colleagues [54] relating to an unusual patient who had exhibited severe complications after XRT; lymphocytes from this patient showed abnormally high levels of residual γ-H2AX foci at 24 h after a CT scan, and their fibroblasts were radiosensitive and clearly DSB repair deficient as indicated by γ-H2AX resolution in vitro. Surprisingly, lymphocytes from this patient irradiated in vitro did not display such a defect in resolving γ-H2AX foci [54]. The combined results of these two studies therefore serve to remind us that we still have much to learn about the

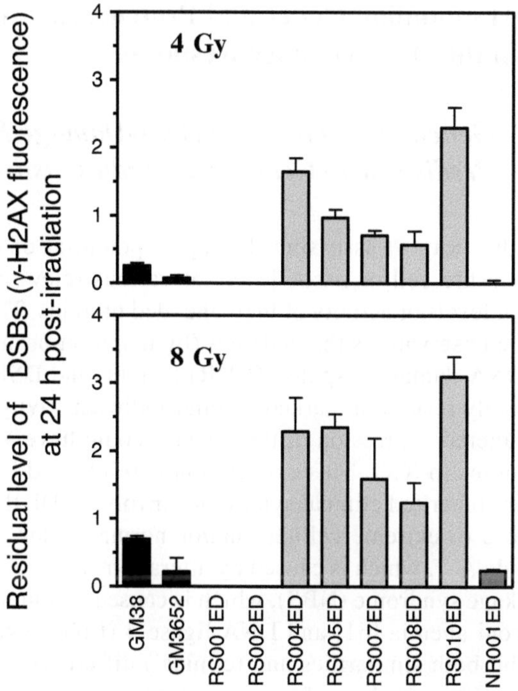

Fig. 4 Residual levels of γ-H2AX (i.e., DSBs) measured at 24 h after exposure of various fibroblast cell cultures to either 4 or 8 Gy of γ-rays. Cell types are as follows: GM38 and GM3652, normal fibroblast strains; RS001/2/4/5/7/10ED, fibroblasts from hypersensitive cancer patients; NR001ED, a fibroblast strain derived from a patient who exhibited no adverse reaction to XRT

complexity of the data generated by the γ-H2AX assay in the context of patient-derived samples.

Although beyond the scope of this chapter, it is interesting to note that early induction of γ-H2AX foci has proven useful in the prediction of dose delivered to XRT patients based on peripheral blood samples [28] and in the identification of individuals with A-T based on the *ex vivo* response of their lymphocytes to 2 Gy of γ rays [45].

In summary, the γ-H2AX assay is extremely sensitive and versatile. Because differences in patient response to XRT undoubtedly have a multifactorial origin, the use of so-called "intermediate end points" based on functional assays such as γ-H2AX IRIF formation and resolution that integrate the responses of multiple proteins is a powerful adjunct to the genomic studies that is discussed later in Sect. 5. Nonetheless, it is fair to say that there is currently no clear consensus regarding the predictive ability of the kinetics of γ-H2AX resolution in the context of the severity of XRT complications. At this point in time there is little information regarding its applicability to predicting tumor responses. One potentially important factor in predicting tumor responses is the cells' p53 status. Olive and colleagues [36] have shown that the correlation between clonogenic survival and γ-H2AX resolution is only apparent when cells are stratified according to their p53 status. We have made essentially the same observation for noncancerous cells [20].

3 Mammalian Genes and Proteins Involved in the DNA-Damage Response

3.1 Genetic Defects in the DNA-Damage Response Predispose to Cell/Tissue Radiosensitivity

Another way of assessing the repair potential of mammalian cells/tissues is to measure cell or tissue levels of various relevant DNA-repair gene mRNAs and/or levels or activity of their encoded proteins. This approach is based largely on the observations that a deficit (by mutation or otherwise) of genes encoding for DNA-damage response (DDR) factors (and DSB repair genes in particular) generally results in marked cellular radiosensitivity, leading to the expectation that altered expression of these genes could have a major impact on cell/tissue responses to XRT. This expectation is reinforced by the existence of a number of rare inherited clinical syndromes involving DDR/DNA-repair genes that predispose to extreme cellular and/or normal tissue radiosensitivity [61]. These include A-T, which is caused by alterations in the *ATM* gene [62, 63], Nijmegen breakage syndrome (NBS), which is caused by mutations in the *NBS1* gene [64], Fanconi anemia [61] and DNA ligase 4 (LIG4) syndrome [65]. A-T and NBS exhibit both similarities and (clinical) differences. Importantly, A-T heterozygotes within a cohort of prostate cancer patients were found to have a radiosensitive phenotype [66], indicating that even a single defective allele can impact on XRT outcome. Another human syndrome similar to A-T, "ataxia telangiectasia-like disorder" or ATLD, is associated with mutation of another DDR gene, *MRE11*.

3.2 The DNA-Damage Response

Exposure of human cells to IR activates a complex response network, collectively known as the DDR, which involves a coordinated series of signaling events that invoke changes in the levels and/or activity of a large number of genes and proteins. Important among these are proteins involved in DNA repair and in cell-cycle checkpoint activation [67, 68]. As discussed in Sect. 3.5, damaged cells can transiently activate cell-cycle checkpoints which allow DNA repair to progress without the complication of critical ongoing DNA-metabolic processes such as DNA synthesis and chromosomal segregation, potentially promoting cell survival. Failure to properly activate such responses in normal cells can lead to genetic instability and cell death or, at the organism level, to the development of cancer if the cell does not die. The important DNA-repair pathways and key proteins required for the repair of IR-induced DNA lesions are outlined in Tables 2 and 3, and are briefly described here.

Table 2 Some important proteins involved in the three major enzymatic pathways for repairing ionizing radiation-induced DNA damage in human cells. For full names of these proteins, see Table 3

Pathway		
Non-homologous end-joining (NHEJ)	Homologous recombination repair (HRR)	Base excision repair (BER)/SSB repair
DNA-PK$_{cs}$, KU70, KU80, XRCC4, LIG4, XLF, PNKP, FEN1, Artemis, APTX, PARP-1, PALF, POLμ, POLλ, MRE11, RAD50, NBS1	RAD51/51B/51C/51D, XRCC2, XRCC3, RAD52, RAD54, RAD55, RAD57, RPA, BRCA1, BRCA2, MRE11, RAD50, NBS1, hMSH1, hMSH6, FEN1, FANC-A/C/D2/E/F/G/L, MUS81-EME1, GEN1, CTP1	NEIL1, NEIL2, NEIL3, hNTH1, FPG, OGG1, OGG2, APE1, PNKP, APTX, LIG1, LIG3, PCNA, FEN1, RFC, POLβ, POLδ, POLε, PARP-1, XRCC1, XPG

3.3 Base Excision Repair

The Base Excision Repair (BER) pathway repairs many types of IR-induced DNA damage; these include various types of base damage and SSBs [69]. Important proteins for repairing base damage include NEIL1, NEIL2, NEIL3, hNTH1, FPG, OGG1 and OGG2; these are examples of the damage-specific DNA glycosylases which initiate the repair of base lesions by cutting the *N*-glycosidic bond connecting the damaged base to the ribose–phosphate backbone, generating an apurinic/apyrimidinic (AP) site. The ribose–phosphate chain at the resulting AP site is then cleaved either by an AP lyase (at the 3′ side) or AP endonuclease (at the 5′ side). The major human AP endonuclease is called APE1 (sometimes called HAP1). Some of the DNA glycosylases also exhibit AP-lyase activity. Next, any "dirty" strand break termini have to be processed to generate "clean" 3′-OH ends before a DNA polymerase can act on them to restore the missing nucleotides, as well as "clean" 3′-OH/5′-phosphate termini so that DNA ligases can seal the break [70]. This requires various deoxyribophosphodiesterases or exonucleases such as PNKP [71, 72], APTX [73], and APE1 itself [74]. The processing of "frank" IR-induced SSBs largely overlaps that of base damage subsequent to the generation of the AP sites.

Further processing of cleaved AP sites involves one of two BER sub-pathways: short-patch or long-patch [75]. In short-patch BER, POLβ replaces a single nucleotide and LIG3 seals the gap. Long-patch BER, which may be invoked to repair more complex lesions [76], involves the removal and resynthesis of up to 15 nucleotides; key participants in this step are PCNA, FEN1, RFC, and POLδ or POLε. LIG1 seals the resulting gap. Not surprisingly, in view of the many coordinated/sequential activities involved, a number of protein-protein interactions are important for BER, at least in part by promoting protein stability. For example, LIG3 interacts with POLβ, PARP-1, and XRCC1. The role of XRCC1 appears to be in the detection and coordination of the processing of SSBs, where it provides a scaffold for assembling the repair complexes [77, 78]; it also interacts with PNKP, stimulating the end-processing activity of the latter [79, 80]. The XPG(ERCC5) protein,

Table 3 Some important radiosensitivity proteins

AIF	Apoptosis inducing factor
APE1	Human AP endonuclease 1
APTX	Aprataxin
ATM	Mutated in ataxia telangiectasia
ATR	ATM and RAD3 related
ATRIP	ATR interacting protein
BAK	BCL-2 homologous antagonist killer
BARD1	BRCA1-associated RING domain protein 1
BAD	BCL-2-associated agonist of cell death
BAX	BCL-2-associated X protein
BCL-2	B-cell lymphoma 2
BCL_{XL}	B-cell lymphoma-extra large
BIK	BCL-2-interacting killer
BIM	BCL2L11, protein from BCL-2 protein family
BRCA1	Breast cancer type 1 susceptibility protein
BRCA2	Breast cancer type 2 susceptibility protein
BLM	The helicase associated with Bloom's syndrome
CDC25/25A/45	CDC45 cell division cycle 25/25A/45-like (*S. cerevisiae*)
CDK1/2	Cyclin-dependent kinase 1/2
CDKN1A	$p21^{WAF1}$/cyclin-dependent kinase inhibitor 1A
CHK2	Checkpoint kinase 2
CK2	Casein kinase 2
CSB	Cockayne syndrome type B (ERCC6)
$DNA\text{-}PK_{cs}$	DNA-dependent protein kinase, catalytic subunit
DR5/TRAIL receptor 2	Tumor necrosis factor receptor superfamily, member 10
ERCC1, etc.	Excision repair cross complementing 1, etc.
FANC-A/C/D2/E/F/G/L	Fanconi anemia proteins A/C/D2/E/F/G/L
FEN1	Flap endonuclease 1
FPG	Formamidopyrimidine-DNA glycosylase
γ-H2AX	Serine-phosphorylated histone H2AX
hMLH1	Human homolog of MutL, 1
hMSH1/2/3/6	human homolog of MutS, 1/2/3/6
hNTH1	Homolog of Escherichia coli endonuclease III
iASPP	Inhibitor of apoptosis stimulating proteins of p53.
JNK	c-Jun N-terminal kinase
LIG1/3/4	DNA ligase I/III/IV
MAP kinase	Mitogen-activated protein kinase
MCL1	Induced myeloid leukemia cell differentiation protein
MDC1	Mediator of DNA damage checkpoint 1
MRE11	Meiotic recombination 11
mTOR	Mammalian target of rapamycin
NBS1	Protein product of the gene mutated in Nijmegen breakage syndrome

(continued)

Table 3 (continued)

NEIL1/2/3	Homologs of E. coli endonuclease VIII 1/2/3
NF-κB	Nuclear factor kappa-light-chain-enhancer of activated B cells
NOXA	Noxa (Latin for damage): a pro-apoptotic member of the BCL-2 protein family
OGG1/2	8-oxoguanine glycosylase 1/2
PALF	PNK/APTX-like factor
PARP-1	Poly(ADP-ribose) polymerase 1
PCNA	Proliferating cell nuclear antigen
PKB/AKT	Protein kinase B
PKC	Protein kinase C
PNKP	Polynucleotide kinase-phosphatase
POLβ/δ/ε/μ/λ	DNA polymerase β/δ/ε/μ/λ
PTEN	Phosphatase and tensin homolog
PUMA	p53 upregulated modulator of apoptosis
RFC	Replication factor C
RPA	Replication protein A
SMAC/DIABLO	Second mitochondria-derived activator of caspases
SMC1/3	Structural maintenance of chromosomes 1/3
SOD2	Mitochondrial iron/manganese superoxide dismutase
XLF	XRCC4-like factor (also known as Cernunnos)
XPA/C/D/F/G	Xeroderma pigmentosum A/C/D/F/G
XRCC1/2/3/4 etc.	X-ray repair cross complementing 1/2/3/4 etc.
9-1-1 complex	RAD9/HUS1/RAD1
53BP1	p53-binding protein 1

which is best characterized for its role in the nucleotide excision repair pathway, also functions as an accessory factor in BER [81].

Although PARP-1 plays a critical role in BER, the nature of this involvement remains to be clearly defined; it may help to recruit repair proteins to DNA ends [82] as well as protecting these ends from nuclease activity [83]. By inhibiting the nuclease-mediated conversion of SSBs to DSBs, PARP-1 may allow time for BER to be completed [84].

3.4 DSB Repair Pathways

Mammalian cells repair IR-induced DSBs mainly through two pathways: NHEJ and HRR. It is generally believed that, outside of S- and G_2-phase, NHEJ is the prevailing mechanism [85]. These pathways share some components; for example, the recognition and early signaling of a DSB in both pathways appears to require the MRE11-RAD50-NBS1 (MRN) complex [86, 87]. MRN contributes in several ways to DSB repair (see below) and activation of cell-cycle checkpoints, as well as transduction of signals [68, 88]. γ-H2AX appears to form at DSBs regardless of whether they are processed by HRR or NHEJ and to play a role in both pathways [89], although its functional role in NHEJ is controversial [85]. Phosphorylation of H2AX induces chromatin remodeling [60] and appears to be essential for recruiting repair proteins and additional chromatin-remodeling proteins [42, 90]. Chromatin remodeling is important for both pathways, initially to facilitate access and assembly of repair complexes, then for restoring the chromatin once the break is rejoined. These processes, which involve posttranslational modification of the tails of histone proteins, such as methylation and phosphorylation, as well as acetylation by enzymes such as the NuA4/TIP60 histone acetyltransferase complex and later their deacetylation by histone deacetylases (e.g., [85, 91, 92]), would intuitively be more important for HRR, in which extensive (kilobase) tracts of DNA are involved in strand exchanges and where γ-H2AX may be more important [52, 85].

3.4.1 Nonhomologous End Joining

NHEJ proteins catalyze the direct rejoining of broken incompatible DNA ends; this requires no or as little as one base pair of sequence homology, and does not involve strand exchanges [85, 93, 94]. The likelihood of an illegitimate rejoining event is therefore high. The NHEJ proteins are unusual in that they can process DNA ends with a diverse range of overhang length, sequence and chemistry. Indeed, the rejoining of the broken ends is probably an iterative process that can proceed via several possible routes [85].

Important participants in NHEJ include the three proteins that comprise the DNA-PK holoenzyme complex, namely the DNA-PK$_{cs}$(XRCC7) catalytic serine/threonine kinase subunit, and the KU70(XRCC6) and KU80(XRCC5) (also known as KU86) proteins that heterodimerize to form the ring-shaped regulatory "KU" subunit. In response to an IR exposure, the highly abundant KU is believed to initiate NHEJ by binding to a DSB, where it stabilizes and aligns the broken termini and protects them from degradation as well as providing a docking point for other proteins [85]. DNA-PK$_{cs}$ is then recruited to generate the active DNA-PK complex. Activated DNA-PK$_{cs}$ is then able to phosphorylate its downstream targets, which include itself and p53. KU70 also interacts with MRE11, an interaction that may recruit the MRN complex to these breaks [95] where it is involved in micro-homology searching [96]. A complex of three other NHEJ proteins, XRCC4, LIG4, and XLF(Cernunnos), performs tail removal and gap filling/ligation functions and is the most flexible DNA ligase known [85, 97]. As with BER, ligation requires the generation of "clean" 3′-OH/5′-phosphate termini by proteins such as PNKP [98], FEN1, MRN, Artemis [99], APTX, and possibly PARP-1 [100]. The major nuclease in NHEJ is the Artemis-DNA-PK$_{cs}$ complex which has an array of nuclease activities, including the ability to endonucleolytically process a variety of types of damaged overhangs to facilitate their ligation. Additional nuclease activity may be provided by PALF (also known as APLF) [85]. Nuclease resection in NHEJ typically occurs over a span of 0–14 base pairs [85]. Other important factors in break rejoining are the various error-prone DNA polymerases, and especially POLμ and POLλ, which bind to KU/DNA complexes [85]. There are many regulatory interactions among the NHEJ proteins, especially involving XRCC4. Recently, an "alternative" NHEJ pathway has been described that appears to be independent of some of the factors that mediate "classical" NHEJ [85].

3.4.2 Homologous Recombination Repair

Unlike NHEJ, HRR requires extensive sequence homology and thus provides a mechanism for error-free repair of a break [101]. HRR is preferred in S- and G_2-phase cells where homologous recombination between sister chromatids can occur efficiently; axiomatically, NHEJ is preferred in G_1/G_0-phase cells. Important HRR proteins include RAD51 and the five RAD51 paralogs (RAD51B, RAD51C, RAD51D, XRCC2, and XRCC3) as well as RAD52, RAD54, RPA, and the BRCA1 and BRCA2(XRCC11) tumor suppressor proteins [102, 103]. RAD51, a DNA-dependent ATPase, catalyses strand exchange between homologous DNA molecules in the presence of the RPA single-stranded DNA-binding protein [104, 105]. RAD51 also interacts with and is modulated by RAD52, RAD54, BRCA2, and p53: for example, interaction with RAD52 enhances its polymerization and homologous pairing/strand exchange activities [105, 106]; phosphorylation of RAD52 post-irradiation is important for this interaction [107]. RAD54, another DNA-dependent ATPase, may function with RAD51 in stabilizing protein–DNA complexes, in remodeling chromatin [108, 109], and in homology searching/strand invasion. The RAD51 paralogs facilitate formation of the polymeric RAD51 filaments. Collectively, RAD51 and its paralogs likely mediate homology searching and strand pairing/exchanges. The MRN complex, discussed above in the context of NHEJ, is involved in HRR in several potential ways, including (1) resecting 5′ DNA ends, (2) removing excess DNA at 3′ flaps, and (3) providing endonuclease, exonuclease, and helicase activities [96]. FEN1 is probably involved in removing 5′ flaps from HRR intermediates [68, 87]. Completion of HRR requires the resynthesis of the deleted DNA sequences by DNA polymerases using the intact homologous sequence for a template and joining of the newly synthesized fragments by DNA ligases.

Additional proteins important for HRR include several other human homologs of the yeast Rad proteins, such as RAD55 and RAD57, as well as the mismatch repair proteins hMSH1 and hMSH6, the Fanconi anemia proteins (FANC-A, FANC-C, FANC-D2, FANC-E, FANC-F, FANC-G(XRCC9), and FANC-L), the MUS81-EME1 endonuclease, and "resolvase" endonucleases such as GEN1 which resolve Holliday junctions. BRCA1- and BRCA2-deficient cells are deficient in HRR but not in NHEJ [110]. Both BRCA1 and BRCA2 associate with the RAD51 complex [111], and indeed the involvement of BRCA2 in DSB repair primarily involves the regulation of RAD51; BRCA1, on the other hand, has a more general role in linking DSB sensing, signaling and effector responses [112, 113].

3.5 *Other DNA-Damage Response Proteins*

The DDR proteins mediate pro-survival responses, such as DNA-damage recognition/repair (discussed above) and the transient activation of cell-cycle checkpoints, as well as cell-death responses such as engagement of the apoptotic and stress-induced premature senescence (SIPS) pathways [2, 114] (see below), as well as less well characterized responses such as autophagy, a conserved stress response regulated by signaling through the mTOR protein [115–117]. Genes and their encoded proteins involved in checkpoint activation, signal transduction, apoptosis, SIPS and autophagy (notably, the components of the mTOR signaling pathway), along with genes whose encoded products respond to early DSB-associated chromatin changes, such as MDC1 and 53BP1, and the E3 ubiquitin ligases RNF8 and RNF168 which are required for the accumulation of DNA repair proteins at DSB sites, are all candidate biomarkers for radiosensitivity testing. Others include CK2 (which can phosphorylate MDC1), cohesins (which are recruited by γ-H2AX to tether the DNA ends during repair), CTP1$^{Sae2/CtIP}$ (a DNA end-processing factor that interacts with MRN in HRR [118]), as well as the γ-H2AX phosphatases and FACT (a heterodimeric histone exchange factor consisting of SSRP1 and SPT16 that exchanges γ-H2AX out of the histone octamer) [85]. There are many others, including sumoylating activities, such as those that act on RAD52 in HRR [119], the 9-1-1 (RAD9/HUS1/RAD1) complex which is involved in sensing some types of bulky DNA lesions but whose role in the response of human cells to IR-induced DSBs is not well defined, and factors such as those recently identified in a broad screen of in vivo substrates for the ATM and ATR kinase activities and confirmed to have a role in the DDR [120].

3.5.1 ATM

The ATM serine-threonine kinase is a major integrator of the cellular response to DSBs [121]. Activation of ATM's kinase activity is dependent on the autophosphorylation and relocation of ATM to the site of the DSB [122]; the recruitment step involves the MRN complex and possibly MDC1 [68, 123, 124]. ATM is also post-translationally modified by other DDR proteins; for example, its acetylation in response to DNA damage is mediated by the TIP60 histone acetyltransferase [125]. The activated ATM molecule can phosphorylate many substrates, including checkpoint regulators and repair proteins such as p53, MDM2, 53BP1, CHK2, BRCA1, BLM, NBS1, H2AX, RAD9, hRAD17, MDC1, SMC1, and c-ABL [68, 120, 126]. These proteins are phosphorylated either directly by ATM or indirectly via the CHK2 checkpoint kinase [121, 127].

3.5.2 p53

An important target of the ATM kinase activity is the p53 tumor-suppressor protein [128, 129]. Following irradiation of cells, wild-type p53 undergoes a series of post-translational modifications by enzymes such as ATM and CHK2 [130, 131] that result in the stabilization/increased levels and activation of p53. Activated p53 regulates many critical responses that follow an IR exposure, such as the coordination of repair and checkpoint activation, either by directly interacting with downstream effector proteins or indirectly by transcriptionally regulating target genes. It also regulates both pro-survival (DNA repair and cell-cycle checkpoint activation) and pro-death (such as apoptosis and SIPS) responses to genomic injury. p53 regulates several DNA-repair pathways, including BER, NHEJ and HRR, by a number of mechanisms (reviewed in [129, 132]).

3.5.3 Cell Cycle Checkpoints

Cells exposed to IR transiently delay their movement through the cell cycle by activating checkpoints in the G_1, S and G_2 phases. Activated wild-type p53 plays a critical role in the G_1 checkpoint by activating transcription of the *p21*^{WAF1} gene; the protein encoded by this gene, the p21^{WAF1} (CDKN1A) cyclin-dependent kinase inhibitor, in turn inhibits the G_1 cyclin-dependent kinases (CDKs), leading to hypo-phosphorylation of the pRb (retinoblastoma) protein and to sequestration of the E2F transcription factors required for progression to S-phase. The S-phase checkpoint, which slows down DNA replication in the presence of DNA lesions, is triggered by parallel ATM-regulated pathways in which activated ATM phosphorylates the checkpoint kinase CHK2 and an alternate pathway involving phosphorylation of NBS1, SMC1, BRCA1, and FANC-D2 [133]. Downstream of CHK2 are CDC25A, CDC45, and cyclin E/CDK2; this "ATM-CHK2-CDC25A axis" suppresses origin firing when IR-induced DNA lesions are present. Downstream from MRN are SMC1 and Tipin (Tim/Timeless-interacting protein); this pathway inhibits the progression of replication forks that encounter a DNA lesion. Both RAD51 and p53 are also required for fork slowing following IR exposure [133]. Important regulators of the G_2 checkpoint include ATM, ATR, CHK1, CHK2, CDC25, CDK1, and 14-3-3σ; its activation involves both p53-dependent and p53-independent mechanisms [134]. A second ATM-dependent G_2 checkpoint has been described which may be important for low dose rate exposures [135], but the role of p53 in this pathway is not clear.

3.5.4 Apoptosis

This genetically regulated form of cell death has two alternative pathways: the *intrinsic* (or "mitochondrial") pathway and the *extrinsic* (or "death receptor") pathway; both of which result in activation of a cascade in which caspase proteins function as either initiators (caspases 2/8/9/10) or effectors/executioners (caspases 3/6/7)

[136, 137]. The mitochondrial pathway is regulated mainly by the BCL-2 family of proteins: pro-apoptotic members include the BAX-like (e.g., BAX and BAK) and the BH3-only (e.g., BIM, BIK, BAD, NOXA, and PUMA) proteins; anti-apoptotic members include BCL-2, BCL_{XL}, and MCL1. Other important proteins are survivin, SMAC/DIABLO and AIF. Important elements of the extrinsic pathway include the death receptors such as FAS/CD95 and DR5/TRAIL receptor 2. Many genes that positively or negatively regulate the apoptotic threshold are regulated by p53. Another pathway activated by IR involves the enzyme acid sphingomyelinase (ASMase) which translocates from lysosomes to the outer layer of cell membrane [138, 139]. ASMase can then hydrolyze sphingomyelin, generating the second messenger ceramide which can trigger apoptosis in some cell types, notably vascular endothelial cells, which may contribute to both normal tissue and tumor responses to XRT.

3.5.5 Stress-Induced Premature Senescence

SIPS, sometimes referred to as accelerated senescence, is another genetically regulated form of cell death that may contribute significantly to normal tissue/tumor responses to XRT [114, 140–142]. It is a favored response of fibroblasts and of many p53 wild-type solid tumor-derived cell lines and manifests as a state of permanent cell cycle arrest, although the cells retain metabolic activity (e.g., [141] and references therein). An important effector of this response is the $p21^{Cip1/Waf1}$ (CDKN1A) protein described above in the context of the G_1 checkpoint. Cells undergoing SIPS following exposure to IR exhibit sustained up-regulation of the *p21* gene/p21 protein in their nucleus (e.g., [141]). p21 activates SIPS by several mechanisms, including CDK inhibition, inhibition of PCNA leading to suppression of DNA synthesis, and alteration of gene expression (reviewed in [143, 144]).

3.5.6 Mitotic Catastrophe/Autophagy

Mitotic catastrophe, in which cells fail to undergo proper mitosis after DNA injury (probably as a result of chromosome mis-segregation and cell fusion), may contribute significantly to clonogenic cell death following an IR exposure [145, 146] and to the response of p53-deficient solid tumors to XRT [114]. Autophagy is another genetically regulated form of programmed cell death in which the cell essentially undergoes self-digestion; it occurs in a variety of human tumor cell lines following exposure to IR and is regulated by the mTOR protein [116, 117, 147].

3.5.7 Signal Transduction

Coordination of the DDR involves components of the intrinsic signal transduction pathways that are utilized to mediate decisions about cell growth and differentiation in response to environmental cues such as those relating to nutrient or oxygen availability [148]. A number of these signaling kinases, including c-JUN, c-FOS, RAS, RAF, NF-κB, MAP kinase, PKC, PKB/AKT, and SAP kinase/JNK, as well as phosphatases such as PTEN that act in these signaling pathways, represent candidate gene products for analysis in the context of predictive biomarkers for XRT outcome.

4 DNA-Repair Gene Transcript/Protein Levels/Activity and Radioresponsiveness

Inactivation of BER genes typically results in only modest, and often no, increase in cellular radiosensitivity, as opposed to DSB-repair defects which typically impart considerable radiosensitization. Also, for BER there can be considerable redundancy of enzymatic function as well as of pathway choice, which probably limits its value for prediction in XRT. Partly for these reasons, only a few studies have examined BER genes/proteins from the perspective of radiosensitivity testing. Most studies have focused on the APE1 AP-endonuclease, and the majority of these have suggested that elevated APE1 levels are predictive or prognostic of poor outcomes. For example, a correlation was found between levels of APE1 and radioresistance in primary pretreatment cervical carcinoma biopsy tissue [149]. High tumor nuclear APE1 levels were significantly associated with poor clinical response among patients with locally advanced squamous-cell carcinoma of the head and neck (SCCHN) treated with radical chemoradiotherapy [150]. APE1 protein levels and activity were significantly higher in adult glioma tissue than in adjacent normal brain [151]; the speculation that this might result in glioma resistance to adjuvant XRT was subsequently confirmed by these authors [152] and extended to medulloblastomas and primitive neuroectodermal tumors treated with XRT and chemotherapy [153]. APE1 protein levels were also abnormally high in germ cell tumors from some testicular cancer patients, and it was again suggested that this might contribute to the relative resistance of these tumors to therapy [154]. APE1 levels were also found to be elevated in prostate cancer [155]. Interestingly, APE1 was among a number of proteins found to be consistently elevated in 3 stably radioresistant prostate cancer cell lines compared to their parental lines [156]; these radioresistant lines were generated by repeated daily exposure to 5 fractions of 2 Gy X-rays. APE1 protein levels in cervical cancer were also higher than in normal cervical tissue, and the survival time of patients with high tumor APE1 levels was significantly shorter than those with low APE1 levels [157]. A negative prognostic significance of elevated APE1 has also been reported for osteosarcoma [158], pancreatic cancers [159] and ovarian and gastroesophageal cancers [160].

In contrast to these studies, APE1 protein levels were not predictive of the radiosensitivity of 11 human tumor and fibroblast cell lines [161] or of patient survival in 88 early stage invasive cervical cancer samples [162]. Sak and colleagues [163] actually reported high levels of APE1 and XRCC1 to be strongly associated with *better* outcome in patients undergoing XRT for muscle-invasive transitional cell carcinoma of the bladder.

Clearly there are studies that do and do not support the general validity of elevated APE1 as a predictive/prognostic marker for poor outcome. Indeed, there are a number of potential confounding factors that might contribute to such inter-study differences. First, APE1 is a multifunctional protein that, in addition to its enzymatic role in BER, has key roles in the activation of several transcription factors including NF- B and wild type p53, which could also contribute to its involvement in XRT outcomes [150, 164]. Alterations in DNA repair activity have also been associated with genetic instability and a more aggressive tumor phenotype, a mechanism that was invoked by Al-Attar and colleagues to interpret their data with ovarian and gastroesophageal cancers [160]. Finally, APE1 expression may be higher in poorly differentiated versus well-differentiated tumors [165]; in fact, the poor response to XRT of bladder tumors with low expression of XRCC1 and APE1 [163] has been suggested to partly reflect the poorly differentiated/more aggressive nature of these tumors, leading to the question of whether APE1/XRCC1 is a prognostic factor relating to tumor aggressiveness or whether it is predictive of response to XRT.

As regards other BER genes, *XRCC1* mRNA levels varied significantly among cell lines derived from human SCCHN, but this parameter was not significantly associated with either cellular radiosensitivity or patient response [166]. Little has been done with DNA glycosylase genes/enzymes in the context of clinical radiosensitivity; studies of other BER factors such as FEN1 and PNKP in the context of clinical radiosensitivity prediction are in progress but the results have not been reported at this time (M. Weinfeld, personal communication).

For reasons noted above, there has been more interest in genes/proteins involved in DSB repair, and especially in the NHEJ pathway which is believed to dominate in human cells. As with BER, the expectation from the DDR perspective is that high expression levels of DSB repair proteins should be associated with radioresistance and hence poorer local control and possibly survival. High levels and/or activity of components of the DNA-PK complex were indeed reported to be predictive of radioresistance in cell lines derived from human SCCHN [167], esophageal cancer [168], and lung carcinoma [169]. Of two small cell lung cancer (SCLC) cell lines established from the same patient, the radioresistant subline exhibited much faster DSB rejoining in the PFGE assay after 20 Gy as well as higher levels of both the DNA-PK$_{cs}$ and RAD51 proteins compared with the more radiosensitive subline [170]. In contrast, there was no correlation between NHEJ component levels/activity and radiosensitivity in a panel of human malignant glioma cell lines [171] or in biopsies from patients with therapy-naïve SCCHN [172].

In regard to clinical tumor responses, elevated levels of DNA-PK components measured pretreatment in cell lines or biopsy material were found to be predictive of poor tumor responsiveness to XRT and/or survival/prognosis in patients with carcinomas of the colon/rectum [173–176], oropharynx/hypopharynx [177], and nasopharynx [178, 179]. In the nasopharyngeal carcinoma study by Lee and colleagues [178], for example, those patients receiving XRT without or with concurrent chemotherapy whose tumors had lower KU70 (but not DNA-PK_{cs}) levels had significantly better local control; however, metastasis-free survival was not improved. Disease-free survival was significantly better among endometrial carcinoma patients whose tumors had lower levels of KU70, but again overall survival was not improved [180]. In patients with non-small cell lung cancer (NSCLC), high tumor to normal tissue expression ratios of *DNA-PK_{cs}*, but not *KU80*, mRNA levels were associated with decreased survival and shorter median survival time [181]. The association for *DNA-PK_{cs}* was stronger in patients who received adjuvant chemotherapy or XRT than in those undergoing surgery alone. Interestingly, tumor *DNA-PK_{cs}* mRNA levels per se were not significantly associated with survival, leading the authors to suggest that tumor to normal tissue ratios may be the better predictor.

Not all of the studies published to date support an association between high levels of DSB repair proteins/transcripts and diminished response to therapy. For example, there was no association between the levels of DNA-PK_{cs}, KU70 or KU80 proteins and tumor radiosensitivity in oral squamous cell carcinoma (OSCC) cell lines and in OSCC patients treated with preoperative XRT; however, DNA-PK levels did increase after IR exposure in a manner that correlated with tumor radioresistance [182]. Similarly, neither DNA-PK activity nor levels of the DNA-PK_{cs}, KU70 or KU80 proteins in glioma cell lines correlated with the clinical response of the patients from which they were derived [171], nor were differences in the levels of DNA-PK_{cs}, KU70, KU80, XRCC4 and NBS1 apparent among specimens from tumor (and normal) tissue types with differing clinical radiosensitivity [183, 184]. A study of progressive esophageal cancers suggested that patients having tumors with *high* expression of DNA-PK_{cs} actually had a better short-term response to chemotherapy-XRT than the low-expressing patients based on tumor regression, although local control was not evaluated [185]. High pretreatment levels of DNA-PK_{cs} and KU80 in combination with low levels of p53 in tonsillar carcinoma biopsies correlated with better locoregional control and survival of patients after XRT [186]. In pretreatment SCCHN tumor biopsies from patients receiving chemoradiotherapy or XRT after induction chemotherapy, tumors in the responder group also had significantly higher *KU70*, *KU80*, and *DNA-PK_{cs}* mRNA levels than those in the non-responder group [187].

One of the few tumor sites where there have been multiple studies that allows some comparison of findings from different research groups is cervical carcinoma. Wilson and colleagues [188] reported a borderline significant correlation between pretreatment KU70 (but *not* KU80) levels and clinical radioresistance (survival) in a panel of 53 cervical carcinoma tumors. Harima et al. [189] also found significantly better clinical response to XRT for KU80-low versus KU80-high tumors; patients with KU80-low tumors also had better survival. In contrast, Beskow and colleagues [190] saw no correlation between response to XRT and DNA-PK_{cs}, KU70 or KU80 levels in biopsies from cervix cancer patients receiving preoperative brachytherapy followed by surgery, nor did the levels of these proteins correlate with long-term survival. These authors [190] noted the obvious discrepancy between their data and the above-mentioned two reports suggesting that low levels of DNA-PK components are predictive of radioresponsiveness in cervical carcinoma. Among the variables that they consider might contribute to this variability are differences between studies in the following: cohort size; methodology (examples might be specific antibodies and sample integrity); tumor volume; tumor stages included in the study cohort (which were certainly quite different among these studies); patient demographics, such as age; treatment plans; intra-tumor heterogeneity; and microenvironmental factors such as hypoxia.

An interesting recent study of cervical cancer patients who had received preoperative XRT and radical surgery showed that tumor DNA-PK_{cs}, KU70 and KU80 protein levels were significantly increased in residual tumors compared to the corresponding primary tumor [191].

As regards normal tissue responses, levels of the NHEJ proteins DNA-PK_{cs}, KU70, KU80, XRCC4, and LIG4 as well as of the RAD51 HRR protein and of the ATM, RAD1 and HUS1 DDR proteins were similar in skin fibroblast cell lines and control cells established from cancer patients with different normal tissue reactions to XRT [192]. Although differences were noted in DNA-PK and XRCC4-LIG4 activity among these cell lines, these were not significantly associated with either the clinical normal tissue response of the patients or with the in vitro radiosensitivity of the cell lines [192]. Loong and colleagues [193] examined lymphoblastoid cell lines derived from 5 patients with late radionecrosis and found that 2 of the lines exhibited cellular radiosensitivity and had a reduced ability to rejoin DSBs in the PFGE assay; they also had six- to tenfold reduced DNA-PK activity in vitro in cell-free extracts despite having normal levels of KU70, KU80, XRCC4 and DNA-PK_{cs} protein.

The role of alterations in BRCA1 and BRCA2 on DNA-repair activity and cancer therapeutics has been discussed elsewhere [194], and is not further considered here.

5 Single Nucleotide Polymorphisms and Response to XRT

5.1 Genetic Polymorphism

As noted in Sect. 3.1, it is fairly well established that interindividual variations in DNA-repair capability arise in part as a result of subtle variations in the genome and that these variations may in turn contribute to the marked differences in sensitivity to normal tissue complications that are seen among cancer patients following XRT [2, 195, 196]. Consequently, the identification of polymorphisms (allelic variants) among individuals in a population has become increasingly important in the context of predicting responses to XRT, with the ultimate expectation of delivering individualized therapy based on validated biomarkers.

The types of variations that occur in the genome include repeating sequences such as short tandem repeats (STRs) and variable number of tandem repeats (VNTRs) as well as single nucleotide polymorphisms or "SNPs," which are usually defined as changes at a single base pair that occur in at least 1% of the population

[197] as opposed to mutations which occur at a frequency below 1%. Although the genome is ~99.9% identical among individuals, the ~0.1% variations (most of which are SNPs) tend to be heritable and stable [198]. For each SNP, a person will be either homozygous for the common allele (wild type), heterozygous for the common/minor allele, or occasionally homozygous for the minor/variant allele. It is estimated that as many as 400,000 SNPs occur in coding regions of genes (so-called cSNPs). Such cSNPs can result in an amino acid substitution (i.e., a missense or non-synonymous event), which may either be innocuous or result in a change in protein function, or they can encode the same amino acid (i.e., a silent or synonymous event). Synonymous SNPs can still have biological consequences, e.g., by altering tRNA binding and mRNA structure/stability. A similar number of SNPs occur in 3′ or 5′ flanking DNA adjacent to the coding sequences; these are called perigenic or "pSNPs" and have the potential to affect the phenotype by altering gene expression or mRNA stability and thereby levels of functional proteins. The remaining and most frequent SNPs are the "rSNPs" that occur randomly in noncoding sequences; these include intronic SNPs, which may influence RNA splicing.

Predisposition to normal tissue toxicity following XRT, i.e., clinical radiosensitivity, is presumed to be a complex phenotype or "quantitative trait" that is polygenic in origin, i.e., it probably reflects the aggregate impact of small alterations in several nonallelic genes [196, 199, 200, 202, 203]. Axiomatically, a polymorphism in a single DNA repair-related gene, or indeed any single marker, should not be able to reliably describe clinical radiosensitivity. Studying multiple SNPs in a panel of candidate "radiosensitivity" genes has thus become the common approach for examining associations with normal tissue complications, and this topic has been the subject of several reviews [199–205]. Typically, candidate gene studies have examined known SNPs available from public-domain databases and have focused on genes involved in DNA repair, cell cycle checkpoint activation, DDR signaling, cell death/apoptosis, oxidative stress, inflammation, and wound healing [201, 202].

5.2 *Polymorphisms in DNA-Repair Genes*

As noted above, many genes in a variety of pathways can influence radiosensitivity. Nonetheless, DNA-repair genes have been prominent in candidate-gene panels for generating SNP signatures for patient radiosensitivity. Although inactivating mutations are uncommon, numerous SNPs in DNA-repair genes have been described [206–208], many of which represent missense events and may be associated with reduced DNA-repair activity [208, 209]. For example, several hundred *XRCC1* SNPs have been reported in public dbSNP databases that might impact on the DNA-repair activity of the XRCC1 BER protein (e.g., [207, 210]; http://www.ncbi.nlm.nih.gov/SNP). Among these are 3 fairly frequent non-synonymous SNPs: codon 194 C>T Arg194Trp; codon 280 G>A Arg280His; and codon 399 G>A Arg399Gln. The Arg399Gln SNP, which has been associated with alterations in DNA-repair proficiency [2, 211], represents an arginine to glutamine change at codon 399 that results from the G>A nucleotide substitution at position 28152 in exon 10. Another widely studied non-synonymous SNP at position 18067 in exon 7 (C>T variant) of the *XRCC3* gene results in the substitution of methionine for threonine at codon 241 (Thr241Met). As was noted in Sect. 3.4.2, the XRCC3 protein is important in the HRR pathway of DSB repair where it is involved in the assembly and stabilization of RAD51 [212].

It is instructive at this point to consider some of the potential pitfalls that might impact on SNP-phenotype association studies. Probably the most important of these is the occurrence of "linkage disequilibrium," whereby two alleles might be co-inherited more commonly than would be predicted based on random assortment; this can result from the occurrence of "haplotype blocks," which are sets of alleles or polymorphisms on a specific chromosomal region that tend to be inherited as a unit (often along lines of ethnicity) [213]. A SNP located with such a haplotype block might therefore be associated with a particular phenotype even though it has no functional involvement therein; rather, as part of a haplotype, it may represent a marker for the phenotype-causing variant(s). Such SNPs are referred to as tagging SNPs or "tag-SNPs." Such a scenario may well be relevant in reported associations between radiosensitivity phenotypes, notably involving the above-mentioned *XRCC1* SNPs [203].

Other potential confounding factors in association studies that might obscure the impact of genetic polymorphisms (as well as inter-study comparisons) include variations or uncertainties in XRT dose or fractionation scheme, target volume, criteria for scoring acute and late toxicities, preexisting medical conditions, biostatistical methods, and the use/nonuse of chemotherapy (e.g., [200, 203, 214, 215]). Another source of variability is in study design, including cohort size and demographics. Some studies have looked at variations within a cohort of consecutive patients; others have undertaken case–control comparisons of reactors versus matching non-reacting controls. Several authors have noted that these studies have typically employed relatively small sample sizes and may be limited by methodological issues, and as such should be regarded as hypothesis generating and in need of validation (e.g., [203, 214]).

5.3 DNA-Repair SNPs as Predictive Factors for Acute and/or Late Normal Tissue Complications Following XRT

In 2004 when we previously reviewed this field [2] a few publications had begun to appear relating SNPs such as *XRCC1* Arg194Trp and Arg399Gln, *XRCC3* Thr241Met, and *APE1* Asp148Glu to clinical radiosensitivity (e.g., [216]). Even at that time, researchers were beginning to simultaneously screen for the effect of alterations in multiple candidate radiosensitivity genes either as individual alleles or in combination [14, 217]. Since then, a considerable body of data has been generated, a systematic review of which is beyond the scope of this chapter; we instead try to illustrate some key points by focusing on the literature for breast and prostate cancer patients where there are several data sets that allow some level of inter-study comparison. In this regard, it is important to remember that some germ-line SNPs that result in a deficit in DNA repair are more prevalent among cancer patients than in the general population [2].

5.3.1 Breast Cancer Patients

As noted above, attributing an underlying mechanistic basis in DNA-repair deficiency for radiosensitivity associations, especially those involving polymorphism of the *XRCC1* BER gene, are, at best, tentative considering the extensive linkage disequilibrium that occurs at this locus. Nonetheless, such information is still useful for developing SNP signatures for radiosensitive patients that might facilitate clinical management decisions. Table 4 contains a summary of such SNP-association studies for breast cancer. In an early hypothesis-generating pilot study, Andreasson and coworkers [218] examined 7 SNPs in the *XRCC1*, *XRCC3*, and *APE1* DNA repair genes as well as in *TGFβ1* (Leu10Pro) and *SOD2* (Val16Ala) in 41 unselected patients receiving XRT post-mastectomy. They reported that several genotypes, namely, *XRCC1* 399-Arg/Arg, *TGFβ1* 10-Pro/Pro and T/T in position –509, and *SOD2* 16-Ala/Val, were significantly associated with an increased risk of skin fibrosis. The *XRCC3* 241-Thr/Thr common genotype correlated with increased risk of both fibrosis and telangiectasia. Combined analysis of multiple SNPs demonstrated that the risk of fibrosis increased with the number of risk alleles. The authors noted the paradox that these *XRCC1* and *XRCC3* alleles had been typically associated with *more efficient* DNA repair. Indeed, when they subsequently retested their findings in a validation cohort of 120 patients receiving similar treatment, the results could not be duplicated insofar as none of the SNPs from the earlier study was significantly associated with skin fibrosis [219]. Furthermore, when a Bonferroni correction for false-positivity was applied to the initial dataset [218], only one genotype—*XRCC3* 241-Thr/Met—retained significance [214]. In a similar study of these same end points in 167 breast cancer patients, Giotopoulos and colleagues [220] found that the *XRCC1* 399-Arg/Gln genotype was associated with late telangiectasia, but not with fibrosis.

Chang-Claude and colleagues observed that the *XRCC1* 399-Gln or *APE1* 148-Glu alleles were associated with fewer acute reactions in normal-weight patients [221]. The same group [222] saw no association of polymorphisms in three genes involved in the HRR pathway for DSB repair—*XRCC3* Thr241Met, *XRCC2* Arg188His, and *NBS1* Glu185Gln—with acute skin toxicity in breast cancer patients receiving XRT after breast-conserving surgery.

Moullan and colleagues [215] also examined the *XRCC1* SNPs in codons 194, 280 and 399 among 254 breast cancer patients who had undergone XRT and found that only the 194-Trp variant allele in exon 6 showed a borderline-significant increase in frequency in reactors over non-reactors. Haplotype analysis at the 3 loci revealed that the combination of the *XRCC1* 194-Trp and 399-Gln variant alleles along with the homozygous common Arg allele at codon 280 was even more strongly

Table 4 Association studies of SNPs in DNA repair/DDR genes and acute or late normal-tissue reactions to XRT for breast cancer patients

Authors	Genes	*N*	Reaction type	Findings
Andreassen et al. 2003 [218]	XRCC1, XRCC3	41	Late skin fibrosis and telangiectasia	XRCC1 399-Arg/Arg common genotype associated with increased risk of fibrosis XRCC3 241-Thr/Thr common genotype associated with increased risk of fibrosis and telangiectasia
Andreassen et al. 2006 [219]	XRCC1, XRCC3, APE1	120	Late skin fibrosis	No correlation between any SNPs and fibrosis
Moullan et al. 2003[215]	XRCC1	254	Acute and/or late	XRCC1 194-Trp and 399-Gln variant alleles together associated with increased reactions
Angèle et al. 2003 [225]	ATM			Homozygous ATM 5557 G>A variant associated with increased reactions
Brem et al. 2006 [223]	XRCC1	247	Acute and/or late	XRCC1 H3 haplotype (common allele at all 4 positions) associated with lower risk of reactions
Chang-Claude et al 2005 [221]	APE1, XRCC1, TP53, CSB	446	Acute moist desquamation	XRCC1 399-Gln or APE1 148-Glu variant alleles associated with decreased risk of acute reactions among normal-weight patients
Popanda et al. 2006 [222]	XRCC3, XRCC2, NBS1	446	Acute skin toxicity	No association of XRCC3 Thr241Met, XRCC2 Arg188His, or NBS1 Glu185Gln SNPs with reactions
Giotopoulos et al. 2007 [220]	XRCC1	167	Late skin fibrosis and telangiectasia	XRCC1 399-Arg/Gln genotype associated with increased risk of telangiectasia but not fibrosis

Suga et al. 2007 [230]	MAD2L2, PTTG1, RAD9A, LIG3	399	Acute	Haplotypes of these 4 DDR-related genes associated with decreased risk of severe acute skin reactions
Chang-Claude et al. 2009 [226]	APE1, XRCC1, XRCC2, XRCC3, XPD, TP53, CDKN1A	409	Late skin toxicity	TP53 72-Arg/Pro or PIN3 variant alleles associated with increased risk of telangiectasia; combination of both variant alleles conferred ~2-fold increased risk
Zschenker et al. 2010 [227]	XRCC1, TGFβ1, XPD, ATM, SOD2, GSTP1	69	Skin fibrosis	Risk of fibrosis nonsignificantly increased for variant XRCC1 (399-Gln) or TGFβ1 (position −509) alleles. Significantly increased risk for all high-risk alleles combined
Sterpone et al. 2010 [228]	XRCC1, OGG1, XRCC3	43	Acute	No association between XRCC1 Arg399Gln, OGG1 Ser326Cys, and XRCC3 Thr241Met SNPs and acute reactions
Mangoni et al. 2010 [229]	XRCC3, MSH2, MSH3	87	Acute skin reaction	XRCC3 241-Met, MSH2 gIVS12-6nt-C, or MSH3 1045-Ala variant alleles associated with increased risk; combination of the XRCC1 194-Trp variant and 399-Arg common allele had high risk

associated with risk of various acute and late reactions. These studies were extended to include analysis of a fourth *XRCC1* SNP, −77 T>C (rs 3213245), in the same cohort [223]. The T>C polymorphism at position −77 is located in the 5′ untranslated region (5′-UTR) and may be associated with decreased promoter activity and thus *XRCC1* expression [224]. The −77C allele by itself was not significantly associated with acute and/or late toxicity. However, inclusion of the −77 T>C genotype in the haplotype analysis indicated that patients with the H3 haplotype (which has the common alleles at all four positions) had a significantly lower risk of adverse reactions. It should be mentioned that interpreting these results is complicated because of the mixed early/late clinical end point used. In this same cohort, patients homozygous for the *ATM* 5557 G>A (Asp1853Asn) variant allele were at significantly increased risk of developing normal tissue reactions [225].

A recent study of SNPs in *APE1*, *XRCC1*, *XRCC2*, *XRCC3*, *XPD*, *TP53*, and *CDKN1A* in 409 breast cancer patients in relation to late skin complications after XRT indicated that patients having either the Arg72Pro or *PIN3* variant *TP53* alleles had a significantly greater risk of telangiectasia, and patients with the *TP53* haplotype having both variant alleles had an almost twofold increased risk [226]. The association of grade 2 or 3 fibrosis and SNPs in the DDR genes *XRCC1* (codon 399), *ATM* (codon 158), and *XPD* (codon 751) as well as in *GSTP1* (codon 105), *SOD2* (codon 16), and *TGFβ1* (position −509) was evaluated in 69 patients receiving XRT [227]. The risk of developing fibrosis tended to be greater among patients with variant *XRCC1* or *TGFβ1* alleles, whereas for *XPD*, *ATM*, *SOD2*, and *GSTP1*, the variant genotype appeared to be associated with a *decreased* risk. Although none of the individual associations was significant, the combination of all of the high-risk alleles was significantly associated with increased fibrosis. No association between the *XRCC1* Arg399Gln, *OGG1* Ser326Cys, and *XRCC3* Thr241Met SNPs and the occurrence of acute reactions to XRT was found among 43 patients, although a reduced in vitro SSB repair ability *was* apparent among those patients who exhibited side effects [228]. In another study involving 87 patients, those having the variant *XRCC3* 241-Met, *hMSH2* gIVS12-6nt-C, or *hMSH3* 1045-Ala alleles were at increased risk of developing severe acute skin reaction following XRT [229]; the combination of the *XRCC1* 194-Trp variant and 399-Arg common allele conferred a high risk of toxicity.

An interesting approach to acute reactions to XRT was reported by the Japanese RadGenomics group using candidate genes identified on the basis of their IR-inducibility in vitro [230]. Among 399 breast cancer patients, 5 haplotypes potentially related to DNA repair—TCC and CCG in the *RAD9A* gene (involved in cell cycle checkpoint activation and DNA repair), GCT in the *LIG3* BER gene, CG in *MAD2L2* (a component of the mitotic spindle assembly checkpoint), and GTTG in *PTTG1* (pituitary tumor transforming gene 1)—were associated with less-severe acute skin reactions.

On the basis of the above studies, it is apparent that association studies for toxicity and SNPs in breast cancer patients are inconsistent and in critical need of validation. The reader is referred to a recent review [204] for additional information about

SNP associations for other DDR genes such as *ATM* and *TP53*, as well as *BRCA1* and *BRCA2* [231, 232].

5.3.2 Prostate Cancer Patients

Interpreting the radiogenomics literature on toxicity in prostate cancer patients (summarized in Table 5) requires some caution because of several potential confounding factors—notably, the presence or absence of conditions such as inflammatory bowel disease, connective tissue diseases, urinary stricture, benign prostatic hyperplasia, diverticular disease, diabetes mellitus, or peritonitis, as well as previous pelvic surgery. Comparing XRT results with brachytherapy and external-beam XRT is also complicated by the need to address the impact of variations in dose, time and fractionation.

Cesaretti et al. [233] examined the association between SNPs in the *ATM* gene and late toxicity among 37 prostate cancer patients treated with ^{125}I brachytherapy. Patients with at least one *ATM* exonic sequence variant were at significantly increased risk for late complications. In a later study of 108 patients [234], these investigators reported that rectal dosimetry was a significant confounding factor in the genetic associations. Having any *ATM* sequence variant was associated with increased risk of late proctitis for patients in whom only a small volume (<1.4 cm^3) of rectum received the full prescription dose. The same authors subsequently reported on *SOD2*, *XRCC1*, and *XRCC3* SNPs in 135 prostate cancer patients who had received brachytherapy with or without supplemental external-beam XRT [235]. Erectile dysfunction was found more often among patients having the *XRCC1* codon 280-G/A genotype compared to the G/G genotype, whereas grade ≥2 late rectal bleeding was more common for patients with the *XRCC3* codon 241-T/C genotype and *SOD2* codon 16-C/T genotype. An important message from these studies is that normal tissue dose-volume effects for prostate cancer are independently important in toxicity outcomes.

Pugh and colleagues [236] performed a case–control study in which they sequenced the exons of *ATM*, *BRCA1*, *XPD*, *H2AX*, *LIG4*, *MDC1*, *MRE11A*, and *RAD50* in 21 reactor and 20 non-reactor prostate brachytherapy patients. Prostate and rectal dosimetry was similar for cases and controls. The case (reactor) cohort contained significantly more patients with at least one *LIG4* coding variant. The synonymous rs28986317 variant in *MDC1* was also more prevalent in the reactor group. Although the cohort sizes were small and the findings are in need of validation, this study does highlight the power of detailed exonic sequencing for discovering novel variants that could prove to be causative rather than tagging alleles.

Two DNA-repair gene SNP-association studies have been performed in prostate cancer patients treated with three-dimensional conformal XRT. We [237] reported on a cohort of 83 patients for whom late toxicity was evaluated. In this pilot study, patients were screened for 49 SNPs in 17 genes including 14 involved in the DDR pathway/DNA repair: *BRCA1*, *BRCA2*, *XRCC1*, *XRCC2*, *XRCC3*, *NBS1*, *RAD51*, *RAD52*, *LIG4*, *ATM*, *hMSH6*, *hMLH1*, *XPD*, and *XPF*. The actuarial dependence of allelic

Table 5 Association studies of SNPs in DNA repair/DDR genes and acute or late normal-tissue reactions to XRT for prostate cancer patients

Authors	Genes	*N*	Reaction type	Findings
Cesaretti et al. 2005 [233]	ATM	37	Late	Possession of at least one ATM exonic sequence variant associated with significantly increased adverse late reactions
Cesaretti et al. 2007 [234]	ATM	108	Late	SNPs in ATM associated with late proctitis among brachytherapy patients who received the full prescription dose to a small volume of the rectum
Burri et al. 2008 [235]	XRCC1, XRCC3, SOD2	135	Late	XRCC1 280-Arg/His genotype associated with erectile dysfunction
				Combination of XRCC3 241-Thr/Met and SOD2 16-Val/Ala genotypes associated with late rectal bleeding
Pugh et al. 2009 [236]	ATM, BRCA1, XPD, H2AX, LIG4, MDC1, MRE11A, RAD50	41	Composite acute and late	Reactor cohort enriched for at least one LIG4 variant SNP; synonymous variant SNP in MDC1 (rs28986317) associated with increased reactions
Damaraju et al. 2006 [237]	ATM, BRCA1, BRCA2, XRCC1, XRCC2, XRCC3, NBS1, RAD51, RAD52, LIG4, hMLH1, hMSH6, XPD, XPF	83	Late	Multivariate analysis indicated significant associations with late reactions for LIG4 T>C Asp568Asp and XPD G>A Asp711Asp
Hall et al. 1998 [238]	ATM	17	Late	Significantly greater frequency of ATM variant in patients with late reactions

Popanda et al. 2009 [201]	XRCC1, APE1, OGG1, XRCC2, XRCC3, NBS1, XPA, ERCC1, XPC, TP53, CDKN1A, MDM2	405	Acute	XRCC1 399-Gln and XRCC3 241-Met variant alleles associated with acute reactions
Suga et al. 2008 [239]	SART1, ID3, EPDR1, PAH, KU70	197	Late	Variants in SART1, ID3, EPDR1, PAH, and KU70 highly predictive for stratifying patients into no reaction versus any reaction groups

variants as well as dose-volume parameters for the bladder and rectum was calculated. Cox multivariate analysis showed significant associations with toxicity for *LIG4* T>C Asp568Asp (CC and CT genotypes) and *XPD* G>A Asp711Asp (AA genotype) variant alleles, as well as rectal D_{30} and mean bladder dose. However, this analysis was not corrected for the statistical impact of multiple comparisons, so it remains in need of validation. Popanda et al. [201] reported on associations between acute toxicity in 405 patients and various SNPs in the *XRCC1*, *APE1*, *OGG1*, *XRCC2*, *XRCC3*, *NBS1*, *XPA*, *ERCC1*, *XPC*, *TP53*, *CDKN1A*, and *MDM2* genes. The *XRCC1* 399-Gln variant genotype appeared to show an association with acute toxicity, as did the *XRCC3* 241-Met variant genotype. No data were reported for late effects.

As noted earlier [230], the RadGenomics group based their candidate gene panel on in vitro IR-inducibility studies. This group reported on 197 prostate cancer patients evaluable for late genitourinary toxicity after carbon-ion XRT [239]. Remarkably, only 28 patients experienced grade 1 toxicity, and only four experienced grade 2 effects, with no grade ≥3 effects. The paucity of moderate late effects and absence of severe late effects together with difficulties inherent in scoring of the milder toxicities is an issue for generalizing these findings, as is the Japanese ethnicity of the patients. However, their 2-stage design identified a group of variants in the *SART1* (involved in cell cycle arrest and pre-mRNA splicing), *ID3* (DNA-binding protein inhibitor 3), *EPDR1* (ependymin related protein 1), *PAH* (phenylalanine hydroxylase), and *KU70* genes that showed high predictive ability. Evaluation of a separate validation cohort suggested reasonable robustness in terms of stratifying patients into no toxicity versus any toxicity groups.

5.3.3 Other Sites

Among other tumor sites receiving some attention in radiogenomics studies of risk of normal tissue complications are head and neck and gynecological cancers. Significant acute and late toxicities frequently occur in SCCHN patients, making this an important site to study. Among 88 SCCHN XRT patients who were screened for 9 SNPs in the DSB repair genes *XRCC3*, *RAD51*, *LIG4*, *KU70*, and *KU80* in the context of acute toxicity, the *XRCC3* codon 722-CT/TT and *KU70* codon 1310-CG/GG genotypes were significantly associated with the development of grade ≥3 dysphagia, but not with mucositis or dermatitis [240]. In a study of 130 patients with SCCHN of the oropharyngeal subsite who were treated with XRT, the 2505-C allele of the *XPF* nucleotide excision repair gene was associated with a reduced dependence on a percutaneous feeding tube (a putative marker for chronic mucosal ulceration and dysphagia) [241]. In a pilot study of 60 nasopharyngeal cancer patients, a significant association with late fibrosis following XRT was observed for SNPs in *TGFβ1* Leu10Pro and *XRCC1* Arg399Gln SNPs (but not *XRCC3* Thr241Met); in both cases the common allele, *TGFβ1* 10-Leu and *XRCC1* 399-Arg, appeared to confer the higher risk [242].

De Ruyck and colleagues [243] examined the association of SNPs in *XRCC1* (Arg194Trp, Arg280His, Arg399Gln, Gln632Gln), *XRCC3* (5′-UTR 4.541A>G,

IVS5-14 17.893A>G, Thr241Met), and *OGG1* (Ser326Cys) with the development of late XRT reactions in 62 women treated for carcinomas of the cervix and endometrium. The *XRCC3* IVS5-14 (17893A>G) AG or AG+GG genotypes were significantly associated with an increased risk of late reactions, whereas the *XRCC1* codon 194-Trp allele showed a significant protective effect [243]. Severe reactions were significantly increased in patients with three or more risk alleles in *XRCC1* and *XRCC3*. These authors also examined in 62 patients with cervical and endometrial tumors and saw no association between the repeat length polymorphisms at microsatellites in *XRCC1*, *XRCC3*, or *KU80* and the incidence of late XRT complications [244].

Azria and colleagues [245] reported that severe normal tissue reactions among XRT patients were associated with their having 4 or more SNPs in the *ATM*, *SOD2*, *XRCC1*, *XRCC3*, *TGFβ1*, and *RAD21* genes.

5.4 Association of DNA Repair SNPs with XRT Efficacy and Treatment Outcome in Cancer Patients

Table 6 lists several studies that have examined the association between SNPs in various genes and tumor radioresponsiveness. These include cancers of the head and neck [246, 247], esophagus [248, 249], lung [250–252], pancreas [159, 253–255], bladder [256], prostate [257], and rectum [258]. Only a few of these studies reported on cohorts treated exclusively with XRT. None of the putative associations with disease outcome end points listed in Table 6 have yet been validated or replicated. In the current era where there is increasing use of concurrent and sequential systemic and local therapies, efforts to develop predictive SNP signatures prior to therapy will probably need to include screening for potential genetic factors that influence the response to all active agents.

It should be noted that these studies examined SNPs either in tumor samples or in normal cells that would report exclusively on germ-line variants. In both approaches, there are many potential confounding factors [204], including biological "inescapables" such as the impact of intratumoral heterogeneity and cancer stem cell populations as well as factors such as sample size, use of clinical versus pathological end points of response, variable use of chemotherapy or surgery, and use/nonuse of a two-stage study design with marker validation.

5.5 Genome-Wide Association Studies and Other High-Throughput Approaches

As noted earlier, all of the radiogenomics studies described above used the candidate gene approach and it must be said have been largely inconclusive or ambiguous. As should be apparent from the previous sections, and as summarized elsewhere [202], even for a single SNP such as *XRCC1* Arg399Gln, different studies have reported significant

Table 6 Association studies of SNPs in DNA repair/DDR genes and tumor radioresponsiveness in patients with various types of cancers

Authors	Cancer site	Genes	*N*	Findings
Carles et al. 2006 [246]	Squamous cell carcinoma of head/neck	XPA, XPC, XPD, ERCC1, XPG, KU80, XRCC1	108	ERCC1, XPG, XPA SNPs significantly associated with time to progression and OS after XRT
Tu et al. 2009 [247]	Squamous cell carcinoma of head/neck	MDM2, TP53	189	Combination of MDM2 nt309 G/G and TP53 codon 72 Arg/Arg associated with poor OS and DFS after surgery plus XRT
Okazaki et al. 2008[253]	Pancreas	ATM, ATR, CHK1, CHK2	119	ATM 60 G>A and CHK1 35 G>A significantly associated with OS; combined genotype effect for up to three deleterious alleles in response to neoadjuvant gemcitabine, XRT
Li et al. 2006 [254]	Pancreas	RECQ1, RAD54L, XRCC1, ATM	92	RECQ1 159 A>C, RAD54L 157 C>T, XRCC1 Arg194Trp, and ATM –77 T>C genotypes significantly associated with OS. Strong combined effect of the 4 genotypes
Li et al. 2006 [255]	Pancreas	RECQ1, RAD54L, XRCC2, XRCC3	378	Significant effect of RECQ1 and RAD54L genotypes on OS. Significantly reduced OS for XRCC2 188-His and XRCC3 17893 A>G (IVS5-14) variant alleles in subgroup analysis. Increasing number of adverse alleles significantly associated with decreased OS

Li et al. 2007 [159]	Pancreas	OGG1, APE1, POLβ, XRCC1, LIG3, LIG4	378	Homozygous variant genotypes of OGG1 G2657A, APE1 Asp148Glu and XRCC1 Arg194Trp (but not Arg399Gln) significantly associated with decreased OS. Homozygous variant genotypes of POLβ associated with significantly increased OS. Combined genotype showed predominant effect of POLβ homozygous variant
Sakano et al. 2006 [256]	Bladder	XPC, XPD, XPG, XRCC1, XRCC3	78	Genotypes with at least one variant allele in XPD or XRCC1 significantly associated with improved cancer-specific survival after cisplatin plus XRT
Warnecke-Eberz et al. 2009 [249]	Esophagus	Akt1, HER-2/neu, ERCC1, FGFR4, GTSP1, MDR1, MGMT, MTHFR, TERT, TS, XRCC1, XRCC3	52	ERCC1 118 C>T predictive for major histopathologic response to neoadjuvant cisplatin/5-fluorouracil/XRT
Wu et al. 2006 [248]	Esophagus	MTHFR, MTR, TS, MDR1, GTSP, MPO, TP53, FAS, FasL, NQO1, XPA, XPC, XPD, XPG, ERCC1, CSB, CCNH, RAD23B	210	XRCC1 399-Gln variant alleles associated with lower chance of pathologic complete response and poor OS after cisplatin/5-fluorouracil/XRT

(continued)

Table 6 (continued)

Authors	Cancer site	Genes	*N*	Findings
Yin et al. 2009 [252]	Lung adenocarcinoma	XRCC1, XPD, ERCC1	257	ERCC1 Asn118Asn CT or TT variant and XRCC1 399-Gln variant genotypes associated with shorter survival among nonsmoking female patients
Yoon et al. 2005 [250]	Non-small cell lung	XRCC1	229	Haplotype pairs other than homozygous TGG had superior OS after XRT
Su et al. 2007 [251]	Non-small cell lung	APE1, ATM, iASPP	230	An A allele (A/T+A/A) at position 67 in iASPP or the APE1 T allele associated with better response to chemoradiotherapy. No association for ATM 60 A>G
Gao et al. 2007 [257]	Prostate	XRCC1	513	Among patients receiving XRT, median survival associated with XRCC1 haplotype based on Arg399Gln and Arg194Trp SNPs.
Cecchin et al. 2011 [258]	Rectum	OGG1, ERCC1, XRCC1, XRCC3	238	OGG1 1245 C>G genotype associated with tumor regression in multivariate analysis. Also significant were ERCC1 8092 C>A, XRCC1 Arg399Gln 28152 G>A, and XRCC3 4541 A>G in the 5′-UTR

OS overall survival, *DFS* disease-free survival

associations between various manifestations of normal tissue toxicity following XRT for which a particular *XRCC1* genotype (either alone or in combination with other SNPs) was predictive for increased risk, decreased risk, or unaltered risk.

However, the advent of critical technologies such as high-throughput microarrays that allow the rapid screening of large numbers of validated SNPs (over one million per chip) without requiring prior knowledge of their function will enable researchers to take a genome-wide association study (GWAS) approach to radiogenomics markers or groups of DNA variations (i.e., haplotypes) by examining tag-SNPs that span the entire genome [200, 259]. In contrast to the candidate-gene approach, GWAS are typically not hypothesis driven and are not biased against intronic/noncoding markers; this is important considering that many SNPs associated with human disease are not cSNPs but rather occur in noncoding regions or introns. A number of potential issues will need to be addressed as radiogenomics progresses from the candidate gene to the GWAS approach, including the fact that sufficiently powered studies will require large numbers of patients [200].

The next few years will also see increasing interest in generating biomarkers of responsiveness to XRT using "next-generation" or "NextGen" DNA sequencing technologies such as "massively parallel sequencing" that provide genomic data in even greater detail than the GWAS approaches [260, 261]. We will also see the application of powerful new systems biology approaches to understanding the effects of IR on cells and of XRT on normal tissues and tumors. For example, Eschrich and colleagues [262] applied systems biology to the discovery of radiosensitivity biomarkers in 48 human cancer cell lines (based on their surviving fraction at 2 Gy). Although none of the genes in the reduced 10-hub network were DSB repair genes per se, a number of DSB repair genes do interact with this network. As noted by the authors, this radiation-biomarker discovery platform could be extremely valuable for the integration of biology into clinical XRT practice.

In addition to SNPs, there are other types of genomic alterations that might be of relevance to radiogenomics. Among these are the copy number variations (CNVs) that are evident between the genomes of individuals and which now rival SNPs with respect to their extent in terms of amount of genomic sequence that they encompass and have been associated with susceptibility to disease. CNVs are widespread and common and represent either inherited or *de novo* duplications or deletions of segments of the genome that can involve from one kilobase to several megabases [263]. Approximately 2000 CNVs have been described to date, but there may be thousands more.

5.6 Large Consortium Genomic Studies

Given the complexity of the phenotype being addressed in radiogenomics studies, it is apparent that very large sample sizes are essential for significant SNP-association studies [202, 203, 214, 264]. For this reason, a number of national and international collaborative networks have been initiated with the intention of recruiting thousands of patients to radiogenomics studies and that link well annotated clinical databases to biospecimens. These resources have been important in the candidate-gene studies undertaken to date and will continue to be so as the GWAS approach is increasingly adopted. They include:

1. Genetic Predictors of Adverse Radiotherapy Effects (Gene-PARE) [265], a multinational initiative headquartered at the Mount Sinai School of Medicine in New York that is banking frozen lymphocytes and DNA from XRT patients linked to a detailed clinical database.
2. Radiogenomics: Assessment of Polymorphisms for Predicting the Effects of Radiotherapy or "RAPPER" [266, 267], a multicentered translational radiogenomics initiative based in the UK that plans to enroll >2,000 breast and prostate cancer patients.
3. The RadGenomics project [268], created in 2001 and centered at the National Institute of Radiological Sciences in Chiba, Japan.
4. GENEtic pathways for the Prediction of the effect of Irradiation (GENEPI) [269] and GENEPI 2 [270], a European consortium, supported by ESTRO, which currently houses more than 4,000 patient samples (DNA, lymphocytes or other normal and tumor tissues) linked to XRT outcome data.
5. International Radiogenomics Consortium [271], created in 2009 to link and stimulate international collaborative efforts.

5.7 Non-repair Factors in Responses to XRT

An important consideration with respect to normal tissue injury after XRT is that genes and pathways that are unrelated to DNA repair may play a major role in the pathogenesis of tissue damage. For example, fibrotic and inflammatory processes are clearly important players in late effects such as fibrosis [59]. Indeed, XRT-induced fibrotic late reactions have been related to abnormal wound healing, possibly leading to the characteristic excessive deposition of extracellular matrix and collagen [272]. Many radiogenomics studies of late effects have therefore included markers such as transforming growth factor-β1 (TGFβ1), a pro-inflammatory fibrogenic cytokine that is thought to induce deposition of collagen and fibronectin [272].

6 Conclusions

Despite major advances in our knowledge of the genetic factors that control cellular radiosensitivity, it is still not possible to predict the clinically observed heterogeneity of response among patients (both normal tissues and tumors) to XRT with adequate precision. The reasons for this are complex. In the context of the candidate gene approach to SNP-typing, clearly there are a multitude of genes that would have to be taken into account in a rigorous screening of radiosensitivity parameters in cells, never mind tissues. The reality is that DNA repair-related parameters even in surrogate cell types studied in vitro may bear a complex relationship to clinical normal tissue toxicity. For example, among 11 normal human fibroblast lines, differences in clonogenic cell survival were found to correlate closely with the level of residual DSBs [273]. Curiously, *KU70* and *KU86* mRNA levels were similar among these cell lines, as was their DNA-PK activity [274]. More recently the levels/activity of the KU70, KU80, XRCC4, and DNA-PK_{cs} proteins in these cells, measured both before and after irradiation, has been found to differ but again did not correlate with the cells' DSB repair capability and in turn radiosensitivity [275]. One thing that these studies strongly reinforce is the value of combining genomic and intermediate phenotypic end points (e.g., see [203]) when undertaking radiogenomics studies.

We are also becoming increasingly cognizant of the fact that our understanding of the basic mechanisms that govern cellular responses to IR remain incomplete, as evidenced for example by the recent realization of the importance of epigenetic factors and the histone code [276] as well as microRNAs [277] in this regard. Indeed, the methylation of lysine 79 of histone H3 was shown to play an important early signaling role in response to IR-induced DSBs [278, 279].

Acknowledgments We are grateful for support by the Alberta Cancer Foundation.

References

1. Emami B, Lyman J, Brown A, Coia L, Goitein M, Munzenrider JE, Shank B, Solin LJ, Wesson M (1991) Tolerance of normal tissue to therapeutic irradiation. Int J Radiat Oncol Biol Phys 15:109–122
2. Murray D, Begg A (2004) DNA repair genes and radiosensitivity. In: Panasci LC, Alaoui-Jamali MA (eds) DNA repair in cancer therapy. Humana Press, Totawa, NJ, pp 211–256
3. Peters LJ (1996) Radiation therapy tolerance limits. For one or for all?—Janeway lecture. Cancer 77:2379–2385
4. Burnet NG, Johansen J, Turesson I, Nyman J, Peacock JH (1998) Describing patients' normal tissue reactions: concerning the possibility of individualising radiotherapy dose prescriptions based on potential predictive assays of normal tissue radiosensitivity. Steering Committee of the BioMed2 European Union Concerted Action Programme on the Development of Predictive Tests of Normal Tissue Response to Radiation Therapy. Int J Cancer 79:606–613
5. West CM (1995) Invited review: intrinsic radiosensitivity as a predictor of patient response to radiotherapy. Br J Radiol 68:827–837
6. Budach W, Classen J, Belka C, Bamberg M (1998) Clinical impact of predictive assays for acute and late radiation morbidity. Strahlenther Onkol 174(suppl 3):20–24
7. Tucker SL, Geara FB, Peters LJ, Brock WA (1996) How much could the radiotherapy dose be altered for individual patients based on a predictive assay of normal-tissue radiosensitivity? Radiother Oncol 38:103–113
8. West CM, Davidson SE, Roberts SA, Hunter RD (1997) The independence of intrinsic radiosensitivity as a prognostic factor for patient response to radiotherapy of carcinoma of the cervix. Br J Cancer 76:1184–1190
9. Girinsky T, Lubin R, Pignon JP, Chavaudra N, Gazeau J, Dubray B, Cosset JM, Socie G, Fertil B (1993) Predictive value of in vitro radiosensitivity parameters in head and neck can-

cers and cervical carcinomas: preliminary correlations with local control and overall survival. Int J Radiat Oncol Biol Phys 25:3–7
10. Olive PL (1998) The role of DNA single- and double-strand breaks in cell killing by ionizing radiation. Radiat Res 150(5 Suppl):S42–S51
11. Badie C, Iliakis G, Foray N, Alsbeih G, Pantellias GE, Okayasu R, Cheong N, Russell NS, Begg AC, Arlett CF, Malaise EP (1995) Defective repair of DNA double-strand breaks and chromosome damage in fibroblasts from a radiosensitive leukemia patient. Cancer Res 55:1232–1234
12. Muller WU, Bauch T, Stuben G, Sack H, Streffer C (2001) Radiation sensitivity of lymphocytes from healthy individuals and cancer patients as measured by the comet assay. Radiat Environ Biophys 40:83–89
13. Muller WU, Bauch T, Streffer C, von Mallek D (2002) Does radiotherapy affect the outcome of the comet assay? Br J Radiol 75:608–614
14. Borgmann K, Roper B, El-Awady R, Brackrock S, Bigalke M, Dörk T, Alberti W, Dikomey E, Dahm-Daphi J (2002) Indicators of late normal tissue response after radiotherapy for head and neck cancer: fibroblasts, lymphocytes, genetics, DNA repair, and chromosome aberrations. Radiother Oncol 64:141–152
15. Marcou Y, D'Andrea A, Jeggo P, Plowman PN (2001) Normal cellular radiosensitivity in an adult Fanconi anaemia patient with marked clinical radiosensitivity. Radiother Oncol 60:75–79
16. Kiltie AE, Ryan AJ, Swindell R, Barber JB, West CM, Magee B, Hendry JH (1999) A correlation between residual radiation-induced DNA double-strand breaks in cultured fibroblasts and late radiotherapy reactions in breast cancer patients. Radiother Oncol 51:55–65
17. Dickson J, Magee B, Stewart A, West CM (2002) Relationship between residual radiation-induced DNA double-strand breaks in cultured fibroblasts and late radiation reactions: a comparison of training and validation cohorts of breast cancer patients. Radiother Oncol 62:321–326
18. El-Awady RA, Mahmoud M, Saleh EM, El-Baky HA, Lotayef M, Dahm-Daphi J, Dikomey E (2005) No correlation between radiosensitivity or double-strand break repair capacity of normal fibroblasts and acute normal tissue reaction after radiotherapy of breast cancer patients. Int J Radiat Biol 81:501–508
19. Wojewódzka M, Buraczewska I, Kruszewski M (2002) A modified neutral comet assay: elimination of lysis at high temperature and validation of the assay with anti-single-stranded DNA antibody. Mutat Res 518:9–20
20. Mirzayans R, Severin D, Murray D (2006) Relationship between DNA double strand break rejoining and cell survival following exposure to ionizing radiation in human fibroblast strains with differing ATM/p53 status: implications for the evaluation of clinical radiosensitivity. Int J Radiat Oncol Biol Phys 66:1498–1505
21. Rothkamm K, Lobrich M (2003) Evidence for a lack of DNA double- strand break repair in human cells exposed to very low x-ray doses. Proc Natl Acad Sci USA 100:5057–5062
22. Kato TA, Nagasawa H, Weil MM, Genik PC, Little JB, Bedford JS (2006) γ-H2AX foci after low-dose-rate irradiation reveal atm haploinsufficiency in mice. Radiat Res 166:47–54
23. Rübe CE, Grudzenski S, Kühne M, Dong X, Rief N, Löbrich M, Rübe C (2008) DNA double-strand break repair of blood lymphocytes and normal tissues analysed in a preclinical mouse model: implications for radiosensitivity testing. Clin Cancer Res 14:6546–6555
24. Kato TA, Nagasawa H, Weil MM, Little JB, Bedford JS (2006) Levels of γ-H2AX foci after low-dose-rate irradiation reveal a DNA DSB rejoining defect in cells from human ATM heterozygotes in two at families and in another apparently normal individual. Radiat Res 166:443–453
25. Meyn MS (1995) Ataxia-telangiectasia and cellular responses to DNA damage. Cancer Res 55:5991–6001
26. Foray N, Arlett CF, Malaise EP (1997) Radiation-induced DNA double-strand breaks and the radiosensitivity of human cells: a closer look. Biochimie 79:567–575

27. Chavaudra N, Bourhis J, Foray N (2004) Quantified relationship between cellular radiosensitivity, DNA repair defects and chromatin relaxation: a study of 19 human tumor cell lines from different origin. Radiother Oncol 73:373–382
28. Sak A, Grehl S, Erichsen P, Engelhard M, Grannass A, Levegrün S, Pöttgen C, Groneberg M, Stuschke M (2007) γ-H2AX foci formation in peripheral blood lymphocytes of tumor patients after local radiotherapy to different sites of the body: dependence on the dose-distribution, irradiated site and time from start of treatment. Int J Radiat Biol 83:639–652
29. Paull TT, Rogakou EP, Yamazaki V, Kirchgessner CU, Gellert M, Bonner WM (2000) A critical role for histone H2AX in recruitment of repair factors to nuclear foci after DNA damage. Curr Biol 10:886–895
30. Löbrich M, Shibata A, Beucher A, Fisher A, Ensminger M, Goodarzi AA, Barton O, Jeggo PA (2010) γH2AX foci analysis for monitoring DNA double-strand break repair: strengths, limitations and optimization. Cell Cycle 9:662–669
31. Pilch DR, Sedelnikova OA, Redon C, Celeste A, Nussenzweig A, Bonner WM (2003) Characteristics of γ-H2AX foci at DNA double-strand breaks sites. Biochem Cell Biol 81:123–129
32. Bekker-Jensen S, Lukas C, Kitagawa R, Melander F, Kastan MB, Bartek J, Lukas J (2006) Spatial organization of the mammalian genome surveillance machinery in response to DNA strand breaks. J Cell Biol 173:195–206
33. Raderschall E, Golub EI, Haaf T (1999) Nuclear foci of mammalian recombination proteins are located at single-stranded DNA regions formed after DNA damage. Proc Natl Acad Sci USA 96:1921–1926
34. Raderschall E, Bazarov A, Cao J, Lurz R, Smith A, Mann W, Ropers HH, Sedivy JM, Golub EI, Fritz E, Haaf T (2002) Formation of higher-order nuclear Rad51 structures is functionally linked to p21 expression and protection from DNA damage-induced apoptosis. J Cell Sci 115:153–164
35. Liu SK, Olive PL, Bristow RG (2008) Biomarkers for DNA DSB inhibitors and radiotherapy clinical trials. Cancer Metastasis Rev 27:445–458
36. Banáth JP, Macphail SH, Olive PL (2004) Radiation sensitivity, H2AX phosphorylation, and kinetics of repair of DNA strand breaks in irradiated cervical cancer cell lines. Cancer Res 64:7144–7149
37. Olive PL, Banáth JP (2004) Phosphorylation of histone H2AX as a measure of radiosensitivity. Int J Radiat Oncol Biol Phys 58:331–335
38. Taneja N, Davis M, Choy JS, Beckett MA, Singh R, Kron SJ, Weichselbaum RR (2004) Histone H2AX phosphorylation as a predictor of radiosensitivity and target for radiotherapy. J Biol Chem 279:2273–2280
39. MacPhail SH, Banáth JP, Yu TY, Chu EH, Lambur H, Olive PL (2003) Expression of phosphorylated histone H2AX in cultured cell lines following exposure to X-rays. Int J Radiat Biol 79:351–358
40. Klokov D, MacPhail SM, Banáth JP, Byrne JP, Olive PL (2006) Phosphorylated histone H2AX in relation to cell survival in tumor cells and xenografts exposed to single and fractionated doses of X-rays. Radiother Oncol 80:223–229
41. Menegakis A, Yaromina A, Eicheler W, Dörfler A, Beuthien-Baumann B, Thames HD, Baumann M, Krause M (2009) Prediction of clonogenic cell survival curves based on the number of residual DNA double strand breaks measured by γH2AX staining. Int J Radiat Biol 85:1032–1041
42. Kao J, Milano MT, Javaheri A, Garofalo MC, Chmura SJ, Weichselbaum RR, Kron SJ (2006) γ-H2AX as a therapeutic target for improving the efficacy of radiation therapy. Curr Cancer Drug Targets 6:197–205
43. Mahrhofer H, Bürger S, Oppitz U, Flentje M, Djuzenova CS (2006) Radiation induced DNA damage and damage repair in human tumor and fibroblast cell lines assessed by histone H2AX phosphorylation. Int J Radiat Oncol Biol Phys 64:573–580
44. Yoshikawa T, Kashino G, Ono K, Watanabe M (2009) Phosphorylated H2AX foci in tumor cells have no correlation with their radiation sensitivities. J Radiat Res 50:151–160

45. Porcedda P, Turinetto V, Brusco A, Cavalieri S, Lantelme E, Orlando L, Ricardi U, Amoroso A, Gregori D, Giachino C (2008) A rapid flow cytometry test based on histone H2AX phosphorylation for the sensitive and specific diagnosis of ataxia telangiectasia. Cytometry A 73:508–516
46. Ismail IH, Wadhra TI, Hammarsten O (2007) An optimized method for detecting γ-H2AX in blood cells reveals a significant interindividual variation in the γ-H2AX response among humans. Nucleic Acids Res 35(5):e36
47. Hamasaki K, Imai K, Nakachi K, Takahashi N, Kodama Y, Kusunoki Y (2007) Short-term culture and γH2AX flow cytometry determine differences in individual radiosensitivity in human peripheral T lymphocytes. Environ Mol Mutagen 48:38–47
48. Bhogal N, Jalali F, Bristow RG (2009) Microscopic imaging of DNA repair foci in irradiated normal tissues. Int J Radiat Biol 85:732–746
49. Pospelova TV, Demidenko ZN, Bukreeva EI, Pospelov VA, Gudkov AV, Blagosklonny MV (2009) Pseudo-DNA damage response in senescent cells. Cell Cycle 8:4112–4118
50. Sedelnikova OA, Bonner WM (2006) γH2AX in cancer cells: a potential biomarker for cancer diagnostics, prediction and recurrence. Cell Cycle 5:2909–2913
51. Bonner WM, Redon CE, Dickey JS, Nakamura AJ, Sedelnikova OA, Solier S, Pommier Y (2008) γH2AX and cancer. Nat Rev Cancer 8:957–967
52. Kinner A, Wu W, Staudt C, Iliakis G (2008) γ-H2AX in recognition and signaling of DNA double-strand breaks in the context of chromatin. Nucleic Acids Res 36:5678–5694
53. Qvarnstrom OF, Simonsson M, Johansson KA, Nyman J, Turesson I (2004) DNA double strand break quantification in skin biopsies. Radiother Oncol 72:311–317
54. Lobrich M, Rief N, Kuhne M, Heckmann M, Fleckenstein J, Rübe C, Uder M (2005) In vivo formation and repair of DNA double-strand breaks after computed tomography examinations. Proc Natl Acad Sci USA 102:8984–8989
55. Rothkamm K, Balroop S, Shekhdar J, Fernie P, Goh V (2007) Leukocyte DNA damage after multi-detector row CT: a quantitative biomarker of low-level radiation exposure. Radiology 242:244–251
56. Simonsson M, Qvarnström F, Nyman J, Johansson KA, Garmo H, Turesson I (2008) Low-dose hypersensitive γH2AX response and infrequent apoptosis in epidermis from radiotherapy patients. Radiother Oncol 88:388–397
57. Kuefner MA, Grudzenski S, Schwab SA, Wiederseiner M, Heckmann M, Bautz W, Lobrich M, Uder M (2009) DNA double-strand breaks and their repair in blood lymphocytes of patients undergoing angiographic procedures. Invest Radiol 44:440–446
58. Olive PL, Banáth JP, Keyes M (2008) Residual γH2AX after irradiation of human lymphocytes and monocytes in vitro and its relation to late effects after prostate brachytherapy. Radiother Oncol 86:336–346
59. Werbrouck J, De Ruyck K, Beels L, Vral A, Van Eijkeren M, De Neve W, Thierens H (2010) Prediction of late normal tissue complications in RT treated gynaecological cancer patients: potential of the γ-H2AX foci assay and association with chromosomal radiosensitivity. Oncol Rep 23:571–578
60. Vasireddy RS, Sprung CN, Cempaka NL, Chao M, McKay MJ (2010) H2AX phosphorylation screen of cells from radiosensitive cancer patients reveals a novel DNA double-strand break repair cellular phenotype. Br J Cancer 102:1511–1518
61. Pollard JM, Gatti RA (2009) Clinical radiation sensitivity with DNA repair disorders: an overview. Int J Radiat Oncol Biol Phys 74:1323–1331
62. Meyn MS (1999) Ataxia-telangiectasia, cancer and the pathobiology of the ATM gene. Clin Genet 55:289–304
63. Gatti R (2001) The inherited basis of human radiosensitivity. Acta Oncol 40:702–711
64. Girard PM, Foray N, Stumm M (2000) Radiosensitivity in Nijmegen Breakage Syndrome cells is attributable to a repair defect and not cell cycle checkpoint defects. Cancer Res 60:4881–4888
65. Girard PM, Kysela B, Härer CJ, Doherty AJ, Jeggo PA (2004) Analysis of DNA ligase IV mutations found in LIG4 syndrome patients: the impact of two linked polymorphisms. Hum Mol Genet 13:2369–2376

66. Hall EJ, Schiff PB, Hanks GE (1998) A preliminary report: frequency of A-T heterozygotes among prostate cancer patients with severe late responses to radiation therapy. Cancer J Sci Am 4:385–389
67. Jackson SP (2002) Sensing and repairing DNA double-strand breaks. Carcinogenesis 23:687–696
68. Kurz EU, Lees-Miller SP (2004) DNA damage induced activation of ATM and ATM-dependent signaling pathways. DNA Repair (Amst) 3:889–900
69. Almeida KH, Sobol RW (2007) A unified view of base excision repair: lesion-dependent protein complexes regulated by post-translational modification. DNA Repair (Amst) 6:695–711
70. Hoeijmakers JHJ (2001) Genome maintenance mechanisms for preventing cancer. Nature 411:366–374
71. Jilani A, Ramotar D, Slack C, Ong C, Yang XM, Scherer SW, Lasko DD (1999) Molecular cloning of the human gene, PNKP, encoding a polynucleotide kinase 3′-phosphatase and evidence for its role in repair of DNA strand breaks caused by oxidative damage. J Biol Chem 274:24176–24186
72. Karimi-Busheri F, Daly G, Robins P, Canas B, Pappin DJ, Sgouros J, Miller GG, Fakhrai H, Davis EM, Le Beau MM, Weinfeld M (1999) Molecular characterization of a human DNA kinase. J Biol Chem 274:24187–24194
73. Takahashi T, Tada M, Igarashi S, Koyama A, Date H, Yokoseki A, Shiga A, Yoshida Y, Tsuji S, Nishizawa M, Onodera O (2007) Aprataxin, causative gene product for EAOH/AOA1, repairs DNA single-strand breaks with damaged 3′-phosphate and 3′-phosphoglycolate ends. Nucleic Acids Res 35:3797–3809
74. Izumi T, Hazra TK, Boldogh I, Tomkinson AE, Park MS, Ikeda S, Mitra S (2000) Requirement for human AP endonuclease 1 for repair of 3′-blocking damage at DNA single-strand breaks induced by reactive oxygen species. Carcinogenesis 21:1329–1334
75. Dianov GL, Sleeth KM, Dianova II, Allinson SL (2003) Repair of abasic sites in DNA. Mutat Res 531:157–163
76. Lindahl T (2001) Keynote: past, present, and future aspects of base excision repair. Prog Nucleic Acid Res Mol Biol 68:xvii–xxx
77. Masson M, Niedergang C, Schreiber V, Muller S, Menissier-de Murcia J, de Murcia G (1998) XRCC1 is specifically associated with poly(ADP-ribose) polymerase and negatively regulates its activity following DNA damage. Mol Cell Biol 18:3563–3571
78. Tomkinson AE, Chen L, Dong Z, Leppard JB, Levin DS, Mackey ZB, Motycka TA (2001) Completion of base excision repair by mammalian DNA ligases. Prog Nucleic Acid Res Mol Biol 68:151–164
79. Whitehouse CJ, Taylor RM, Thistlethwaite A, Zhang H, Karimi-Busheri F, Lasko DD, Weinfeld M, Caldecott KW (2001) XRCC1 stimulates human polynucleotide kinase activity at damaged DNA termini and accelerates DNA single-strand break repair. Cell 104:107–117
80. Mani RS, Fanta M, Karimi-Busheri F, Silver E, Virgen CA, Caldecott KW, Cass CE, Weinfeld M (2007) XRCC1 stimulates polynucleotide kinase by enhancing its damage discrimination and displacement from DNA repair intermediates. J Biol Chem 282:28004–28013
81. Bessho T (1999) Nucleotide excision repair 3′ endonuclease XPG stimulates the activity of base excision repair enzyme thymine glycol DNA glycosylase. Nucleic Acids Res 27:979–983
82. Okano S, Lan L, Caldecott KW, Mori T, Yasui A (2003) Spatial and temporal cellular responses to single-strand breaks in human cells. Mol Cell Biol 23:3974–3981
83. Satoh MS, Lindahl T (1992) Role of poly(ADP-ribose) formation in DNA repair. Nature 356:356–358
84. Woodhouse BC, Dianova II, Parsons JL, Dianov GL (2008) Poly(ADP-ribose) polymerase-1 modulates DNA repair capacity and prevents formation of DNA double strand breaks. DNA Repair (Amst) 7:932–940
85. Lieber MR (2010) The mechanism of double-strand DNA break repair by the nonhomologous DNA end-joining pathway. Annu Rev Biochem 79:181–211

86. Kuschel B, Auranen A, McBride S, Novik KL, Antoniou A, Lipscombe JM, Day NE, Easton DF, Ponder BA, Pharoah PD, Dunning A (2002) Variants in DNA double-strand break repair genes and breast cancer susceptibility. Hum Mol Genet 11:1399–1407
87. Valerie K, Povirk LF (2003) Regulation and mechanisms of mammalian double-strand break repair. Oncogene 22:5792–5812
88. Petrini JH, Stracker TH (2003) The cellular response to DNA double-strand breaks: defining the sensors and mediators. Trends Cell Biol 13:458–462
89. Shrivastav M, De Haro LP, Nickoloff JA (2008) Regulation of DNA double-strand break repair pathway choice. Cell Res 18:134–147
90. Downs JA (2007) Chromatin structure and DNA double-strand break responses in cancer progression and therapy. Oncogene 26:7765–7772
91. Downs JA, Allard S, Jobin-Robitaille O, Javaheri A, Auger A, Bouchard N, Kron SJ, Jackson SP, Côté J (2004) Binding of chromatin-modifying activities to phosphorylated histone H2A at DNA damage sites. Mol Cell 16:979–990
92. Utley RT, Lacoste N, Jobin-Robitaille O, Allard S, Côté J (2005) Regulation of NuA4 histone acetyltransferase activity in transcription and DNA repair by phosphorylation of histone H4. Mol Cell Biol 25:8179–8190
93. Lieber MR (2008) The mechanism of human nonhomologous DNA end joining. J Biol Chem 283:1–5
94. Weterings E, Chen DJ (2008) The endless tale of non-homologous end-joining. Cell Res 18:114–124
95. Goedecke W, Eijpe M, Offenberg HH, van Aalderen M, Heyting C (1999) Mre11 and Ku70 interact in somatic cells, but are differentially expressed in early meiosis. Nat Genet 23:194–198
96. Paull TT, Gellert M (2000) A mechanistic basis for Mre11-directed DNA joining at microhomologies. Proc Natl Acad Sci USA 97:6409–6414
97. Jeggo PA (1998) Identification of genes involved in repair of DNA double-strand breaks in mammalian cells. Radiat Res 150(5 Suppl):S80–S91
98. Koch CA, Agyei R, Galicia S, Metalnikov P, O'Donnell P, Starostine A, Weinfeld M, Durocher D (2004) Xrcc4 physically links DNA end processing by polynucleotide kinase to DNA ligation by DNA ligase IV. EMBO J 23:3874–3885
99. Jeggo PA, Lobrich M (2005) Artemis links ATM to double strand break rejoining. Cell Cycle 4:359–862
100. Ruscetti T, Lehnert BE, Halbrook J, Le Trong H, Hoekstra MF, Chen DJ, Peterson SR (1998) Stimulation of the DNA-dependent protein kinase by poly(ADP-ribose) polymerase. J Biol Chem 273:14461–14467
101. Li X, Heyer WD (2008) Homologous recombination in DNA repair and DNA damage tolerance. Cell Res 18:99–113
102. Zhang J, Powell SN (2005) The role of the BRCA1 tumor suppressor in DNA double-strand break repair. Mol Cancer Res 3:531–539
103. Zhuang J, Zhang J, Willers H, Wang H, Chung JH, van Gent DC, Hallahan DE, Powell SN, Xia F (2006) Checkpoint kinase 2-mediated phosphorylation of BRCA1 regulates the fidelity of nonhomologous end-joining. Cancer Res 66:1401–1408
104. Baumann P, Benson FE, West SC (1996) Human Rad51 protein promotes ATP-dependent homologous pairing and strand transfer reactions in vitro. Cell 87:757–766
105. Benson FE, Baumann P, West SC (1998) Synergistic actions of Rad51 and Rad52 in recombination and DNA repair. Nature 391:401–404
106. McIlwraith MJ, Van Dyck E, Masson JY, Stasiak AZ, Stasiak A, West SC (2000) Reconstitution of the strand invasion step of double-strand break repair using human Rad51 Rad52 and RPA proteins. J Mol Biol 304:151–164
107. Chen G, Yuan SS, Liu W, Xu Y, Trujillo K, Song B, Cong F, Goff SP, Wu Y, Arlinghaus R, Baltimore D, Gasser PJ, Park MS, Sung P, Lee EY (1999) Radiation-induced assembly of Rad51 and Rad52 recombination complex requires ATM and c-Abl. J Biol Chem 274:12748–12752

108. Kingston RE, Bunker CA, Imbalzano AN (1996) Repression and activation by multiprotein complexes that alter chromatin structure. Genes Dev 10:905–920
109. Swagemakers SM, Essers J, de Wit J, Hoeijmakers JH, Kanaar R (1998) The human RAD54 recombinational DNA repair protein is a double-stranded DNA-dependent ATPase. J Biol Chem 273:28292–28297
110. McCabe N, Turner NC, Lord CJ, Kluzek K, Bialkowska A, Swift S, Giavara S, O'Connor MJ, Tutt AN, Zdzienicka MZ, Smith GC, Ashworth A (2006) Deficiency in the repair of DNA damage by homologous recombination and sensitivity to poly(ADP-ribose) polymerase inhibition. Cancer Res 66:8109–8115
111. Chen Y, Lee WH, Chew HK (1999) Emerging roles of BRCA1 in transcriptional regulation and DNA repair. J Cell Physiol 181:385–392
112. Rassool FV, Tomkinson AE (2010) Targeting abnormal DNA double strand break repair in cancer. Cell Mol Life Sci 67:3699–3710, Epub ahead of print
113. Nagaraju G, Scully R (2007) Minding the gap: the underground functions of BRCA1 and BRCA2 at stalled replication forks. DNA Repair (Amst) 6:1018–1031
114. Eriksson D, Stigbrand T (2010) Radiation-induced cell death mechanisms. Tumour Biol 31:363–372
115. Hait WN, Jin S, Yang JM (2006) A matter of life or death (or both): understanding autophagy in cancer. Clin Cancer Res 12:1961–1965
116. Ito H, Daido S, Kanzawa T, Kondo S, Kondo Y (2005) Radiation-induced autophagy is associated with LC3 and its inhibition sensitizes malignant glioma cells. Int J Oncol 26:1401–1410
117. Paglin S, Yahalom J (2006) Pathways that regulate autophagy and their role in mediating tumor response to treatment. Autophagy 2:291–293
118. Williams RS, Dodson GE, Limbo O, Yamada Y, Williams JS, Guenther G, Classen S, Glover JN, Iwasaki H, Russell P, Tainer JA (2009) Nbs1 flexibly tethers Ctp1 and Mre11-Rad50 to coordinate DNA double-strand break processing and repair. Cell 139:87–99
119. Altmannova V, Eckert-Boulet N, Arneric M, Kolesar P, Chaloupkova R, Damborsky J, Sung P, Zhao X, Lisby M, Krejci L (2010) Rad52 SUMOylation affects the efficiency of the DNA repair. Nucleic Acids Res 38:4708–4721
120. Matsuoka S, Ballif BA, Smogorzewska A, McDonald ER 3rd, Hurov KE, Luo J, Bakalarski CE, Zhao Z, Solimini N, Lerenthal Y, Shiloh Y, Gygi SP, Elledge SJ (2007) ATM and ATR substrate analysis reveals extensive protein networks responsive to DNA damage. Science 316:1160–1166
121. Shiloh Y (2003) ATM: ready, set, go. Cell Cycle 2:116–117
122. Bakkenist CJ, Kastan MB (2003) DNA damage activates ATM through intermolecular autophosphorylation and dimer dissociation. Nature 421:499–506
123. Uziel T, Lerenthal Y, Moyal L, Andegeko Y, Mittelman L, Shiloh Y (2003) Requirement of the MRN complex for ATM activation by DNA damage. EMBO J 22:5612–5621
124. Lee JH, Paull TT (2004) Direct activation of the ATM protein kinase by the Mre11/Rad50/Nbs1 complex. Science 304:93–96
125. Sun Y, Jiang X, Chen S, Fernandes N, Price BD (2005) A role for the Tip60 histone acetyltransferase in the acetylation and activation of ATM. Proc Natl Acad Sci USA 102:13182–13187
126. Niida H, Nakanishi M (2006) DNA damage checkpoints in mammals. Mutagenesis 21:3–9
127. Bartek J, Lukas J (2003) Chk1 and Chk2 kinases in checkpoint control and cancer. Cancer Cell 3:421–429
128. Kastan MB, Lim DS, Kim ST, Yang D (2001) ATM–a key determinant of multiple cellular responses to irradiation. Acta Oncol 40:686–688
129. Sengupta S, Harris CC (2005) p53: traffic cop at the crossroads of DNA repair and recombination. Nat Rev Mol Cell Biol 6:44–55
130. Appella E, Anderson CW (2001) Post-translational modifications and activation of p53 by genotoxic stresses. Eur J Biochem 268:2764–2772

131. Brooks CL, Gu W (2003) Ubiquitination, phosphorylation and acetylation: the molecular basis for p53 regulation. Curr Opin Cell Biol 15:164–171
132. Murray D, Mirzayans R (2007) Role of p53 in the repair of ionizing radiation-induced DNA damage. In: Landseer BR (ed) New research on DNA repair. Nova, Hauppauge, New York, pp 325–373
133. Willis N, Rhind N (2009) Regulation of DNA replication by the S-phase DNA damage checkpoint. Cell Div 4:13
134. Taylor WR, Stark GR (2001) Regulation of the G2/M transition by p53. Oncogene 20:1803–1815
135. Xu B, Kim ST, Lim DS, Kastan MB (2002) Two molecularly distinct G(2)/M checkpoints are induced by ionizing irradiation. Mol Cell Biol 22:1049–1059
136. Vangestel C, Van de Wiele C, Mees G, Peeters M (2009) Forcing cancer cells to commit suicide. Cancer Biother Radiopharm 24:395–407
137. Jin Z, El-Deiry WS (2005) Overview of cell death signaling pathways. Cancer Biol Ther 4:139–163
138. Kolesnick R, Fuks Z (2003) Radiation and ceramide-induced apoptosis. Oncogene 22:5897–5906
139. Corre I, Niaudet C, Paris F (2010) Plasma membrane signaling induced by ionizing radiation. Mutat Res 704:61–67
140. Suzuki M, Boothman DA (2008) Stress-induced premature senescence (SIPS)—influence of SIPS on radiotherapy. J Radiat Res 49:105–112
141. Mirzayans R, Scott A, Cameron M, Murray D (2005) Induction of accelerated senescence by γ radiation in human solid tumor-derived cell lines expressing wild-type TP53. Radiat Res 163:53–62
142. Gewirtz DA, Holt SE, Elmore LW (2008) Accelerated senescence: an emerging role in tumor cell response to chemotherapy and radiation. Biochem Pharmacol 76:947–957
143. Mirzayans R, Murray D (2007) Cellular senescence: implications for cancer therapy. In: Garvey RB (ed) New research on cell aging. Nova Science Publishers, Inc., Hauppauge, New York, pp 1–64
144. Roninson IB, Broude EV, Chang BD (2001) If not apoptosis, then what? Treatment-induced senescence and mitotic catastrophe in tumor cells. Drug Resist Updat 4:303–313
145. Jonathan EC, Bernhard EJ, McKenna WG (1999) How does radiation kill cells? Curr Opin Chem Biol 3:77–83
146. Brown JM, Wouters BG (2001) Apoptosis: mediator or mode of cell killing by anticancer agents? Drug Resist Updat 4:135–135
147. Daido S, Yamamoto A, Fujiwara K, Sawaya R, Kondo S, Kondo Y (2005) Inhibition of the DNA-dependent protein kinase catalytic subunit radiosensitizes malignant glioma cells by inducing autophagy. Cancer Res 65:4368–4375
148. Schmidt-Ullrich RK, Dent P, Grant S (2000) Signal transduction and cellular radiation responses. Radiat Res 153:245–257
149. Herring CJ, West CM, Wilks DP, Davidson SE, Hunter RD, Berry P, Forster G, MacKinnon J, Rafferty JA, Elder RH, Hendry JH, Margison GP (1998) Levels of the DNA repair enzyme human apurinic/apyrimidinic endonuclease (APE1, APEX, Ref-1) are associated with the intrinsic radiosensitivity of cervical cancers. Br J Cancer 78:1128–1133
150. Koukourakis MI, Giatromanolaki A, Kakolyris S, Sivridis E, Georgoulias V, Funtzilas G, Hickson ID, Gatter KC, Harris AL (2001) Nuclear expression of human apurinic/apyrimidinic endonuclease (HAP1/Ref-1) in head-and-neck cancer is associated with resistance to chemoradiotherapy and poor outcome. Int J Radiat Oncol Biol Phys 50:27–36
151. Bobola MS, Blank A, Berger MS, Stevens BA, Silber JR (2001) Apurinic/apyrimidinic endonuclease activity is elevated in human adult gliomas. Clin Cancer Res 7:3510–3518
152. Bobola MS, Emond MJ, Blank A, Meade EH, Kolstoe DD, Berger MS, Rostomily RC, Silbergeld DL, Spence AM, Silber JR (2004) Apurinic endonuclease activity in adult gliomas and time to tumor progression after alkylating agent-based chemotherapy and after radiotherapy. Clin Cancer Res 10:7875–7883

153. Bobola MS, Finn LS, Ellenbogen RG, Geyer JR, Berger MS, Braga JM, Meade EH, Gross ME, Silber JR (2005) Apurinic/apyrimidinic endonuclease activity is associated with response to radiation and chemotherapy in medulloblastoma and primitive neuroectodermal tumors. Clin Cancer Res 11:7405–7414
154. Robertson KA, Bullock HA, Xu Y, Tritt R, Zimmerman E, Ulbright TM, Foster RS, Einhorn LH, Kelley MR (2001) Altered expression of Ape1/ref-1 in germ cell tumors and overexpression in NT2 cells confers resistance to bleomycin and radiation. Cancer Res 61:2220–2225
155. Kelley MR, Cheng L, Foster R, Tritt R, Jiang J, Broshears J, Koch M (2001) Elevated and altered expression of the multifunctional DNA base excision repair and redox enzyme Ape1/ref-1 in prostate cancer. Clin Cancer Res 7:824–830
156. Skvortsova I, Skvortsov S, Stasyk T, Raju U, Popper BA, Schiestl B, von Guggenberg E, Neher A, Bonn GK, Huber LA, Lukas P (2008) Intracellular signaling pathways regulating radioresistance of human prostate carcinoma cells. Proteomics 8:4521–4533
157. Qing Y, Wang D, Lei X, Xiang DB, Li MX, Li ZP, Shan JL (2009) The expression of APE1 and its correlation with prognostic significance after 252Cf radiotherapy in cervical cancer. Sichuan Da Xue Xue Bao Yi Xue Ban 40:125–128
158. Wang D, Luo M, Kelley MR (2004) Human apurinic endonuclease 1 (APE1) expression and prognostic significance in osteosarcoma: enhanced sensitivity of osteosarcoma to DNA damaging agents using silencing RNA APE1 expression inhibition. Mol Cancer Ther 3:679–686
159. Li D, Li Y, Jiao L, Chang DZ, Beinart G, Wolff RA, Evans DB, Hassan MM, Abbruzzese JL (2007) Effects of base excision repair gene polymorphisms on pancreatic cancer survival. Int J Cancer 120:1748–1754
160. Al-Attar A, Gossage L, Fareed KR, Shehata M, Mohammed M, Zaitoun AM, Soomro I, Lobo DN, Abbotts R, Chan S, Madhusudan S (2010) Human apurinic/apyrimidinic endonuclease (APE1) is a prognostic factor in ovarian, gastro-oesophageal and pancreatico-biliary cancers. Br J Cancer 102:704–709
161. Herring CJ, Deans B, Elder RH, Rafferty JA, MacKinnon J, Barzilay G, Hickson ID, Hendry JH, Margison GP (1999) Expression levels of the DNA repair enzyme HAP1 do not correlate with the radiosensitivities of human or HAP1-transfected rat cell lines. Br J Cancer 80:940–945
162. Schindl M, Oberhuber G, Pichlbauer EG, Obermair A, Birner P, Kelley MR (2001) DNA repair-redox enzyme apurinic endonuclease in cervical cancer: evaluation of redox control of HIF-1α and prognostic significance. Int J Oncol 19:799–802
163. Sak SC, Harnden P, Johnston CF, Paul AB, Kiltie AE (2005) APE1 and XRCC1 protein expression levels predict cancer-specific survival following radical radiotherapy in bladder cancer. Clin Cancer Res 11:6205–6211
164. Tell G, Quadrifoglio F, Tiribelli C, Kelley MR (2009) The many functions of APE1/Ref-1: not only a DNA repair enzyme. Antioxid Redox Signal 11:601–620
165. Di Maso V, Avellini C, Crocè LS, Rosso N, Quadrifoglio F, Cesaratto L, Codarin E, Bedogni G, Beltrami CA, Tell G, Tiribelli C (2007) Subcellular localization of APE1/Ref-1 in human hepatocellular carcinoma: possible prognostic significance. Mol Med 13:89–96
166. Dunphy EJ, Beckett MA, Thompson LH, Weichselbaum RR (1992) Expression of the polymorphic human DNA repair gene XRCC1 does not correlate with radiosensitivity in the cells of human head and neck tumor cell lines. Radiat Res 130:166–170
167. Polischouk AG, Cedervall B, Ljungquist S, Flygare J, Hellgren D, Grénman R, Lewensohn R (1999) DNA double-strand break repair, DNA-PK, and DNA ligases in two human squamous carcinoma cell lines with different radiosensitivity. Int J Radiat Oncol Biol Phys 43:191–198
168. Zhao HJ, Hosoi Y, Miyachi H, Ishii K, Yoshida M, Nemoto K, Takai Y, Yamada S, Suzuki N, Ono T (2000) DNA-dependent protein kinase activity correlates with Ku70 expression and radiation sensitivity in esophageal cancer cell lines. Clin Cancer Res 6:1073–1078
169. Sirzen F, Nilsson A, Zhivotovsky B, Lewensohn R (1999) DNA-dependent protein kinase content and activity in lung carcinoma cell lines: correlation with intrinsic radiosensitivity. Eur J Cancer 35:111–116

170. Hansen LT, Lundin C, Helleday T, Poulsen HS, Sørensen CS, Petersen LN, Spang-Thomsen M (2003) DNA repair rate and etoposide (VP16) resistance of tumor cell subpopulations derived from a single human small cell lung cancer. Lung Cancer 40:157–164
171. Allalunis-Turner MJ, Lintott LG, Barron GM, Day RS, Lees-Miller SP (1995) Lack of correlation between DNA-dependent protein kinase activity and tumor cell radiosensitivity. Cancer Res 55:5200–5202
172. Bjork-Eriksson T, West C, Nilsson A, Magnusson B, Svensson M, Karlsson E, Slevin N, Lewensohn R, Mercke C (1999) The immunohistochemical expression of DNA-PKcs and Ku (p70/p80) in head and neck cancers: relationships with radiosensitivity. Int J Radiat Oncol Biol Phys 45:1005–1010
173. Komuro Y, Watanabe T, Hosoi Y, Matsumoto Y, Nakagawa K, Tsuno N, Kazama S, Kitayama J, Suzuki N, Nagawa H (2002) The expression pattern of Ku correlates with tumor radiosensitivity and disease free survival in patients with rectal carcinoma. Cancer 95:1199–1205
174. Komuro Y, Watanabe T, Hosoi Y, Matsumoto Y, Nakagawa K, Saito S, Ishihara S, Kazama S, Tsuno N, Kitayama J, Suzuki N, Tsurita G, Muto T, Nagawa H (2003) Prediction of tumor radiosensitivity in rectal carcinoma based on p53 and Ku70 expression. J Exp Clin Cancer Res 22:223–228
175. Komuro Y, Watanabe T, Hosoi Y, Matsumoto Y, Nakagawa K, Suzuki N, Nagawa H (2005) Prognostic significance of Ku70 protein expression in patients with advanced colorectal cancer. Hepatogastroenterology 52:995–998
176. Komuro Y, Watanabe T, Tsurita G, Muto T, Nagawa H (2005) Evaluating the combination of molecular prognostic factors in tumor radiosensitivity in rectal cancer. Hepatogastroenterology 52:666–671
177. Sakata K, Matsumoto Y, Satoh M, Oouchi A, Nagakura H, Koito K, Hosoi Y, Hareyama M, Suzuki N (2001) Clinical studies of immunohistochemical staining of DNA-dependent protein kinase in oropharyngeal and hypopharyngeal carcinomas. Radiat Med 19:93–97
178. Lee SW, Cho KJ, Park JH, Kim SY, Nam SY, Lee BJ, Kim SB, Choi SH, Kim JH, Ahn SD, Shin SS, Choi EK, Yu E (2005) Expressions of Ku70 and DNA-PKcs as prognostic indicators of local control in nasopharyngeal carcinoma. Int J Radiat Oncol Biol Phys 62:1451–1457
179. Yan SS, Liu L, Liu ZG, Zeng MS, Song LB, Xia YF (2008) Expression and clinical significance of DNA-PKcs in nasopharyngeal carcinoma. Ai Zheng 27:979–983
180. Saygili U, Gorkay IB, Koyuncuoglu M, Gol M, Uslu T, Erten O (2004) The relationship between expression of Ku70 and survival in irradiated patients with endometrial carcinoma. Gynecol Oncol 95:518–522
181. Xing J, Wu X, Vaporciyan AA, Spitz MR, Gu J (2008) Prognostic significance of ataxia-telangiectasia mutated, DNA-dependent protein kinase catalytic subunit, and Ku heterodimeric regulatory complex 86-kD subunit expression in patients with nonsmall cell lung cancer. Cancer 112:2756–2764
182. Shintani S, Mihara M, Li C, Nakahara Y, Hino S, Nakashiro K, Hamakawa H (2003) Up-regulation of DNA-dependent protein kinase correlates with radiation resistance in oral squamous cell carcinoma. Cancer Sci 94:894–900
183. Sakata K, Matsumoto Y, Tauchi H, Satoh M, Oouchi A, Nagakura H, Koito K, Hosoi Y, Suzuki N, Komatsu K, Hareyama M (2001) Expression of genes involved in repair of DNA double-strand breaks in normal and tumor tissues. Int J Radiat Oncol Biol Phys 49:161–167
184. Sakata K, Someya M, Matsumoto Y, Hareyama M (2007) Ability to repair DNA double-strand breaks related to cancer susceptibility and radiosensitivity. Radiat Med 25:433–438
185. Noguchi T, Shibata T, Fumoto S, Uchida Y, Mueller W, Takeno S (2002) DNA-PKcs expression in esophageal cancer as a predictor for chemoradiation therapeutic sensitivity. Ann Surg Oncol 9:1017–1022
186. Friesland S, Kanter-Lewensohn L, Tell R, Munck-Wikland E, Lewensohn R, Nilsson A (2003) Expression of Ku86 confers favorable outcome of tonsillar carcinoma treated with radiotherapy. Head Neck 25:313–321
187. Pavón MA, Parreño M, León X, Sancho FJ, Céspedes MV, Casanova I, Lopez-Pousa A, Mangues MA, Quer M, Barnadas A, Mangues R (2008) Ku70 predicts response and primary

tumor recurrence after therapy in locally advanced head and neck cancer. Int J Cancer 123:1068–1079
188. Wilson CR, Davidson SE, Margison GP, Jackson SP, Hendry JH, West CM (2000) Expression of Ku70 correlates with survival in carcinoma of the cervix. Br J Cancer 83:1702–1706
189. Harima Y, Sawada S, Miyazaki Y, Kin K, Ishihara H, Imamura M, Sougawa M, Shikata N, Ohnishi T (2003) Expression of Ku80 in cervical cancer correlates with response to radiotherapy and survival. Am J Clin Oncol 26:e80–e85
190. Beskow C, Kanter L, Holgersson A, Nilsson B, Frankendal B, Avall-Lundqvist E, Lewensohn R (2006) Expression of DNA damage response proteins and complete remission after radiotherapy of stage IB-IIA of cervical cancer. Br J Cancer 94:1683–1689
191. Beskow C, Skikuniene J, Holgersson A, Nilsson B, Lewensohn R, Kanter L, Viktorsson K (2009) Radioresistant cervical cancer shows upregulation of the NHEJ proteins DNA-PKcs, Ku70 and Ku86. Br J Cancer 101:816–821
192. Carlomagno F, Burnet NG, Turesson I, Nyman J, Peacock JH, Dunning AM, Ponder BA, Jackson SP (2000) Comparison of DNA repair protein expression and activities between human fibroblast cell lines with different radiosensitivities. Int J Cancer 85:845–849
193. Loong SL, Korzh S, Price A (2004) Reduced DNA-dependent protein kinase activity in two cell lines derived from adult cancer patients with late radionecrosis. Oncogene 23:5562–5566
194. Xia F, Powell SN (2002) The molecular basis of radiosensitivity and chemosensitivity in the treatment of breast cancer. Semin Radiat Oncol 12:296–304
195. Turesson I, Nyman J, Holmberg E, Odén A (1996) Prognostic factors for acute and late skin reactions in radiotherapy patients. Int J Radiat Oncol Biol Phys 36:1065–1075
196. Andreassen CN, Alsner J, Overgaard J (2002) Does variability in normal tissue reactions after radiotherapy have a genetic basis—where and how to look for it? Radiother Oncol 64:131–140
197. Brookes AJ (1999) The essence of SNPs. Gene 234:177–186
198. Stoneking M (2001) Single nucleotide polymorphisms. From the evolutionary past. Nature 409:821–822
199. Andreassen CN (2005) Can risk of radiotherapy-induced normal tissue complications be predicted from genetic profiles? Acta Oncol 44:801–815
200. Barnett GC, West CM, Dunning AM, Elliott RM, Coles CE, Pharoah PD, Burnet NG (2009) Normal tissue reactions to radiotherapy: towards tailoring treatment dose by genotype. Nat Rev Cancer 9:134–142
201. Popanda O, Marquardt JU, Chang-Claude J, Schmezer P (2009) Genetic variation in normal tissue toxicity induced by ionizing radiation. Mutat Res 667:58–69
202. Andreassen CN, Alsner J (2009) Genetic variants and normal tissue toxicity after radiotherapy: a systematic review. Radiother Oncol 92:299–309
203. Alsner J, Andreassen CN, Overgaard J (2008) Genetic markers for prediction of normal tissue toxicity after radiotherapy. Semin Radiat Oncol 18:126–135
204. Parliament MB, Murray D (2010) Single nucleotide polymorphisms of DNA repair genes as predictors of radioresponse. Semin Radiat Oncol 20:232–240
205. Fernet M, Hall J (2004) Genetic biomarkers of therapeutic radiation sensitivity. DNA Repair (Amst) 3:1237–1243
206. Mohrenweiser H, Jones I (1998) Variation in DNA repair is a factor in cancer susceptibility: a paradigm for the promises and perils of individual and population risk estimation? Mutat Res 400:15–24
207. Shen R, Jones I, Mohrenweiser H (1998) Nonconservative amino acid substitution variants exist at polymorphic frequency in DNA repair genes in healthy humans. Cancer Res 58:604–608
208. Mohrenweiser HW, Xi T, Vazquez-Matias J, Jones IM (2002) Identification of 127 amino acid substitution variants in screening 37 DNA repair genes in humans. Cancer Epidemiol Biomark Prev 11:1054–1064

209. Goode EL, Ulrich CM, Potter JD (2002) Polymorphisms in DNA repair genes and associations with cancer risk. Cancer Epidemiol Biomark Prev 11:1513–1530
210. Yin M, Tan D, Wei Q (2009) Genetic variants of the XRCC1 gene and susceptibility to esophageal cancer: a meta-analysis. Int J Clin Exp Med 2:26–35
211. Cornetta T, Festa F, Testa A, Cozzi R (2006) DNA damage repair and genetic polymorphisms: assessment of individual sensitivity and repair capacity. Int J Radiat Oncol Biol Phys 66:537–545
212. Bishop D, Ear U, Bhattacharyya A, Calderone C, Beckett M, Weichselbaum R (1998) XRCC3 is required for assembly of Rad51 complexes in vivo. J Biol Chem 273:21482–21488
213. Gabriel SB, Schaffner SF, Nguyen H, Moore JM, Roy J, Blumenstiel B, Higgins J, DeFelice M, Lochner A, Faggart M, Liu-Cordero SN, Rotimi C, Adeyemo A, Cooper R, Ward R, Lander ES, Daly MJ, Altshuler D (2002) The structure of haplotype blocks in the human genome. Science 296:2225–2229
214. Bentzen SM (2008) From cellular to high-throughput predictive assays in radiation oncology: challenges and opportunities. Semin Radiat Oncol 18:75–88
215. Moullan N, Cox DG, Angèle S, Romestaing P, Gérard JP, Hall J (2003) Polymorphisms in the DNA repair gene XRCC1, breast cancer risk, and response to radiotherapy. Cancer Epidemiol Biomark Prev 12:1168–1174
216. Hu JJ, Smith TR, Miller MS, Mohrenweiser HW, Golden A, Case LD (2001) Amino acid substitution variants of APE1 and XRCC1 genes associated with ionizing radiation sensitivity. Carcinogenesis 22:917–922
217. Rogers PB, Plowman PN, Harris SJ, Arlett CF (2000) Four radiation hypersensitivity cases and their implications for clinical radiotherapy. Radiother Oncol 57:143–154
218. Andreassen CN, Alsner J, Overgaard M, Overgaard J (2003) Prediction of normal tissue radiosensitivity from polymorphisms in candidate genes. Radiother Oncol 69:127–135
219. Andreassen CN, Alsner J, Overgaard M, Sørensen FB, Overgaard J (2006) Risk of radiation-induced subcutaneous fibrosis in relation to single nucleotide polymorphisms in TGFB1, SOD2, XRCC1, XRCC3, APEX and ATM–a study based on DNA from formalin fixed paraffin embedded tissue samples. Int J Radiat Biol 82:577–586
220. Giotopoulos G, Symonds RP, Foweraker K, Griffin M, Peat I, Osman A, Plumb M (2007) The late radiotherapy normal tissue injury phenotypes of telangiectasia, fibrosis and atrophy in breast cancer patients have distinct genotype-dependent causes. Br J Cancer 96:1001–1007
221. Chang-Claude J, Popanda O, Tan XL, Kropp S, Helmbold I, von Fournier D, Haase W, Sautter-Bihl ML, Wenz F, Schmezer P, Ambrosone CB (2005) Association between polymorphisms in the DNA repair genes, XRCC1, APE1, and XPD and acute side effects of radiotherapy in breast cancer patients. Clin Cancer Res 11:4802–4809
222. Popanda O, Tan XL, Ambrosone CB, Kropp S, Helmbold I, von Fournier D, Haase W, Sautter-Bihl ML, Wenz F, Schmezer P, Chang-Claude J (2006) Genetic polymorphisms in the DNA double-strand break repair genes XRCC3, XRCC2, and NBS1 are not associated with acute side effects of radiotherapy in breast cancer patients. Cancer Epidemiol Biomark Prev 15:1048–1050
223. Brem R, Cox DG, Chapot B, Moullan N, Romestaing P, Gérard JP, Pisani P, Hall J (2006) The XRCC1-77 T->C variant: haplotypes, breast cancer risk, response to radiotherapy and the cellular response to DNA damage. Carcinogenesis 27:2469–2474
224. Hao B, Miao X, Li Y, Zhang X, Sun T, Liang G, Zhao Y, Zhou Y, Wang H, Chen X, Zhang L, Tan W, Wei Q, Lin D, He F (2006) A novel T-77C polymorphism in DNA repair gene XRCC1 contributes to diminished promoter activity and increased risk of non-small cell lung cancer. Oncogene 25:3613–3620
225. Angèle S, Romestaing P, Moullan N, Vuillaume M, Chapot B, Friesen M, Jongmans W, Cox DG, Pisani P, Gérard JP, Hall J (2003) ATM haplotypes and cellular response to DNA damage: association with breast cancer risk and clinical radiosensitivity. Cancer Res 63:8717–8725

226. Chang-Claude J, Ambrosone CB, Lilla C, Kropp S, Helmbold I, von Fournier D, Haase W, Sautter-Bihl ML, Wenz F, Schmezer P, Popanda O (2009) Genetic polymorphisms in DNA repair and damage response genes and late normal tissue complications of radiotherapy for breast cancer. Br J Cancer 100:1680–1686
227. Zschenker O, Raabe A, Boeckelmann IK, Borstelmann S, Szymczak S, Wellek S, Rades D, Hoeller U, Ziegler A, Dikomey E, Borgmann K (2010) Association of single nucleotide polymorphisms in ATM, GSTP1, SOD2, TGFB1, XPD and XRCC1 with clinical and cellular radiosensitivity. Radiother Oncol 97:26–32
228. Sterpone S, Cornetta T, Padua L, Mastellone V, Giammarino D, Testa A, Tirindelli D, Cozzi R, Donato V (2010) DNA repair capacity and acute radiotherapy adverse effects in Italian breast cancer patients. Mutat Res 684:43–48
229. Mangoni M, Bisanzi S, Carozzi F, Sani C, Biti G, Livi L, Barletta E, Costantini AS, Gorini G (2010) Association between genetic polymorphisms in the XRCC1, XRCC3, XPD, GSTM1, GSTT1, MSH2, MLH1, MSH3, and MGMT genes and radiosensitivity in breast cancer patients. Int J Radiat Oncol Biol Phys 81:52–58, Epub ahead of print
230. Suga T, Ishikawa A, Kohda M, Otsuka Y, Yamada S, Yamamoto N, Shibamoto Y, Ogawa Y, Nomura K, Sho K, Omura M, Sekiguchi K, Kikuchi Y, Michikawa Y, Noda S, Sagara M, Ohashi J, Yoshinaga S, Mizoe J, Tsujii H, Iwakawa M, Imai T (2007) Haplotype-based analysis of genes associated with risk of adverse skin reactions after radiotherapy in breast cancer patients. Int J Radiat Oncol Biol Phys 69:685–693
231. Pierce LJ, Strawderman M, Narod SA, Oliviotto I, Eisen A, Dawson L, Gaffney D, Solin LJ, Nixon A, Garber J, Berg C, Isaacs C, Heimann R, Olopade OI, Haffty B, Weber BL (2000) Effect of radiotherapy after breast-conserving treatment in women with breast cancer and germline BRCA1/2 mutations. J Clin Oncol 18:3360–3369
232. Shanley S, McReynolds K, Ardern-Jones A, Ahern R, Fernando I, Yarnold J, Evans G, Eccles D, Hodgson S, Ashley S, Ashcroft L, Tutt A, Bancroft E, Short S, Gui G (2006) Late toxicity is not increased in BRCA1/BRCA2 mutation carriers undergoing breast radiotherapy in the United Kingdom. Clin Cancer Res 12:7025–7032
233. Cesaretti JA, Stock RG, Lehrer S, Atencio DA, Bernstein JL, Stone NN, Wallenstein S, Green S, Loeb K, Kollmeier M, Smith M, Rosenstein BS (2005) ATM sequence variants are predictive of adverse radiotherapy response among patients treated for prostate cancer. Int J Radiat Oncol Biol Phys 61:196–202
234. Cesaretti JA, Stock RG, Atencio DP, Peters SA, Peters CA, Burri RJ, Stone NN, Rosenstein BS (2007) A genetically determined dose-volume histogram predicts for rectal bleeding among patients treated with prostate brachytherapy. Int J Radiat Oncol Biol Phys 68:1410–1416
235. Burri RJ, Stock RG, Cesaretti JA, Atencio DP, Peters S, Peters CA, Fan G, Stone NN, Ostrer H, Rosenstein BS (2008) Association of single nucleotide polymorphisms in SOD2, XRCC1 and XRCC3 with susceptibility for the development of adverse effects resulting from radiotherapy for prostate cancer. Radiat Res 170:49–59
236. Pugh TJ, Keyes M, Barclay L, Delaney A, Krzywinski M, Thomas D, Novik K, Yang C, Agranovich A, McKenzie M, Morris WJ, Olive PL, Marra MA, Moore RA (2009) Sequence variant discovery in DNA repair genes from radiosensitive and radiotolerant prostate brachytherapy patients. Clin Cancer Res 15:5008–5016
237. Damaraju S, Murray D, Dufour J, Carandang D, Myrehaug S, Fallone G, Field C, Greiner R, Hanson J, Cass CE, Parliament M (2006) Association of DNA repair and steroid metabolism gene polymorphisms with clinical late toxicity in patients treated with conformal radiotherapy for prostate cancer. Clin Cancer Res 12:2545–2554
238. Hall EJ, Schiff PB, Hanks GE, Brenner DJ, Russo J, Chen J, Sawant SG, Pandita TK (1998) A preliminary report: frequency of A-T heterozygotes among prostate cancer patients with severe late responses to radiation therapy. Cancer J Sci Am 4:385–389
239. Suga T, Iwakawa M, Tsuji H, Ishikawa H, Oda E, Noda S, Otsuka Y, Ishikawa A, Ishikawa K, Shimazaki J, Mizoe JE, Tsujii H, Imai T (2008) Influence of multiple genetic polymorphisms on genitourinary morbidity after carbon ion radiotherapy for prostate cancer. Int J Radiat Oncol Biol Phys 72:808–813

240. Werbrouck J, De Ruyck K, Duprez F, Veldeman L, Claes K, Van Eijkeren M, Boterberg T, Willems P, Vral A, De Neve W, Thierens H (2009) Acute normal tissue reactions in head-and-neck cancer patients treated with IMRT: influence of dose and association with genetic polymorphisms in DNA DSB repair genes. Int J Radiat Oncol Biol Phys 73:1187–1195
241. Kornguth DG, Garden AS, Zheng Y, Dahlstrom KR, Wei Q, Sturgis EM (2005) Gastrostomy in oropharyngeal cancer patients with ERCC4 (XPF) germline variants. Int J Radiat Oncol Biol Phys 62:665–671
242. Alsbeih G, Al-Harbi N, Al-Hadyan K, El-Sebaie M, Al-Rajhi N (2010) Association between normal tissue complications after radiotherapy and polymorphic variations in TGFB1 and XRCC1 genes. Radiat Res 173:505–511
243. De Ruyck K, Van Eijkeren M, Claes K, Morthier R, De Paepe A, Vral A, De Ridder L, Thierens H (2005) Radiation-induced damage to normal tissues after radiotherapy in patients treated for gynecologic tumors: association with single nucleotide polymorphisms in XRCC1, XRCC3, and OGG1 genes and in vitro chromosomal radiosensitivity in lymphocytes. Int J Radiat Oncol Biol Phys 62:1140–1149
244. De Ruyck K, Wilding CS, Van Eijkeren M, Morthier R, Tawn EJ, Thierens H (2005) Microsatellite polymorphisms in DNA repair genes XRCC1, XRCC3 and XRCC5 in patients with gynecological tumors: association with late clinical radiosensitivity and cancer incidence. Radiat Res 164:237–244
245. Azria D, Ozsahin M, Kramar A, Peters S, Atencio DP, Crompton NE, Mornex F, Pèlegrin A, Dubois JB, Mirimanoff RO, Rosenstein BS (2008) Single nucleotide polymorphisms, apoptosis, and the development of severe late adverse effects after radiotherapy. Clin Cancer Res 14:6284–6288
246. Carles J, Monzo M, Amat M, Jansa S, Artells R, Navarro A, Foro P, Alameda F, Gayete A, Gel B, Miguel M, Albanell J, Fabregat X (2006) Single-nucleotide polymorphisms in base excision repair, nucleotide excision repair, and double strand break genes as markers for response to radiotherapy in patients with Stage I to II head-and-neck cancer. Int J Radiat Oncol Biol Phys 66:1022–1030
247. Tu HF, Chen HW, Kao SY, Lin SC, Liu CJ, Chang KW (2008) MDM2 SNP 309 and p53 codon 72 polymorphisms are associated with the outcome of oral carcinoma patients receiving postoperative irradiation. Radiother Oncol 87:243–252
248. Wu X, Gu J, Wu TT, Swisher SG, Liao Z, Correa AM, Liu J, Etzel CJ, Amos CI, Huang M, Chiang SS, Milas L, Hittelman WN, Ajani JA (2006) Genetic variations in radiation and chemotherapy drug action pathways predict clinical outcomes in esophageal cancer. J Clin Oncol 24:3789–3798
249. Warnecke-Eberz U, Vallböhmer D, Alakus H, Kütting F, Lurje G, Bollschweiler E, Wienand-Dorweiler A, Drebber U, Hölscher AH, Metzger R (2009) ERCC1 and XRCC1 gene polymorphisms predict response to neoadjuvant radiochemotherapy in esophageal cancer. J Gastrointest Surg 13:1411–1421
250. Yoon SM, Hong YC, Park HJ, Lee JE, Kim SY, Kim JH, Lee SW, Park SY, Lee JS, Choi EK (2005) The polymorphism and haplotypes of XRCC1 and survival of non-small-cell lung cancer after radiotherapy. Int J Radiat Oncol Biol Phys 63:885–891
251. Su D, Ma S, Liu P, Jiang Z, Lv W, Zhang Y, Deng Q, Smith S, Yu H (2007) Genetic polymorphisms and treatment response in advanced non-small cell lung cancer. Lung Cancer 56:281–288
252. Yin Z, Zhou B, He Q, Li M, Guan P, Li X, Cui Z, Xue X, Su M, Ma R, Bai W, Xia S, Jiang Y, Xu S, Lv Y, Li X (2009) Association between polymorphisms in DNA repair genes and survival of non-smoking female patients with lung adenocarcinoma. BMC Cancer 9:439
253. Okazaki T, Jiao L, Chang P, Evans DB, Abbruzzese JL, Li D (2008) Single-nucleotide polymorphisms of DNA damage response genes are associated with overall survival in patients with pancreatic cancer. Clin Cancer Res 14:2042–2048
254. Li D, Frazier M, Evans DB, Hess KR, Crane CH, Jiao L, Abbruzzese JL (2006) Single nucleotide polymorphisms of RecQ1, RAD54L, and ATM genes are associated with reduced survival of pancreatic cancer. J Clin Oncol 24:1720–1728

255. Li D, Liu H, Jiao L, Chang DZ, Beinart G, Wolff RA, Evans DB, Hassan MM, Abbruzzese JL (2006) Significant effect of homologous recombination DNA repair gene polymorphisms on pancreatic cancer survival. Cancer Res 66:3323–3330
256. Sakano S, Wada T, Matsumoto H, Sugiyama S, Inoue R, Eguchi S, Ito H, Ohmi C, Matsuyama H, Naito K (2006) Single nucleotide polymorphisms in DNA repair genes might be prognostic factors in muscle-invasive bladder cancer patients treated with chemoradiotherapy. Br J Cancer 95:561–570
257. Gao R, Price DK, Dahut WL, Reed E, Figg WD (2010) Genetic polymorphisms in XRCC1 associated with radiation therapy in prostate cancer. Cancer Biol Ther 10:13–18
258. Cecchin E, Agostini M, Pucciarelli S, De Paoli A, Canzonieri V, Sigon R, De Mattia E, Friso ML, Biason P, Visentin M, Nitti D, Toffoli G (2011) Tumor response is predicted by patient genetic profile in rectal cancer patients treated with neo-adjuvant chemo-radiotherapy, Pharmacogenomics J 11:214–226
259. Kerns SL, Ostrer H, Stock R, Li W, Moore J, Pearlman A, Campbell C, Shao Y, Stone N, Kusnetz L, Rosenstein BS (2010) Genome-wide association study to identify single nucleotide polymorphisms (SNPs) associated with the development of erectile dysfunction in African-American men after radiotherapy for prostate cancer. Int J Radiat Oncol Biol Phys 78:1292–1300
260. Hutchison CA 3rd (2007) DNA sequencing: bench to bedside and beyond. Nucleic Acids Res 35:6227–6237
261. Das AK, Bell MH, Nirodi CS, Story MD, Minna JD (2010) Radiogenomics predicting tumor responses to radiotherapy in lung cancer. Semin Radiat Oncol 20:149–155
262. Eschrich S, Zhang H, Zhao H, Boulware D, Lee JH, Bloom G, Torres-Roca JF (2009) Systems biology modeling of the radiation sensitivity network: a biomarker discovery platform. Int J Radiat Oncol Biol Phys 75:497–505
263. Redon R, Ishikawa S, Fitch KR, Feuk L, Perry GH, Andrews TD, Fiegler H, Shapero MH, Carson AR, Chen W, Cho EK, Dallaire S, Freeman JL, González JR, Gratacòs M, Huang J, Kalaitzopoulos D, Komura D, MacDonald JR, Marshall CR, Mei R, Montgomery L, Nishimura K, Okamura K, Shen F, Somerville MJ, Tchinda J, Valsesia A, Woodwark C, Yang F, Zhang J, Zerjal T, Zhang J, Armengol L, Conrad DF, Estivill X, Tyler-Smith C, Carter NP, Aburatani H, Lee C, Jones KW, Scherer SW, Hurles ME (2006) Global variation in copy number in the human genome. Nature 444:444–454
264. West CM, McKay MJ, Hölscher T, Baumann M, Stratford IJ, Bristow RG, Iwakawa M, Imai T, Zingde SM, Anscher MS, Bourhis J, Begg AC, Haustermans K, Bentzen SM, Hendry JH (2005) Molecular markers predicting radiotherapy response: report and recommendations from an International Atomic Energy Agency technical meeting. Int J Radiat Oncol Biol Phys 62:1264–1273
265. Ho AY, Atencio DP, Peters S, Stock RG, Formenti SC, Cesaretti JA, Green S, Haffty B, Drumea K, Leitzin L, Kuten A, Azria D, Ozsahin M, Overgaard J, Andreassen CN, Trop CS, Park J, Rosenstein BS (2006) Genetic predictors of adverse radiotherapy effects: the Gene-PARE project. Int J Radiat Oncol Biol Phys 65:646–655
266. Burnet NG, Elliott RM, Dunning A, West CM (2006) Radiosensitivity, radiogenomics and RAPPER. Clin Oncol (R Coll Radiol) 18:525–528
267. http://science.cancerresearchuk.org/research/who-and-what-we-fund/browse-by-location/cambridge/university-of-cambridge/grants/neil-burnet-4764-rapper-radiogenomics-assessment-of-polymorphisms
268. Iwakawa M, Imai T, Harada Y, Ban S, Michikawa Y, Saegusa K, Sagara M, Tsuji A, Noda S, Ishikawa A (2002) RadGenomics project. Nippon Igaku Hoshasen Gakkai Zasshi 62:484–489
269. Baumann M, Hölscher T, Begg AC (2003) Towards genetic prediction of radiation responses: ESTRO's GENEPI project. Radiother Oncol 69:121–125
270. http://www.istworld.org/ProjectDetails.aspx?ProjectId=4e08539ae75e47c48698715be3e7f2a1

271. West C, Rosenstein BS, Alsner J, Azria D, Barnett G, Begg A, Bentzen S, Burnet N, Chang-Claude J, Chuang E, Coles C, De Ruyck K, De Ruysscher D, Dunning A, Elliott R, Fachal L, Hall J, Haustermans K, Herskind C, Hoelscher T, Imai T, Iwakawa M, Jones D, Kulich C, EQUAL-ESTRO, Langendijk JH, O'Neils P, Ozsahin M, Parliament M, Polanski A, Rosenstein B, Seminara D, Symonds P, Talbot C, Thierens H, Vega A, West C, Yarnold J. (2010) Establishment of a Radiogenomics Consortium. Int J Radiat Oncol Biol Phys 76:1295–1296
272. Bentzen SM (2006) Preventing or reducing late side effects of radiation therapy: radiobiology meets molecular pathology. Nat Rev Cancer 6:702–713
273. Dikomey E, Brammer I, Johansen J, Bentzen SM, Overgaard J (2000) Relationship between DNA double-strand breaks, cell killing, and fibrosis studied in confluent skin fibroblasts derived from breast cancer patients. Int J Radiat Oncol Biol Phys 46:481–490
274. Kasten U, Plottner N, Johansen J, Overgaard J, Dikomey E (1999) Ku70/80 gene expression and DNA-dependent protein kinase (DNA-PK) activity do not correlate with double-strand break (dsb) repair capacity and cellular radiosensitivity in normal human fibroblasts. Br J Cancer 79:1037–1041
275. Kasten-Pisula U, Vronskaja S, Overgaard J, Dikomey E (2008) In normal human fibroblasts variation in DSB repair capacity cannot be ascribed to radiation-induced changes in the localisation, expression or activity of major NHEJ proteins. Radiother Oncol 86:321–328
276. Cosgrove MS, Boeke JD, Wolberger C (2004) Regulated nucleosome mobility and the histone code. Nat Struct Mol Biol 11:1037–1043
277. Hummel R, Hussey DJ, Haier J (2010) MicroRNAs: predictors and modifiers of chemo- and radiotherapy in different tumour types. Eur J Cancer 46:298–311
278. Stucki M, Jackson SP (2004) Tudor domains track down DNA breaks. Nat Cell Biol 6:1150–1152
279. Huyen Y, Zgheib O, Ditullio RA Jr, Gorgoulis VG, Zacharatos P, Petty TJ, Sheston EA, Mellert HS, Stavridi ES, Halazonetis TD (2004) Methylated lysine 79 of histone H3 targets 53BP1 to DNA double-strand breaks. Nature 432:406–411

Important Roles of ERCC1 in DNA Repair and Targeted Therapy

Jennifer J. Hu, Joseph de la Garza, Sathish K. Srinivasan, Ann Alyssa Kurian, and Feng Gong

J.J. Hu (✉)
Sylvester Comprehensive Cancer Center, University of Miami Miller School of Medicine, 1120 NW 14 ST. CRB 1511, Miami, FL 33136, USA

Interdepartmental Program in Human Genetics and Genomics, University of Miami Miller School of Medicine, 1120 NW 14 ST. CRB 1511, Miami, FL 33136, USA

Department of Epidemiology and Public Health, University of Miami Miller School of Medicine, 1120 NW 14 ST. CRB 1511, Miami, FL 33136, USA

Biochemistry and Molecular Biology, University of Miami Miller School of Medicine, 1120 NW 14 ST. CRB 1511, Miami, FL 33136, USA
e-mail: jhu@med.miami.edu

J. de la Garza
Sylvester Comprehensive Cancer Center, University of Miami Miller School of Medicine, 1120 NW 14 ST. CRB 1511, Miami, FL 33136, USA

Department of Obstetrics and Gynecology, University of Miami Miller School of Medicine, 1120 NW 14 ST. CRB 1511, Miami, FL 33136, USA

S.K. Srinivasan
Interdepartmental Program in Human Genetics and Genomics, University of Miami Miller School of Medicine, 1120 NW 14 ST. CRB 1511, Miami, FL 33136, USA

A.A. Kurian • J. J. Hu (✉)
Sylvester Comprehensive Cancer Center, University of Miami Miller School of Medicine, 1120 NW 14 ST. CRB 1511, Miami, FL 33136, USA
e-mail: jhu@med.miami.edu

F. Gong
Sylvester Comprehensive Cancer Center, University of Miami Miller School of Medicine, 1120 NW 14 ST. CRB 1511, Miami, FL 33136, USA

Biochemistry and Molecular Biology, University of Miami Miller School of Medicine, 1120 NW 14 ST. CRB 1511, Miami, FL 33136, USA

L. Panasci et al. (eds.), *Advances in DNA Repair in Cancer Therapy*, Cancer Drug Discovery and Development 72, DOI 10.1007/978-1-4614-4741-2_5,

© Springer Science+Business Media, LLC 2013

1 Functions of ERCC1 in Multiple DNA Repair Pathways

The ERCC1 protein (excision repair cross-complementing rodent repair deficiency, complementation group 1) forms a heterodimer with the Xeroderma pigmentosum group F (XPF) endonuclease (also known as ERCC4), and the heterodimeric endonuclease catalyzes the 5′ incision in the process of excising the DNA lesion. The ERCC1–XPF heterodimer has an important role in genome maintenance. While most of the DNA repair proteins function only in a specific repair pathway, ERCC1–XPF is involved in multiple DNA repair pathways and telomere maintenance, making this heterodimer not only an attractive therapeutic target, but also a biomarker to predict treatment outcome.

The classical role of the ERCC1–XPF heterodimer lies in its involvement in the nucleotide excision repair (NER) pathway. NER has been extensively studied and the core mechanism is relatively well understood. It consists of three main steps: (1) lesion detection, (2) dual incision to remove an oligonucleotide containing the lesion, and (3) repair synthesis. ERCC1–XPF complex and the xeroderma pigmentosum group G (XPG) endonuclease are responsible for the dual incision step to release the lesion-containing oligonucleotide. XPG cuts 3′ to the damaged base, while ERCC1–XPF incises DNA 5′ to a lesion. Only XPF contains the nuclease domain of the ERCC1–XPF complex, but it requires ERCC1 for subsequent nuclease activity [1]. ERCC1 is essential for heterodimer positioning, as the central domain of ERCC1 binds with maximal affinity to single-stranded overhangs 15 nucleotides or longer with a preference for 5′ overhangs [2]. XPA appears to have a role in damage verification and is also necessary to load and position the ERCC1–XPF complex correctly onto the damaged DNA in order to start the incision process [3].

NER is one of the most versatile of all DNA repair mechanisms dealing with many different kinds of DNA damage that occur in the form of bulky adducts [4]. Typical substrates for NER include UV-induced photoadducts such as cyclobutane pyrimidine dimers (CPDs) and (6-4) photoproducts [5], intrastrand cross-links, and bulky chemical adducts. Of note, NER is an important pathway in the repair of intrastrand cross-links and bulky adducts that are induced by chemotherapy.

Unlike other NER factors, ERCC1–XPF heterodimer is also involved in multiple repair pathways, including double strand break repair (DSB) and interstrand cross-link repair (ICL). Homologous recombination (HR) is regarded as being an error-free process to repair DSB. A template, usually a sister chromatid, is needed to carry out the repair event [6]. It was proposed that ERCC1 was required for removal of long nonhomologous tails from 3′-OH ends of invading strands during targeted homologous recombination in Chinese hamster ovary cells [7].

For the repair of ICLs, taking the incision activities of the ERCC1–XPF heterodimer into account, it was proposed that a Y structure near the damage is first formed, for example, during DNA replication. ERCC1–XPF then cleaves at the 3′ side of one arm of the ICL and then makes an additional incision at the 5′ side. The replication fork collapses and recombination and NER events can take place to complete the ICL repair [8]. Fisher and coworkers found that the heterodimer does not only cut 5′ of the psoralen lesion but also cuts 3′ of the ICL, resulting in a DSB near the cross-linked site. Therefore, ERCC1–XPF appears to be involved in unhooking of ICLs [9].

Furthermore, the ERCC1–XPF complex also plays roles in telomere maintenance where it interacts with the telomere binding protein 2 (TRF2). At telomeric ends, ERCC1–XPF appears to be required for degrading 3′ G-rich overhangs when TRF2 function is inhibited [10]. In addition, over-expression of TRF2 in mouse keratinocytes led to XPF-dependent telomere loss, increased DNA damage, premature ageing and cancer [11]. Considering the important roles of ERCC1 in multiple DNA repair pathways and telomere maintenance, it may serve as a potential therapeutic target to enhance treatment efficacy.

2 ERCC1 and Clinical Outcome of Lung Cancer

Lung cancer is the leading cause of cancer deaths in American men and women; there are estimated 221,130 new cases and 156,940 deaths in 2011 [12]. It can be classified into two subtypes, non-small cell lung cancer (NSCLC) and small cell lung cancer (SCLC). The standard treatment for lung cancer includes platinum-based chemotherapy in combination with other non-platinum based regimens.

ERCC1 single nucleotide polymorphisms (SNPs), mRNA expression and protein expression levels have been associated with the sensitivity to chemotherapy with potential application as biomarkers for predicting treatment response and survival in lung cancer patients. We have summarized the results of several relatively larger studies evaluating the prognostic value of ERCC1-based biomarkers in lung cancer (Table 1). The two common *ERCC1* SNPs, *C354T* and *C8092A* were investigated in 115 NSCLC patients and the results showed a significant association between both SNPs and response to platinum based chemotherapy [13]. In 158 never-smokers with adenocarcinoma, *ERCC1 8092AA* genotype was associated with better response to gemcitabine/cisplatin [14].

ERCC1 mRNA and protein expression levels were measured predominantly from formalin fixed and paraffin embedded tumor tissues of NSCLC patients. It was reported that high levels of ERCC1 mRNA were significantly associated with poor overall survival (OS) in patients treated with platinum-based chemotherapy [15]. In a different study, patients with ERCC1 negative protein status showed increased survival as compared to those patients with ERCC1 positive tumors [16]. Two studies reported that lower ERCC1 protein levels were significantly associated with an increased response rate to platinum-based chemotherapy [17, 18]. In a randomized multicenter Phase III trial, patients with adenocarcinoma and negative ERCC1 protein expression had greater cisplatin sensitivity [19].

In a recent review, the association among ERCC1 SNPs, expression levels, and treatment response in SCLC was also discussed [20]. It was concluded that molecular markers based on ERCC1 might not be used for predicting treatment response in

Table 1 ERCC1 and clinical outcome of lung cancer

References	ERCC1 variable	Population	Sample size	Therapy	Outcomes
[13]	*ERCC1* SNPs *C354T* and *C8092A*	NSCLC	115	Cisplatin- or platinum-based chemotherapy	SNPs 354 C to T and SNP 8092 C to A was associated with increased or decreased treatment response
[14]	*ERCC1* SNP *C8092A and XRCC1*	Never-smokers with lung adenocarcinoma; stages IIIB/IV	158	Gefitinib or gemcitabine plus cisplatin	Patients with 8092CA genotype
[15]	ERCC1 mRNA expression	NSCLC	100	Platinum-based doublet chemotherapy	Higher expression was associated with poor overall survival (OS)
[16]	ERCC1 protein expression	NSCLC	163	Carboplatin and gemcitabine	Negative ERCC1 was associated with better survival in men but not in women
[18]	ERCC1 protein expression	NSCLC; stages IIIB/IV	170	Carboplatin and gemcitabine	Lower ERCC1 was associated with better response
[19]	ERCC1 protein expression	NSCLC	443 (264 with ERCC1 data)	Platinum-based doublet and triplet chemotherapy	Negative ERCC1 interacts with adenocarcinomas in longer survival
[17]	ERCC1 protein expression	NSCLC; International Adjuvant Lung Trial	784	Adjuvant chemotherapy with cisplatin plus vindesine or vinblastine or vinorelbine or etoposide	Patients with low ERCC1 benefit from adjuvant chemotherapy; ERCC1 was marginally prognostic for survival

SCLC patients due to a lack of clinical evidence from larger studies. Therefore, it is important for investigators to focus on studying response to chemotherapy with various molecular markers in large clinical trials of SCLC.

3 ERCC1 and Clinical Outcome of Ovarian Cancer

It is estimated that in the year of 2011, ovarian cancer will be newly diagnosed in 21,990 women and result in 15,460 deaths [12]. The standard therapy for women with ovarian cancer is primary aggressive cytoreductive surgery followed with adjuvant platinum and taxane-based chemotherapy [21, 22]. Despite efforts to provide optimal surgical management and improve adjuvant therapies, ovarian cancer remains the deadliest gynecologic cancer. Approximately 60–80 % of women diagnosed with ovarian cancer will respond to initial therapy and enter clinical remission. Another subset of women will demonstrate clinical progression during treatment and will have platinum-refractory disease. Some women who have a clinical response initially will subsequently have disease recurrence within the first 6 months following completion of therapy and demonstrate platinum resistance [23].

The success of chemotherapy is limited by the ability of an agent to target and kill cancer cells. Platinum resistance is a primary mechanism responsible for the limitations in treatment and limited survival of ovarian cancer. Up-regulation of several DNA repair pathways may contribute to platinum resistance and ultimate failure of therapy in the clinical setting, such as DNA mismatch repair, base excision repair, and NER. ERCC1–XPF heterodimers are DNA repair proteins essential for both NER and DNA cross-link repair pathways. Thus, ERCC1–XPF heterodimers are absolutely required for the repair of all types of platinum induced DNA damage [23]. Numerous studies have been conducted with aims to discover the clinical significance of ERCC1 and the NER pathway in ovarian cancer, among them are those listed in Table 2.

In a small study of 60 epithelial ovarian cancer (EOC) cases who received initial cytoreductive surgery, followed by six cycles of platinum based chemotherapy, there was an association between the *ERCC1* SNP *118 CT/TT* genotypes and platinum resistance; however, there was no significant association with overall survival in this cohort [24]. In 178 EOC cases who received platinum-based chemotherapy, patients with higher ERCC1 expression or the CC genotype may benefit from platinum plus paclitaxel, while low ERCC1 or the *C/T* or *T/T* genotype may respond well to platinum without paclitaxel [25].

Two latter studies continued to investigate *ERCC1* SNPs *C118T* and *C8092A*. In 2008, the Gynecologic Oncology Group (GOG)-172 clinical trial comparing paclitaxel/cisplatin IV to paclitaxel/cisplatin IV/IP demonstrated that again that codon 118 SNP was not associated with overall survival (OS) or progression free survival (PFS). However, the codon 8092 *CA/AA* genotypes were associated with worse PFS and OS [26]. In contrast, subsequent investigation within the GOG-182 protocol, which is a randomized trial of carboplatin, paclitaxel, gemcitabine, pegylated liposomal

Table 2 ERCC1 and clinical outcome of ovarian cancer

References	ERCC1 variable	Patient population	Sample size	Therapy	Outcomes
[24]	*ERCC1* SNP *C118T*	Adequately staged EOC, all stages	60	Carboplatin and paclitaxel	SNP 118 *CT/TT* genotypes can predict platinum resistance but not OS
[25]	*ERCC1* SNP *C118T*	Stage I–IV EOC	178	Platinum-based chemotherapy with cisplatin or carboplatin IV +/– paclitaxel IV	Higher ERCC1 or the *CC* genotype may benefit from platinum + paclitaxel, while low ERCC1 or the C/T or T/T genotype may respond well to platinum without paclitaxel
[26]	*ERCC1* SNPs *C118T* and *C8092A*	Optimally reduced, stage III EOC (GOG 172)	233	Paclitaxel and cisplatin IV or paclitaxel and cisplatin	SNP 118 was not associated with OS and PFS. The codon 8092 *CA/AA* genotypes were associated with worse PFS and OS
[23]	*ERCC1* SNPs *C118T* and *C8092A*	Optimally or suboptimally reduced patients with stage III–IV EOC (GOG 182)	280	Carboplatin, paclitaxel IV, gemcitabine, pegylated liposomal doxorubicin, and topotecan	SNP 118 *CT/TT* genotypes were associated with better survival in patients treated with platinum and paclitaxel based chemotherapy
[27]	ERCC1 mRNA in leukocytes	Optimally reduced, stage III EOC (GOG 158)	170	Paclitaxel and cisplatin IV or paclitaxel and carboplatin	ERCC1 in leukocytes is not associated with PFS or OS
[29]	ERCC1 protein expression	Chemotherapy naïve patients with stage I–IV EOC	101	Carboplatin and paclitaxel IV	Platinum resistance found in 75 % of tumors with positive ERCC1 protein expression
[28]	ERCC1 protein expression	High-grade serous adenocarcinoma stage III–IV	77	Platinum based chemotherapy with cisplatin or carboplatin	ERCC1-negative tumors have better OS

EOC Epithelial ovarian cancer, *OS* overall survival, *PFS* progression free survival doxorubicin, and topotecan in various regimens IV, identified that SNP 118 *CT/TT* genotypes were associated with better survival in patients treated with platinum and paclitaxel-based chemotherapy [23]. Differences in route of administration, type of regimen, stage of disease, race, stage of residual disease, study designs, and end points may explain, at least in part, the conflicting results between studies.

In addition to SNPs, *ERCC1* gene expression was evaluated using mRNA isolated from peripheral blood leukocytes as a surrogate marker in GOG-158 clinical trial, which is a randomized clinical trial of 170 women with EOC comparing paclitaxel with carboplatin vs. cisplatin as initial adjuvant chemotherapy. There was no association between ERCC1 expression and PFS or OS. However, tumor tissue was not available in this study to assess direct *ERCC1* expression, and the authors stated that mRNA expression of *ERCC1* in peripheral leukocytes may not be a reliable surrogate marker in understanding the influence of the *ERCC1* gene and its effect on chemotherapy resistance and patient survival [27].

Two other studies evaluated ERCC1 protein expression to understand the specific genetic foci responsible or representative of the influence that ERCC1–XPF heterodimers have upon clinical relevance in ovarian cancer [28, 29]. The first study performed immunohistochemistry for ERCC1 protein in chemotherapy naïve patients with stage I–IV EOC and then followed outcomes after they received six cycles of platinum based chemotherapy. They demonstrated that platinum resistance was present in 75 % of tumors with positive ERCC1 protein expression [29]. Similarly, the results from the second study also showed that ERCC1 protein level was associated with OS [28].

With limited literature, it is still unclear the exact clinical usefulness of ERCC1 in EOC. It has been identified that the ERCC1–XPF heterodimer is an integral part of the NER pathway, and that this pathway affects the response to platinum-based chemotherapy in EOC. However, the appropriate target for ERCC1 evaluation which will allow for integration of ERCC1 in treatment planning is yet to be clearly identified. Further research is necessary to identify pathways in which ERCC1 can alter components of the pathway resulting in an impact on effective tailored therapy for patients with EOC.

4 ERCC1 and Clinical Outcome of Other Cancers

ERCC1 genotype has been shown to be an effective biomarker to predict the efficacy of chemotherapy for various types of cancer and chemotherapy regimens (Table 3). The results from two studies showed that the ERCC1 118 SNP was associated with efficacy of oxaliplatin-based chemotherapy in patients with metastatic colorectal cancer [30, 31]. In a phase II clinical trial of variations of FOLFOX treatment regimens, the ERCC1 118 CC genotype was correlated with a longer PFS [30]. In the second study of 113 patients with metastatic colorectal cancer, analysis for correlation between genotype and clinical response showed that patients with polymorphism C/T in ERCC1-118 showed higher ERCC1 mRNA concentrations

Table 3 ERCC1 and clinical outcome of multiple cancers

References	ERCC1	Patient population	Sample Size	Therapy	Outcomes
[30]	*ERCC1* SNP *C118T*	Metastatic colorectal cancer	118	FOLFOX	118 C/C genotype was associated with better PFS
[31]	*ERCC1* SNP *C118T*	Metastatic colorectal cancer	113	Oxaliplatin and 5 fluorouracil, FOLFOX4 and xelox regimens	118 C/T genotype displayed higher ERCC1 mRNA and more resistance to oxaliplatin
[34]	ERCC1 mRNA expression	Locally advanced bladder cancer	108	1. Cisplatin + methotrexate 2. Methotrexate + vinblastine + epirubicin + cisplatin	Higher ERCC1 mRNA expression was associated with reduced PFS for both groups
[32]	ERCC1 protein expression	Gastroesophageal cancer	142	Epirubicin + cisplatin + 5-FU, or epirubicin + cisplatin + capecitabine before surgery	ERCC1 expression was associated with poor disease-specific ($P = 0.020$) and OS
[33]	ERCC1 protein expression	Pancreatic adenocarcinoma	95	Pancreaticoduodenectomy with adjuvant chemotherapy	Higher ERCC1 expression was associated with lower PFS and OS in 73 patients who underwent adjuvant chemotherapy

and more resistance to oxaliplatin treatment. Also, a combination of the C/C SNP in *ERCC1*-118 and Arg/Arg in XRCC1-399 correlated to better treatment response than either one of those polymorphisms alone [31].

A study of 142 cases with gastroesophageal cancer treatment with platinum-based chemotherapy drugs aimed to correlate the expression of several genes involved in DNA repair, including ERCC1, with the clinicopathological outcomes of treatment [32]. This study used tumor regression grade (TGR) as a means of analyzing clinicopathological response. TGR1 is complete regression and TGR2 shows the presence of rare residual cancer cells scattered through the fibrosis. TGR3 showed a slightly greater number of residual cells than TGR2. TRG4 showed residual cancer outgrowing fibrosis and TGR5 showed the absence of any regression. Results showed that positive ERCC1 expression correlated with negative treatment response (TGR4 or 5) and ERCC1 negative cases showed significantly greater median disease specific survival than ERCC1 positive cases and significantly greater overall survival than ERCC1 positive cases. The results suggest that tumor regression and ERCC1 nuclear protein expression are promising predictive markers in gastroesophageal cancer patients receiving neo-adjuvant platinum-based chemotherapy [32].

ERCC1 has been shown to be a potential prognostic marker for disease progression in pancreatic adenocarcinoma [33]. There was differential ERCC1 expression in pancreatic adenocarcinoma using immunohistochemistry and it was assessed as a prognostic marker for disease progression. Ninety-five pancreatic adenocarcinoma patients who underwent pancreaticoduodenectomy with available tissue samples were used ERCC1 expression analysis, PFS, and OS. The study results showed that pancreatic adenocarcinoma displayed differential levels of ERCC1 and higher levels of ERCC1 expression were associated with lower PFS and OS [33].

ERCC1 expression levels was evaluated in 108 bladder cancer patients participating in a Phase III clinical trial of an adjuvant cisplatin based chemotherapy including methotrexate [34]. The study sought to determine the usefulness of ERCC1 expression as a predictive biomarker and whether the effect as a biomarker varies with the type of chemotherapy used. Patients in the study received either CM regimen consisting of cisplatin and methotrexate, or M-VEC regimen consisting of methotrexate, vinblastine, epirubicin, and cisplatin. Results showed that lower ERCC1 levels correlated with greater PFS and higher ERCC1 levels correlated with disease progression [34]. No significant difference was found between effects of ERCC1 levels in each chemotherapy regimen independently.

5 ERCC1–XPF as Cancer Therapeutic Target

It has been well established that DNA repair pathways can enable tumor cells to survive DNA damage that is induced by chemotherapeutic or radiation treatments; therefore, inhibitors of DNA repair pathways might prove efficacious when used in combination with DNA-damaging chemotherapeutic drugs. In addition, deficient

DNA repair pathways that arise during carcinogenesis can drive some cancer cells reliant on limited DNA repair pathways for survival. Therefore, DNA repair inhibitors to target these pathways in such tumors could prove useful as single-agent therapies with selective efficacy and fewer side effects. In addition, DNA repair inhibitors can also be used in combination with DNA-damaging anticancer agents to increase the efficiency of the cancer treatment by inhibiting DNA repair pathway(s) critical for removing toxic DNA damages.

Earlier clinical studies suggest that platinum-DNA adduct may be an important biomarker for the biological effect of platinum-based chemotherapy. DNA repair particularly NER plays an important role in treatment response and resistance to platinum-based chemotherapeutic agents. One critical gene within NER pathway is the *ERCC1* gene. Data exist in multiple human cancer sites that *ERCC1* SNPs and/or expression may serve as useful biomarkers in predicting clinical outcome when platinum-based chemotherapy is utilized. Furthermore, the ERCC1–XPF complex is also involved in HR of DSBR, ICL repair, and telomere maintenance. Therefore, the ERCC1–XPF complex makes an attractive biomarker in predicting clinical outcome in cancer patients as well as a novel treatment target in chemo- and/or radiation sensitization.

Resistance to several chemotherapy drugs has been previously correlated with the over expression of both the ERCC1 and XPF proteins. These proteins form a heterodimeric endonuclease complex, which is recruited to DNA through a secondary interaction between ERCC1 and the XPA protein. Although ERCC1 is a potential anticancer drug target, it does not have intrinsic enzymatic activity. Therefore, modulation of ERCC1 might be less desirable than understanding the clinical relevance of protein–protein interactions within the NER pathway or between ERCC1 and other repair pathways, as potential targets for improving the efficacy of chemotherapy drugs.

ERCC1 expression can be suppressed by emodin, a tyrosine kinase inhibitor, which is a natural anthraquinone derivative isolated from the roots and rhizomes of numerous plants [35]. In addition to suppressing ERCC1 expression, the cytotoxicity to capecitabine can be enhanced by emodin by down-regulating the expression of Rad51 and up-regulation of thymidine phosphorylase expression [36]. In human tongue cancer cells, emodin treatment induced DNA damage and inhibited DNA repair-associated gene expression, including ATM, ATR, 14-3-3sigma, BRCA1, DNA-PK, and MGMT [37]. This inhibitory effect is supported by the previous observation that epidermal growth factor up-regulate ERCC1 through MAPK (ERK1/2) signaling [38]. Moreover, inhibition of HER2–PI3K–AKT signal pathway down-regulates ERCC1 that may contribute to the synergism between trastuzumab and chemotherapy [39].

Three marine-derived NER inhibitors, trabectedin (Et743; Yondelis), PM01183, and PM00104 showed enhanced activity toward cisplatin- and oxaliplatin-resistant ovarian carcinoma cells or Ewing's sarcoma cell lines [40, 41]. In addition to its function as a DNA repair inhibitor due to drug-related DSBs and adduct formation, trabectedin treatment also results in perturbation in the transcription of inducible genes, such as the multidrug resistance gene MDR1.

ERCC1–XPF involved in different repair pathways through specific protein–protein interactions and selective disruption of these interactions can influence different repair pathways separately. UCN-01 (7-hydroxystaurosporine) has been developed as an anticancer agent that potentiates cisplatin and carboplatin toxicity (demonstrated in preclinical models and a phase I clinical trial, respectively), which has been shown to interfere with the interaction of ERCC1 and XPA [42]. The XPA-binding domain of ERCC1 is required for NER only but not other DNA repair pathways [43]. It is not clear whether this specificity may have limitation considering other repair pathways may serve as the backup mechanisms to remove chemotherapeutic agent-induced DNA damages.

6 Conclusions and Future Prospective

The ERCC1–XPF complex has been evaluated as a biomarker in predicting clinical outcome as well as a potential treatment target. In multiple cancer sites (e.g., lung, ovarian, colorectal, etc.) higher levels of ERCC1 mRNA/proteins correlate with poor clinical outcome and resistance to platinum-based chemotherapies. The *ERCC1* SNP at codon 118 leads to a C:T substitution that may influence mRNA/protein levels, DNA repair capacity, and treatment response. The upstream promoter region (around 410 base pairs) to the *ERCC1* initiation site contains a variety of transcription factor binding sites for GATA-1, p53, AP-1, c-Jun, JunB, ER-α, and NF-κB1; they may contribute to ERCC1 expression regulation and resistance to cisplatin.

The potential application of inhibitors of ERCC1 expression or protein–protein interactions in cancer therapy is starting to become apparent. Several approaches have shown promising clinical applications. First, ERCC1 expression can be suppressed by emodin, a tyrosine kinase inhibitor to enhance sensitivities to different chemotherapeutic agents. Second, three marine-derived NER inhibitors, trabectedin (Et743; Yondelis), PM01183, and PM00104 showed enhanced activity toward cisplatin- and oxaliplatin-resistant ovarian carcinoma cells or Ewing's sarcoma cell lines. Third, UCN-01 has been developed as an anticancer agent by targeting the interaction between ERCC1 and XPA to potentiate cisplatin and carboplatin toxicity.

Selective inhibitions of DNA repair pathways have the potential to sensitize chemotherapeutic drugs; synthetic lethality is another potential application of DNA repair inhibitors in tumor with defective DNA repair for selective tumor cell killing. Considering the involvement of ERCC1 in multiple DNA repair pathways critical for repairing DNA damages induced by a variety of chemotherapeutic agents used for human cancers, future development of ERCC1 inhibitors as single or combination treatment may have a great impact on designing new and more effective cancer therapies.

References

1. Tripsianes K et al (2005) The structure of the human ERCC1/XPF interaction domains reveals a complementary role for the two proteins in nucleotide excision repair. Structure 13(12):1849–1858
2. Tsodikov OV et al (2005) Crystal structure and DNA binding functions of ERCC1, a subunit of the DNA structure-specific endonuclease XPF-ERCC1. Proc Natl Acad Sci USA 102(32):11236–11241
3. Enzlin JH, Scharer OD (2002) The active site of the DNA repair endonuclease XPF-ERCC1 forms a highly conserved nuclease motif. EMBO J 21(8):2045–2053
4. de Boer J, Hoeijmakers JH (2000) Nucleotide excision repair and human syndromes. Carcinogenesis 21(3):453–460
5. Cleaver JE, Lam ET, Revet I (2009) Disorders of nucleotide excision repair: the genetic and molecular basis of heterogeneity. Nat Rev Genet 10(11):756–768
6. Sargent RG et al (1997) Recombination-dependent deletion formation in mammalian cells deficient in the nucleotide excision repair gene ERCC1. Proc Natl Acad Sci USA 94(24):13122–13127
7. Adair GM et al (2000) Role of ERCC1 in removal of long non-homologous tails during targeted homologous recombination. EMBO J 19(20):5552–5561
8. Kuraoka I et al (2000) Repair of an interstrand DNA cross-link initiated by ERCC1-XPF repair/recombination nuclease. J Biol Chem 275(34):26632–26636
9. Fisher LA, Bessho M, Bessho T (2008) Processing of a psoralen DNA interstrand cross-link by XPF-ERCC1 complex in vitro. J Biol Chem 283(3):1275–1281
10. Zhu XD et al (2003) ERCC1/XPF removes the 3′ overhang from uncapped telomeres and represses formation of telomeric DNA-containing double minute chromosomes. Mol Cell 12(6):1489–1498
11. Munoz P et al (2005) XPF nuclease-dependent telomere loss and increased DNA damage in mice overexpressing TRF2 result in premature aging and cancer. Nat Genet 37(10):1063–1071
12. Siegel R et al (2011) Cancer statistics, 2011: the impact of eliminating socioeconomic and racial disparities on premature cancer deaths. CA Cancer J Clin 61(4):212–236
13. Li F et al (2010) Association between polymorphisms of ERCC1 and XPD and clinical response to platinum-based chemotherapy in advanced non-small cell lung cancer. Am J Clin Oncol 33(5):489–494
14. Han JY et al (2011) DNA repair gene polymorphisms and benefit from gefitinib in never-smokers with lung adenocarcinoma. Cancer 117(14):3201–3208
15. Ren S et al (2010) High-level mRNA of excision repair cross-complementation group 1 gene is associated with poor outcome of platinum-based doublet chemotherapy of advanced nonsmall cell lung cancer patients. Cancer Invest 28(10):1078–1083
16. Holm B et al (2009) Different impact of excision repair cross-complementation group 1 on survival in male and female patients with inoperable non-small-cell lung cancer treated with carboplatin and gemcitabine. J Clin Oncol 27(26):4254–4259
17. Bepler G et al (2011) ERCC1 and RRM1 in the international adjuvant lung trial by automated quantitative in situ analysis. Am J Pathol 178(1):69–78
18. Reynolds C et al (2009) Randomized phase III trial of gemcitabine-based chemotherapy with in situ RRM1 and ERCC1 protein levels for response prediction in non-small-cell lung cancer. J Clin Oncol 27(34):5808–5815
19. Vilmar AC, Santoni-Rugiu E, Sorensen JB (2010) ERCC1 and histopathology in advanced NSCLC patients randomized in a large multicenter phase III trial. Ann Oncol 21(9):1817–1824
20. Knez L et al (2011) Predictive value of multidrug resistance proteins, topoisomerases II and ERCC1 in small cell lung cancer: a systematic review. Lung Cancer 72(3):271–279
21. Hoskins WJ et al (1994) The effect of diameter of largest residual disease on survival after primary cytoreductive surgery in patients with suboptimal residual epithelial ovarian carcinoma. Am J Obstet Gynecol 170(4):974–979, discussion 979–980

22. Ozols RF et al (2003) Phase III trial of carboplatin and paclitaxel compared with cisplatin and paclitaxel in patients with optimally resected stage III ovarian cancer: a Gynecologic Oncology Group study. J Clin Oncol 21(17):3194–3200
23. Krivak TC et al (2011) Single nucleotide polypmorphisms in ERCC1 are associated with disease progression, and survival in patients with advanced stage ovarian and primary peritoneal carcinoma; a Gynecologic Oncology Group study. Gynecol Oncol 122(1):121–126
24. Kang S et al (2006) Association between excision repair cross-complementation group 1 polymorphism and clinical outcome of platinum-based chemotherapy in patients with epithelial ovarian cancer. Exp Mol Med 38(3):320–324
25. Smith S et al (2007) ERCC1 genotype and phenotype in epithelial ovarian cancer identify patients likely to benefit from paclitaxel treatment in addition to platinum-based therapy. J Clin Oncol 25(33):5172–5179
26. Krivak TC et al (2008) Relationship between ERCC1 polymorphisms, disease progression, and survival in the Gynecologic Oncology Group Phase III Trial of intraperitoneal versus intravenous cisplatin and paclitaxel for stage III epithelial ovarian cancer. J Clin Oncol 26(21):3598–3606
27. Darcy KM, Tian C, Reed E (2007) A Gynecologic Oncology Group study of platinum-DNA adducts and excision repair cross-complementation group 1 expression in optimal, stage III epithelial ovarian cancer treated with platinum-taxane chemotherapy. Cancer Res 67(9):4474–4481
28. Scheil-Bertram S et al (2010) Excision repair cross-complementation group 1 protein overexpression as a predictor of poor survival for high-grade serous ovarian adenocarcinoma. Gynecol Oncol 119(2):325–331
29. Steffensen KD, Waldstrom M, Jakobsen A (2009) The relationship of platinum resistance and ERCC1 protein expression in epithelial ovarian cancer. Int J Gynecol Cancer 19(5):820–825
30. Chua W et al (2009) Molecular markers of response and toxicity to FOLFOX chemotherapy in metastatic colorectal cancer. Br J Cancer 101(6):998–1004
31. Liang J et al (2010) The combination of ERCC1 and XRCC1 gene polymorphisms better predicts clinical outcome to oxaliplatin-based chemotherapy in metastatic colorectal cancer. Cancer Chemother Pharmacol 66(3):493–500
32. Fareed KR et al (2010) Tumour regression and ERCC1 nuclear protein expression predict clinical outcome in patients with gastro-oesophageal cancer treated with neoadjuvant chemotherapy. Br J Cancer 102(11):1600–1607
33. Maithel SK et al (2011) Differential expression of ERCC1 in pancreas adenocarcinoma: high tumor expression is associated with earlier recurrence and shortened survival after resection. Ann Surg Oncol 18:2699–2705
34. Hoffmann AC et al (2010) MDR1 and ERCC1 expression predict outcome of patients with locally advanced bladder cancer receiving adjuvant chemotherapy. Neoplasia 12(8):628–636
35. Ko JC et al (2010) Emodin enhances cisplatin-induced cytotoxicity via down-regulation of ERCC1 and inactivation of ERK1/2. Lung Cancer 69(2):155–164
36. Ko JC et al (2011) Modulation of Rad51, ERCC1, and thymidine phosphorylase by emodin result in synergistic cytotoxic effect in combination with capecitabine. Biochem Pharmacol 81(5):680–690
37. Chen YY et al (2010) Emodin, aloe-emodin and rhein induced DNA damage and inhibited DNA repair gene expression in SCC-4 human tongue cancer cells. Anticancer Res 30(3):945–951
38. Yacoub A et al (2003) Epidermal growth factor and ionizing radiation up-regulate the DNA repair genes XRCC1 and ERCC1 in DU145 and LNCaP prostate carcinoma through MAPK signaling. Radiat Res 159(4):439–452
39. Li XL et al (2010) The sequence-dependent cytotoxic effect of trastuzumab in combination with 5-Fluorouracil or cisplatin on gastric cancer cell lines. Cancer Invest 28(10):1038–1047
40. Guirouilh-Barbat J, Antony S, Pommier Y (2009) Zalypsis (PM00104) is a potent inducer of gamma-H2AX foci and reveals the importance of the C ring of trabectedin for transcription-coupled repair inhibition. Mol Cancer Ther 8(7):2007–2014

41. Soares DG et al (2011) Trabectedin and its C subunit modified analogue PM01183 attenuate nucleotide excision repair and show activity toward platinum-resistant cells. Mol Cancer Ther 10(8):1481–1489
42. Barakat KH et al (2009) Characterization of an inhibitory dynamic pharmacophore for the ERCC1-XPA interaction using a combined molecular dynamics and virtual screening approach. J Mol Graph Model 28(2):113–130
43. Orelli B et al (2010) The XPA-binding domain of ERCC1 is required for nucleotide excision repair but not other DNA repair pathways. J Biol Chem 285(6):3705–3712

The Role of BRCA1 and BRCA2 in Anticancer Drug Therapy

Andrew Y. Shuen, Jorge S. Reis-Filho, and William D. Foulkes

1 Introduction

The genome is under constant assault from both endogenous and exogenous sources such as reactive oxygen species and ionizing radiation capable of inducing a wide array of mutagenic changes [1]. To maintain genomic integrity cells have evolved elegant mechanisms to recognize DNA damage, arrest the cell cycle, and activate specific repair pathways. One of the most cytotoxic lesions that a cell must contend with is a double-strand break (DSB) because even a single unrepaired DSB is capable of inducing cell death [2]. To repair a DSB, cells have at least four mechanisms at their disposal: homologous recombination (HR), single-strand annealing (SSA), nonhomologous end-joining (NHEJ), and microhomology-mediated end joining (MMEJ) (Fig. 1) [3]. HR relies on the sister chromatid as a template to fill in damaged or missing DNA, restoring the chromosome to its original condition. In cells with competent DNA repair mechanisms, HR is the preferred pathway of repair

A.Y. Shuen
Department of Medical Genetics, McGill University Health Centre,
Montreal, QC, Canada H3G 1A4

J.S. Reis-Filho
Molecular Pathology Team, The Breakthrough Breast Cancer Research Centre,
The Institute of Cancer Research, London SW3 6JB, UK

W.D. Foulkes (✉)
Department of Medical Genetics, McGill University Health Centre,
Montreal, QC, Canada H3G 1A4

Program in Cancer Genetics, Departments of Oncology and Human Genetics,
McGill University,Montreal, QC, Canada H2W 1S6

Department of Medical Genetics, Lady Davis Institute, Jewish General Hospital,
Montreal, QC, Canada H3T 1E2
e-mail: william.foulkes@mcgill.ca

L. Panasci et al. (eds.), *Advances in DNA Repair in Cancer Therapy*,
Cancer Drug Discovery and Development 72, DOI 10.1007/978-1-4614-4741-2_6,

© Springer Science+Business Media, LLC 2013

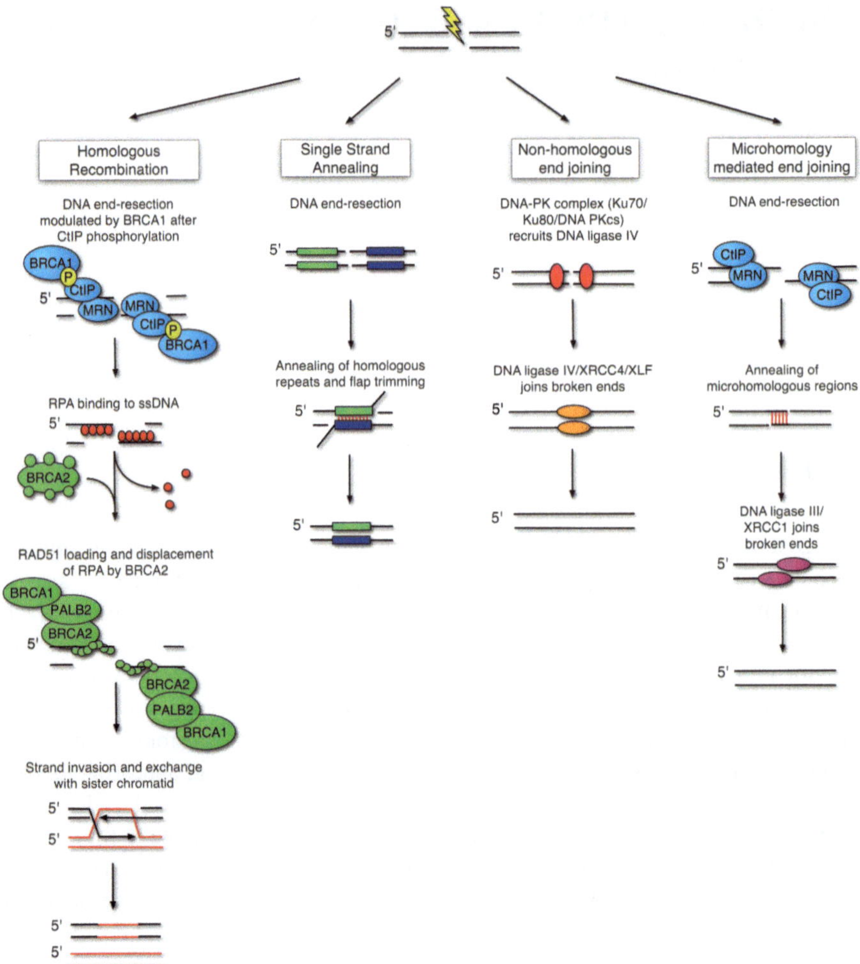

Fig. 1 Schema describing DNA repair pathways following a double-strand break (DSB). Homologous recombination is the preferred pathway during S and G2 phases of the cell cycle and is considered an error-free pathway. NHEJ, MMEJ, and SSA, on the other hand, are thought to be error-prone pathways because they introduce deletions at broken-ends and may promiscuously ligate nonadjacent ends creating gross chromosomal aberrations. *XRCC1/4* X-ray repair complementing defective repair in Chinese hamster cells 1/4. *DNA-PK* DNA dependent protein kinase catalytic subunit. *MRN* Mre11-Rad50-Nbs1

during the S and G2 phase of the cell cycle when the sister chromatid is available [4]. SSA, a variant of HR that is thought to play a minor role in the repair DSBs, utilizes homologous repeats surrounding a DSB to anneal the broken ends resulting in the deletion of the intervening sequence. In contrast, NHEJ and MMEJ both operate throughout the cell cycle and directly ligate two ends of a DSB; however, MMEJ always introduces small deletions at broken ends to produce a region of microhomology to facilitate ligation [5]. The important point to note is that HR is considered an error-free pathway whereas SSA, NHEJ, and MMEJ are error-prone because they can create gross chromosomal aberrations if ligation occurs incorrectly—potentially leading to neoplastic transformation [1].

BRCA1 and BRCA2 are tumor suppressors essential for the faithful repair of DSBs by HR [6]. However, BRCA1 also participates in other cellular functions important in maintaining genomic integrity including the assembly of the mitotic spindle [7], centrosome duplication [8], cell-cycle control [9–14], chromatin remodeling at sites of DSBs [15, 16], and DNA decantenation [17]. In contrast, the role of BRCA2 is primarily to regulate RAD51 filament formation, which is a critical step in catalyzing strand invasion and homologous recombination (Fig. 1).

Cells lacking BRCA1 or BRCA2 are unable to repair DSBs by HR and must resort to more error-prone pathways such as MMEJ and SSA. These cells display gross chromosomal rearrangements such as large deletions, translocations, and fusions during successive rounds of cell division [18]. While the vast majority of these lesions result in cell death, the genetic instability caused by loss of competent HR leads to a dramatically increased number of genetic alterations, which provide a rich background for Darwinian forces to act at the level of the tumor microenvironment, promoting the emergence of multiple clones, some of which have the capability to divide autonomously and metastasize [19]. The importance of BRCA genes in maintaining genomic integrity is underscored by patients who harbor germline mutations in *BRCA1* or *BRCA2* and have a markedly increased predisposition to develop, among others, breast and ovarian cancers [20].

Since the discovery of *BRCA1* and *BRCA2* more than 15 years ago [21, 22], understanding their function has been of primary importance and much progress has been made. In this review, we summarize the role BRCA1 and BRCA2 play in homologous recombination and how this knowledge can be utilized to target tumors deficient in this cellular pathway in hereditary as well as sporadic cancers.

2 Structure and Function of BRCA1

BRCA1 is composed of 1,863 amino acids and contains three functionally important domains (Fig. 2). At its amino terminal is a RING-finger domain with E3 ubiquitin ligase activity (Box 1). It is normally found in association with its heterodimeric protein partner BARD1 (which is itself a RING E3 ubiquitin ligase). This interaction stabilizes the complex, preventing its degradation [23] and enhances its E3 ligase function [24]. In addition, the ubiquitin ligase activity of BRCA1 is activated upon two post-modificational processes: auto-ubiquitination [25] and SUMOylation [26, 27]. It is not yet clear how the ubiquitin ligase activity of BRCA1 is increased; however, two possible scenarios can be envisaged. One is that ubiquitination or SUMOylation directly alters the conformation of the RING-finger domain increasing enzymatic activity. A second possibility could be that posttranslational

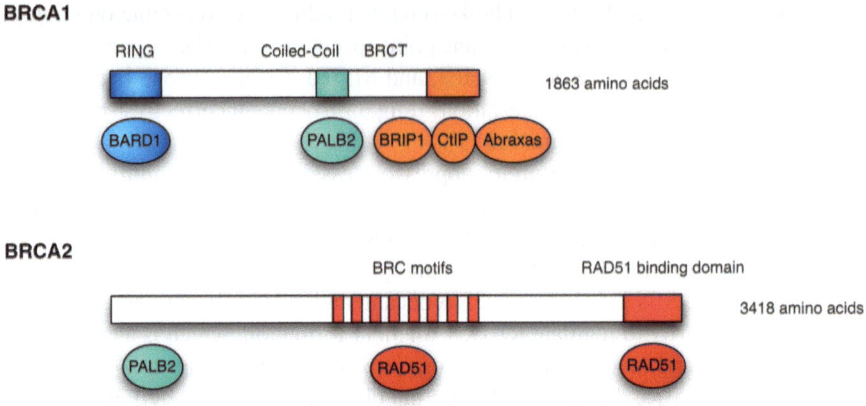

Fig. 2 Functional domains and interacting partners of human BRCA1 and BRCA2 proteins. Only domains (*listed above*) and protein partners (*drawn below*) that were discussed in this review are described. Proteins are color-coded with its corresponding interacting domain

modifications increase affinity for the E2 conjugating enzyme UbcH5a, accelerating ubiquitin transfer. BRCA1 has been shown to ubiquitinate various proteins including histones (H2A, H2AX, and H2B) [25, 28], CtIP [29], γ-tubulin [8], nucleophosmin [30], RNA polymerase II [31, 32], and ERα [33]. How ubiquitination of these target proteins modifies their function is unclear; however, germ-line mutations derived from patients with breast cancer that abolish RING finger ligase activity are observed to result in checkpoint deregulation and sensitivity to ionizing radiation [34, 35]. Strikingly these effects are independent of homologous recombination [36]. To reconcile this apparent paradox, Zhu et al. propose that BRCA1 acts in vivo to regulate expression of satellite DNA that is normally silenced by ubiquitination of H2A and that overexpression of satellite transcripts is linked to genomic instability [28]. However, the function of satellite transcripts and how its aberrant expression leads to tumor development are currently unknown.

At the carboxyl end of BRCA1 are tandem BRCT domains which contain a phosphate binding core providing an interface for phosphorylated proteins [37, 38]. Phosphorylation, mediated primarily by the kinases ATM and CHEK2, is an important spatiotemporal regulator of proteins involved in check-point control and DNA repair. The tandem BRCT domain of BRCA1 helps localize it to nuclear foci by binding to different phosphorylated intermediates including Abraxas [10, 14, 39], CtIP [40], and BRIP1 [13]. These protein complexes form three distinct entities during HR and each has important functions at sites of DNA breakage. For instance BRCA1 in association with Abraxas and RAP80 has been shown to regulate the G2-M checkpoint. When BRCA1 is bound to CtIP coupled with MRN, however, it regulates DNA end-resection diverting the pathway away from MMEJ towards HR (Fig. 2) [41].

BRCA1 also plays a more central role in HR. BRCA1 contains a coiled-coil domain present near the carboxyl terminal which binds PALB2 (Partner and Localizer of BRCA2) [42, 43]. PABL2 physically bridges BRCA1 to BRCA2. This complex in turn mediates the final enzymatic step of RAD51 assembly, and strand exchange between homologous chromosomes (described below).

Box 1

Ubiquitination is a posttranslational modification in which ubiquitin, a small peptide molecule of 76 amino acids, covalently tags larger proteins. This process requires the sequential coupling of three enzymatic reactions: an E1 activating enzyme, an E2 conjugating enzyme and an E3 ligase. The unique combination of a diverse array of E2 and E3 enzymes allows specific proteins to be targeted for ubiquitination. In a similar process, SUMOylation involves the tagging of larger proteins by a small ubiquitin-like modifier (SUMO) using a different but parallel enzymatic cascade consisting of E1, E2 and E3 enzymes. Modification of a protein by ubiquitin or SUMO can alter its conformation or modify its surface to allow or prohibit protein interactions.

3 Structure and Function of BRCA2

Although bearing similar names, BRCA2 is structurally and functionally distinct from BRCA1. It is a much larger (3,418 amino acids) protein containing eight BRC motifs, which enable binding to RAD51 [44, 45] and another distinct RAD51-binding domain at its terminal end [46, 47] (Fig. 1). Full-length human BRCA2 had never been purified to sufficient quantities due to its large molecular size. As such, its functions could only be derived from studying BRCA2 orthologues and smaller protein fragments. Recently, however, three teams using different approaches have managed to obtain purified full-length human BRCA2, providing an unprecedented in vitro analysis of its molecular functions [48–50]. All three papers were able to demonstrate that BRCA2 mediates loading of RAD51 onto ssDNA while displacing RPA (a protein that binds ssDNA preventing secondary DNA structures from forming). In addition, Jensen et al. and Thorslund et al. show that BRCA2 prevents RAD51 association to dsDNA, which would inhibit HR, and favors RAD51 association to ssDNA or dsDNA with ssDNA tails. Jensen et al. and Liu et al. demonstrate that BRCA2 inhibits RAD51 hydrolysis of ATP, which stabilizes the nucleoprotein filament. Much more is still to be learned about BRCA2. For example, there is direct evidence to demonstrate that, despite being evolutionary conserved domains, not all BRC motifs are required for competent HR to be elicited in the presence of DNA DSBs, suggesting that they may have alternate or modulatory roles in HR [51]. Understanding how its interacting partners—such as PALB2 and BRCA1—affect BRCA2 function is also unclear. With full-length BRCA2 at hand characterization of its complex molecular functions will be more readily answered.

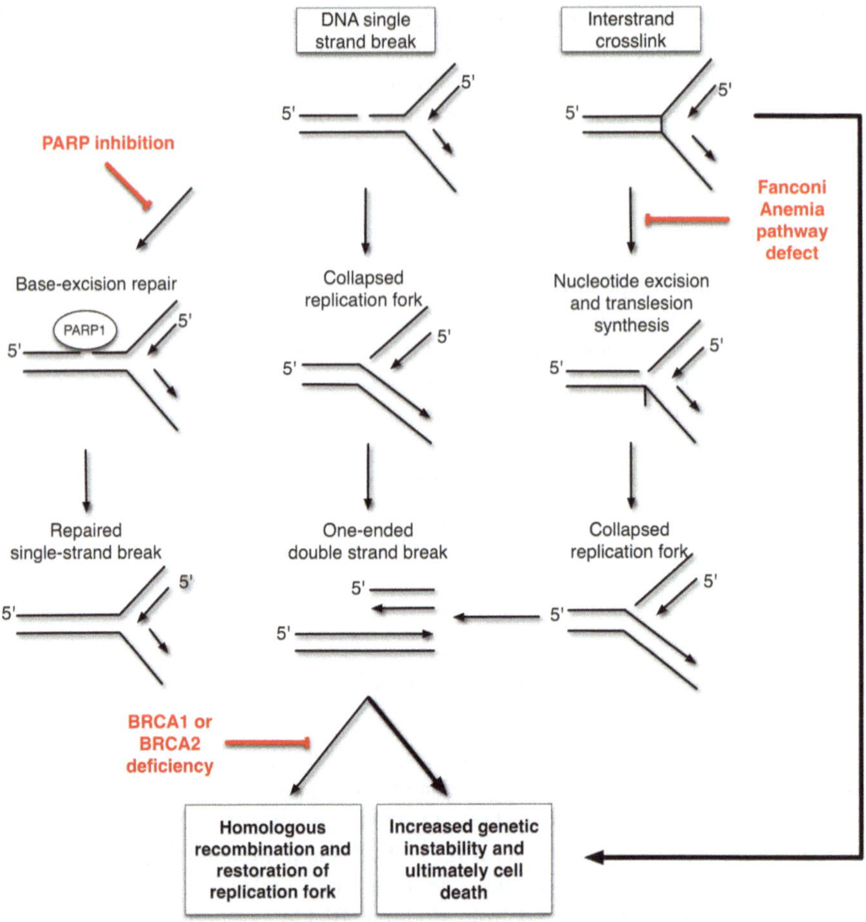

Fig. 3 Schema describing DNA repair pathways following a single-strand break and interstrand cross-link (ICL). A single-strand break is normally repaired by base-excision repair (BER). If PARP is inhibited, however, BER is defective and a double-strand break is induced during the S phase of the cell cycle requiring homologous recombination (HR) to mediate repair and regenerate the replication fork. Tumor cells unable to properly repair DNA damage by both BER and HR, resort to more error prone mechanisms such as NHEJ, MMEJ and SSA, which induces genomic instability and ultimately cell death. Interstrand cross-links require both intact Fanconi anemia (FA) and HR pathways to mediate its repair. A defect in any one of these pathways leads to chromosome breakage and cell death

4 Fanconi Anemia and Homologous Repair

An interstrand cross-link (ICL) is another highly cytotoxic lesion that prevents separation of complementary strands of DNA during replication. A specialized pathway is necessary to recognize and remove a cross-link but in so doing, a DSB is generated, requiring the HR machinery to complete the repair (Fig. 3) [52].

Patients with defects in ICL repair develop a rare genetic condition known as Fanconi anemia (FA), characterized by aplastic anemia, multiple congenital defects, susceptibility to both hematologic and solid malignancies, and sensitivity to ICL agents such as platinum drugs and mitomycin C. It is a heterogeneous disease caused by defects (either by recessive or X-linked mutations) in 1 of 13 genes, three of which—BRIP1, PALB2, and BRCA2—are also proteins involved in HR, providing further evidence that these two pathways are closely interrelated. See chapter "Repair of DNA Interstrand Cross-links Produced by Cancer Chemotherapeutic Drugs" for a detailed review of ICL repair.

5 Targeting BRCA1 and BRCA2-Mutated Tumors

Nearly all ovarian carcinomas and most breast cancers derived from patients with germ-line *BRCA1* and *BRCA2* mutations have lost their remaining wild-type allele and thus the ability to repair DSBs by HR [53, 54]. These cancers instead rely on complementary pathways such as NHEJ, to maintain some degree of genomic stability. By contrast, "healthy" cells with only one functional BRCA gene still have an intact HR pathway, a biological difference that can be exploited. In cells with a defective HR pathway, agents that introduce ICLs, such as mitomycin C and platinum drugs, are not effectively repaired and induce cell death, while those still capable of repairing DSBs by HR are relatively spared. Early data from platinum-based regimens on carriers of BRCA1 mutations have suggested some efficacy in treating breast cancer in the neoadjuvant setting [55]; however, stronger data will be needed before its clinical use in treating BRCA-associated cancers can be routinely proposed [56]. Anthracyclines, which intercalate DNA causing DSBs, may also be effective [57, 58]; however, reports from cell and clinical data are conflicting [55, 59]. Based on in vitro data, taxanes (traditionally used in the treatment of sporadic breast and ovarian cancer) may be of lesser benefit in hereditary cancers, at least for those lacking BRCA1 [60, 61]. Nevertheless, available clinical data do not support this view.

A novel class of drugs called PARP (Poly ADP-ribose polymerase) inhibitors was developed to target cells deficient in HR pathways [62, 63]. By inhibiting PARP, base-excision repair (BER) is impaired leading to the accumulation of unrepaired single-strand breaks, which during S phase lead to stalling and/or collapse of replication forks, and eventually degenerate into DSBs (Fig. 3). While normal cells have the capacity to compensate for PARP inhibitor-mediated loss of BER via HR, cells without the means to repair the damage by HR have to resort to error prone mechanisms (i.e., SSA, NHEJ, and MMEJ) to repair the DNA DSBs. These observations have led to the development of synthetic lethal approaches to target BRCA1 and BRCA2 deficient cancers [62, 63]. An extended phase I study [64] and two phase 2 clinical trials in *BRCA1* and *BRCA2* carriers have shown promise in the treatment of both metastatic breast [65] and ovarian cancers [66]. Further clinical trials are underway to evaluate whether PARP inhibitors act synergistically in combination with other chemotherapeutic agents such as cisplatin.

A caveat to cisplatin and PARP therapy is the cancer's inevitable progression towards drug resistance. Studies have described the mechanism of drug resistance in *BRCA1*- and *BRCA2*-mutant tumors as intragenic deletions and secondary mutations induced by error-prone repair pathways such as NHEJ and SSA that restore an open-reading frame resulting in the expression of a functional BRCA1 or BRCA2 protein [51, 67–69]. It would seem that while loss of BRCA1 or BRCA2 is advantageous early in the progression of tumor development, the presence of BRCA1 or BRCA2 in its later stages may have little if any effect on tumor viability. In addition, it is thought that mutations are stochastic events and therefore the larger the tumor population the greater the likelihood that a revertant mutation will arise. Taken together, this would suggest that treatment with cisplatin or PARP inhibitors in the very early stages of cancer would have the greatest chance of eliminating disease, while treatment beyond a certain stage of development will likely end in relapse.

6 Targeting Sporadic Cancers Lacking Homologous Recombination

Do sporadic cancers harbor defects in HR and FA pathways, and if they do, would targeting them with cisplatin and PARP inhibitors be effective? A logical first step in answering this question would be to determine whether BRCA1 and BRCA2 are mutated in sporadic cases. About 20% of high grade serous ovarian carcinomas [70] and a similar percentage in triple-negative breast cancers (TNBC) [71] have germ line or somatic mutations in BRCA1/2; however, BRCA1/2 are also found to be down-regulated by other means such as epigenetic silencing [72–76] and transcriptional repression [77, 78]. In the latter example, the hypothesized role of *EMSY* amplification and BRCA2 suppression has been called into question as it appears that EMSY amplification in cancer cell lines is not associated with impaired HR function or increased sensitivity to cisplatin or PARP inhibition [79]. It has been previously suggested that the consequences of early BRCA1 deficiency dictate tumor lineage and phenotype [80] and that cell phenotype or "BRCAness" may be used as a surrogate marker for an underlying *BRCA1* mutation [81]. Cells with a BRCA2 deficiency, however, seem not to follow a particular lineage, which is reflected by a lack of an association for *BRCA2*-associated tumors to a histopathologic phenotype that distinguishes them from sporadic cancers.

BRCA1 deficient breast cancers are characteristically "triple-negative" meaning they lack estrogen and progesterone receptors and do not over-express HER2 [82]; a tendency that could be explained by a haploinsufficiency of BRCA1 leading to a failure of luminal-progenitor cells to differentiate [83] thus creating a comparatively larger pool of basal-like stem cells that have the potential to give rise to a triple-negative phenotype [84, 85]. The unique biology of BRCA1 may underlie the phenotype seen in sporadic TNBCs providing the rational for clinical trials targeting TNBC with cisplatin [86] and PARP inhibitors [87]. However, promising results in a phase 2 trial for the novel therapeutic drug iniparib [87], a previously ascribed PARP inhibitor, failed to meet clinical outcomes in a subsequent phase 3 trial (http://en.sanofi-aventis.com/research_innovation/rd_key_figures/rd_key_figures.asp). Although the mechanism by which iniparib achieves its antitumor effects is unclear, its failure in the phase 3 trial may be due to the plasticity by which BRCA1 is down-regulated allowing tumor cells to more readily reactivate BRCA1 function leading to earlier resistance. Another possibility could be because TNBC is a convergent phenotype of a heterogeneous disease with only a small subgroup having an underlying BRCA1 defect. More reliable methods at predicting HR and FA function are being sought such as gene expression profiling [88] and radiation-induced RAD51 foci formation [89]; however, it is expected that next-generation sequencing technologies may ultimately prove to be the "gold standard" in the prediction of the ability to repair DNA. A genomic landscape not only characterizes all the mutations found within HR and FA related genes, but also describes the genetic signature of HR dysfunction. A comprehensive understanding of tumor biology however will rely on more than just genomic data. As a testament to the rapid advances made in sequencing technology and bioinformatics, a recent paper demonstrated the monumental task of analyzing 466 tumors across different platforms, integrating copy number variation, exomic, epigenomic, transcriptomic and proteomic data, providing a comprehensive understanding cancer drivers and drugable targets for the major breast cancer subtypes [90].

7 Conclusion

Studying the molecular pathways underlying hereditary breast and ovarian cancers has elucidated the processes that drive tumor progression, processes that are also common to sporadic cancers. Novel therapies are available to target cells defective in HR and FA pathways; however, determining which tumors have an underlying HR and FA defect is complex with no single method capable of providing a complete picture. As we begin to enter the genomic age, next-generation sequencing should allow full molecular characterization of cancer architecture and function including the tumor's ability to respond to DNA damage—setting the stage for personalized medicine.

References

1. Jackson SP, Bartek J (2009) The DNA-damage response in human biology and disease. Nature 461:1071–1078
2. Bennett CB, Lewis AL, Baldwin KK, Resnick MA (1993) Lethality induced by a single site-specific double-strand break in a dispensable yeast plasmid. Proc Natl Acad Sci USA 90:5613–5617
3. Helleday T, Lo J, van Gent DC, Engelward BP (2007) DNA double-strand break repair: from mechanistic understanding to cancer treatment. DNA Repair (Amst) 6:923–935
4. Rothkamm K, Krüger I, Thompson LH, Löbrich M (2003) Pathways of DNA double-strand break repair during the mammalian cell cycle. Mol Cell Biol 23:5706–5715
5. McVey M, Lee SE (2008) MMEJ repair of double-strand breaks (director's cut): deleted sequences and alternative endings. Trends Genet 24:529–538

6. Narod SA, Foulkes WD (2004) BRCA1 and BRCA2: 1994 and beyond. Nat Rev Cancer 4:665–676
7. Joukov V, Groen AC, Prokhorova T, Gerson R, White E, Rodriguez A, Walter JC, Livingston DM (2006) The BRCA1/BARD1 heterodimer modulates Ran-dependent mitotic spindle assembly. Cell 127:539–552
8. Starita LM, Machida Y, Sankaran S, Elias JE, Griffin K, Schlegel BP, Gygi SP, Parvin JD (2004) BRCA1-dependent ubiquitination of gamma-tubulin regulates centrosome number. Mol Cell Biol 24:8457–8466
9. Kumaraswamy E, Shiekhattar R (2007) Activation of BRCA1/BRCA2-associated helicase BACH1 is required for timely progression through S phase. Mol Cell Biol 27:6733–6741
10. Liu Z, Wu J, Yu X (2007) CCDC98 targets BRCA1 to DNA damage sites. Nat Struct Mol Biol 14:716–720
11. Kim H, Chen J, Yu X (2007) Ubiquitin-binding protein RAP80 mediates BRCA1-dependent DNA damage response. Science 316:1202–1205
12. Sobhian B, Shao G, Lilli DR, Culhane AC, Moreau LA, Xia B, Livingston DM, Greenberg RA (2007) RAP80 targets BRCA1 to specific ubiquitin structures at DNA damage sites. Science 316:1198–1202
13. Cantor SB, Bell DW, Ganesan S, Kass EM, Drapkin R, Grossman S, Wahrer DC, Sgroi DC, Lane WS, Haber DA et al (2001) BACH1, a novel helicase-like protein, interacts directly with BRCA1 and contributes to its DNA repair function. Cell 105:149–160
14. Wang B, Matsuoka S, Ballif BA, Zhang D, Smogorzewska A, Gygi SP, Elledge SJ (2007) Abraxas and RAP80 form a BRCA1 protein complex required for the DNA damage response. Science 316:1194–1198
15. Yarden RI, Brody LC (1999) BRCA1 interacts with components of the histone deacetylase complex. Proc Natl Acad Sci USA 96:4983–4988
16. Bochar DA, Wang L, Beniya H, Kinev A, Xue Y, Lane WS, Wang W, Kashanchi F, Shiekhattar R (2000) BRCA1 is associated with a human SWI/SNF-related complex: linking chromatin remodeling to breast cancer. Cell 102:257–265
17. Lou Z, Minter-Dykhouse K, Chen J (2005) BRCA1 participates in DNA decatenation. Nat Struct Mol Biol 12:589–593
18. Venkitaraman AR (2002) Cancer susceptibility and the functions of BRCA1 and BRCA2. Cell 108:171–182
19. Ashworth A, Lord CJ, Reis-Filho JS (2011) Genetic Interactions in Cancer Progression and Treatment. Cell 145:30–38
20. Foulkes WD (2008) Inherited susceptibility to common cancers. N Engl J Med 359:2143–2153
21. Miki Y, Swensen J, Shattuck-Eidens D, Futreal PA, Harshman K, Tavtigian S, Liu Q, Cochran C, Bennett LM, Ding W (1994) A strong candidate for the breast and ovarian cancer susceptibility gene BRCA1. Science 266:66–71
22. Wooster R, Bignell G, Lancaster J, Swift S, Seal S, Mangion J, Collins N, Gregory S, Gumbs C, Micklem G (1995) Identification of the breast cancer susceptibility gene BRCA2. Nature 378:789–792
23. Joukov V, Chen J, Fox EA, Green JB, Livingston DM (2001) Functional communication between endogenous BRCA1 and its partner, BARD1, during Xenopus laevis development. Proc Natl Acad Sci USA 98:12078–12083
24. Xia Y, Pao GM, Chen H-W, Verma IM, Hunter T (2003) Enhancement of BRCA1 E3 ubiquitin ligase activity through direct interaction with the BARD1 protein. J Biol Chem 278:5255–5263
25. Mallery DL, Vandenberg CJ, Hiom K (2002) Activation of the E3 ligase function of the BRCA1/BARD1 complex by polyubiquitin chains. EMBO J 21:6755–6762
26. Morris JR, Boutell C, Keppler M, Densham R, Weekes D, Alamshah A, Butler L, Galanty Y, Pangon L, Kiuchi T et al (2009) The SUMO modification pathway is involved in the BRCA1 response to genotoxic stress. Nature 462:886–890
27. Galanty Y, Belotserkovskaya R, Coates J, Polo S, Miller KM, Jackson SP (2009) Mammalian SUMO E3-ligases PIAS1 and PIAS4 promote responses to DNA double-strand breaks. Nature 462:935–939

28. Zhu Q, Pao GM, Huynh AM, Suh H, Tonnu N, Nederlof PM, Gage FH, Verma IM (2011) BRCA1 tumour suppression occurs via heterochromatin-mediated silencing. Nature 477:179–184
29. Yu X (2006) BRCA1 ubiquitinates its phosphorylation-dependent binding partner CtIP. Genes Dev 20:1721–1726
30. Sato K, Hayami R, Wu W, Nishikawa T, Nishikawa H, Okuda Y, Ogata H, Fukuda M, Ohta T (2004) Nucleophosmin/B23 is a candidate substrate for the BRCA1-BARD1 ubiquitin ligase. J Biol Chem 279:30919–30922
31. Wu W, Nishikawa H, Hayami R, Sato K, Honda A, Aratani S, Nakajima T, Fukuda M, Ohta T (2007) BRCA1 ubiquitinates RPB8 in response to DNA damage. Cancer Res 67:951–958
32. Starita LM, Horwitz AA, Keogh M-C, Ishioka C, Parvin JD, Chiba N (2005) BRCA1/BARD1 ubiquitinate phosphorylated RNA polymerase II. J Biol Chem 280:24498–24505
33. Eakin CM, Maccoss MJ, Finney GL, Klevit RE (2007) Estrogen receptor alpha is a putative substrate for the BRCA1 ubiquitin ligase. Proc Natl Acad Sci USA 104:5794–5799
34. Hashizume R, Fukuda M, Maeda I, Nishikawa H, Oyake D, Yabuki Y, Ogata H, Ohta T (2001) The RING heterodimer BRCA1-BARD1 is a ubiquitin ligase inactivated by a breast cancer-derived mutation. J Biol Chem 276:14537–14540
35. Ruffner H, Joazeiro CA, Hemmati D, Hunter T, Verma IM (2001) Cancer-predisposing mutations within the RING domain of BRCA1: loss of ubiquitin protein ligase activity and protection from radiation hypersensitivity. Proc Natl Acad Sci USA 98:5134–5139
36. Reid LJ, Shakya R, Modi AP, Lokshin M, Cheng J-T, Jasin M, Baer R, Ludwig T (2008) E3 ligase activity of BRCA1 is not essential for mammalian cell viability or homology-directed repair of double-strand DNA breaks. Proc Natl Acad Sci USA 105:20876–20881
37. Yu X (2003) The BRCT domain is a phospho-protein binding domain. Science 302:639–642
38. Manke IA, Lowery DM, Nguyen A, Yaffe MB (2003) BRCT repeats as phosphopeptide-binding modules involved in protein targeting. Science 302:636–639
39. Kim H, Huang J, Chen J (2007) CCDC98 is a BRCA1-BRCT domain-binding protein involved in the DNA damage response. Nat Struct Mol Biol 14:710–715
40. Yu X, Wu LC, Bowcock AM, Aronheim A, Baer R (1998) The C-terminal (BRCT) domains of BRCA1 interact in vivo with CtIP, a protein implicated in the CtBP pathway of transcriptional repression. J Biol Chem 273:25388–25392
41. Yun MH, Hiom K (2009) CtIP-BRCA1 modulates the choice of DNA double-strand-break repair pathway throughout the cell cycle. Nature 459:460–463
42. Sy SM-H, Huen MSY, Chen J (2009) PALB2 is an integral component of the BRCA complex required for homologous recombination repair. Proc Natl Acad Sci USA 106:7155–7160
43. Zhang F, Fan Q, Ren K, Andreassen PR (2009) PALB2 functionally connects the breast cancer susceptibility proteins BRCA1 and BRCA2. Mol Cancer Res 7:1110–1118
44. Wong AK, Pero R, Ormonde PA, Tavtigian SV, Bartel PL (1997) RAD51 interacts with the evolutionarily conserved BRC motifs in the human breast cancer susceptibility gene brca2. J Biol Chem 272:31941–31944
45. Galkin VE, Esashi F, Yu X, Yang S, West SC, Egelman EH (2005) BRCA2 BRC motifs bind RAD51-DNA filaments. Proc Natl Acad Sci USA 102:8537–8542
46. Davies OR, Pellegrini L (2007) Interaction with the BRCA2 C terminus protects RAD51-DNA filaments from disassembly by BRC repeats. Nat Struct Mol Biol 14:475–483
47. Esashi F, Galkin VE, Yu X, Egelman EH, West SC (2007) Stabilization of RAD51 nucleoprotein filaments by the C-terminal region of BRCA2. Nat Struct Mol Biol 14:468–474
48. Jensen RB, Carreira A, Kowalczykowski SC (2010) Purified human BRCA2 stimulates RAD51-mediated recombination. Nature 467:678–683
49. Thorslund T, McIlwraith MJ, Compton SA, Lekomtsev S, Petronczki M, Griffith JD, West SC (2010) The breast cancer tumor suppressor BRCA2 promotes the specific targeting of RAD51 to single-stranded DNA. Nat Struct Mol Biol 17:1263–1265
50. Liu J, Doty T, Gibson B, Heyer W-D (2010) Human BRCA2 protein promotes RAD51 filament formation on RPA-covered single-stranded DNA. Nat Struct Mol Biol 17:1260–1262
51. Edwards SL, Brough R, Lord CJ, Natrajan R, Vatcheva R, Levine DA, Boyd J, Reis-Filho JS, Ashworth A (2008) Resistance to therapy caused by intragenic deletion in BRCA2. Nature 451:1111–1115

52. D'Andrea AD (2010) Susceptibility pathways in Fanconi's anemia and breast cancer. N Engl J Med 362:1909–1919
53. Berchuck A, Heron KA, Carney ME, Lancaster JM, Fraser EG, Vinson VL, Deffenbaugh AM, Miron A, Marks JR, Futreal PA et al (1998) Frequency of germline and somatic BRCA1 mutations in ovarian cancer. Clin Cancer Res 4:2433–2437
54. Osorio A, de la Hoya M, Rodríguez-López R, Martínez-Ramírez A, Cazorla A, Granizo JJ, Esteller M, Rivas C, Caldés T, Benítez J (2002) Loss of heterozygosity analysis at the BRCA loci in tumor samples from patients with familial breast cancer. Int J Cancer 99:305–309
55. Byrski T, Gronwald J, Huzarski T, Grzybowska E, Budryk M, Stawicka M, Mierzwa T, Szwiec M, Wiśniowski R, Siolek M et al (2010) Pathologic complete response rates in young women with BRCA1-positive breast cancers after neoadjuvant chemotherapy. J Clin Oncol 28:375–379
56. Carey LA (2010) Targeted chemotherapy? Platinum in BRCA1-dysfunctional breast cancer. J Clin Oncol 28:361–363
57. Chappuis PO, Goffin J, Wong N, Perret C, Ghadirian P, Tonin PN, Foulkes WD (2002) A significant response to neoadjuvant chemotherapy in BRCA1/2 related breast cancer. J Med Genet 39:608–610
58. Brodie SG, Xu X, Qiao W, Li WM, Cao L, Deng CX (2001) Multiple genetic changes are associated with mammary tumorigenesis in Brca1 conditional knockout mice. Oncogene 20:7514–7523
59. Tassone P, Tagliaferri P, Perricelli A, Blotta S, Quaresima B, Martelli ML, Goel A, Barbieri V, Costanzo F, Boland CR et al (2003) BRCA1 expression modulates chemosensitivity of BRCA1-defective HCC1937 human breast cancer cells. Br J Cancer 88:1285–1291
60. Lafarge S, Sylvain V, Ferrara M, Bignon YJ (2001) Inhibition of BRCA1 leads to increased chemoresistance to microtubule-interfering agents, an effect that involves the JNK pathway. Oncogene 20:6597–6606
61. Quinn JE, Kennedy RD, Mullan PB, Gilmore PM, Carty M, Johnston PG, Harkin DP (2003) BRCA1 functions as a differential modulator of chemotherapy-induced apoptosis. Cancer Res 63:6221–6228
62. Farmer H, McCabe N, Lord CJ, Tutt ANJ, Johnson DA, Richardson TB, Santarosa M, Dillon KJ, Hickson I, Knights C et al (2005) Targeting the DNA repair defect in BRCA mutant cells as a therapeutic strategy. Nature 434:917–921
63. Bryant HE, Schultz N, Thomas HD, Parker KM, Flower D, Lopez E, Kyle S, Meuth M, Curtin NJ, Helleday T (2005) Specific killing of BRCA2-deficient tumours with inhibitors of poly(ADP-ribose) polymerase. Nature 434:913–917
64. Fong P, Boss D, Yap T, Tutt A, Wu P (2009) Inhibition of poly(ADP-ribose) polymerase in tumors from BRCA mutation carriers. N Engl J Med 361(2):123–134
65. Tutt A, Robson M, Garber JE, Domchek SM, Audeh MW, Weitzel JN, Friedlander M, Arun B, Loman N, Schmutzler RK et al (2010) Oral poly(ADP-ribose) polymerase inhibitor olaparib in patients with BRCA1 or BRCA2 mutations and advanced breast cancer: a proof-of-concept trial. Lancet 376:235–244
66. Audeh MW, Carmichael J, Penson RT, Friedlander M, Powell B, Bell-McGuinn KM, Scott C, Weitzel JN, Oaknin A, Loman N et al (2010) Oral poly(ADP-ribose) polymerase inhibitor olaparib in patients with BRCA1 or BRCA2 mutations and recurrent ovarian cancer: a proof-of-concept trial. Lancet 376:245–251
67. Swisher EM, Sakai W, Karlan BY, Wurz K, Urban N, Taniguchi T (2008) Secondary BRCA1 mutations in BRCA1-mutated ovarian carcinomas with platinum resistance. Cancer Res 68:2581–2586
68. Norquist B, Wurz KA, Pennil CC, Garcia R, Gross J, Sakai W, Karlan BY, Taniguchi T, Swisher EM (2011) Secondary somatic mutations restoring BRCA1/2 predict chemotherapy resistance in hereditary ovarian carcinomas. J Clin Oncol 29:3008–3015
69. Sakai W, Swisher EM, Karlan BY, Agarwal MK, Higgins J, Friedman C, Villegas E, Jacquemont C, Farrugia DJ, Couch FJ et al (2008) Secondary mutations as a mechanism of cisplatin resistance in BRCA2-mutated cancers. Nature 451:1116–1120
70. Cancer Genome Atlas Research Network (2011) Integrated genomic analyses of ovarian carcinoma. Nature 474:609–615

71. Gonzalez-Angulo AM, Timms KM, Liu S, Chen H, Litton JK, Potter J, Lanchbury JS, Stemke-Hale K, Hennessy BT, Arun BK et al (2011) Incidence and outcome of BRCA mutations in unselected patients with triple receptor-negative breast cancer. Clin Cancer Res 17:1082–1089
72. Hilton JL, Geisler JP, Rathe JA, Hattermann-Zogg MA, DeYoung B, Buller RE (2002) Inactivation of BRCA1 and BRCA2 in ovarian cancer. J Natl Cancer Inst 94:1396–1406
73. Esteller M, Silva JM, Dominguez G, Bonilla F, Matias-Guiu X, Lerma E, Bussaglia E, Prat J, Harkes IC, Repasky EA et al (2000) Promoter hypermethylation and BRCA1 inactivation in sporadic breast and ovarian tumors. J Natl Cancer Inst 92:564–569
74. Catteau A, Harris WH, Xu CF, Solomon E (1999) Methylation of the BRCA1 promoter region in sporadic breast and ovarian cancer: correlation with disease characteristics. Oncogene 18:1957–1965
75. Baldwin RL, Nemeth E, Tran H, Shvartsman H, Cass I, Narod S, Karlan BY (2000) BRCA1 promoter region hypermethylation in ovarian carcinoma: a population-based study. Cancer Res 60:5329–5333
76. Geisler JP, Hatterman-Zogg MA, Rathe JA, Buller RE (2002) Frequency of BRCA1 dysfunction in ovarian cancer. J Natl Cancer Inst 94:61–67
77. Beger C, Pierce LN, Kruger M, Marcusson EG, Robbins JM, Welcsh P, Welch PJ, Welte K, King MC, Barber JR et al (2001) Identification of Id4 as a regulator of BRCA1 expression by using a ribozyme-library-based inverse genomics approach. Proc Natl Acad Sci USA 98:130–135
78. Hughes-Davies L, Huntsman D, Ruas M, Fuks F, Bye J, Chin S-F, Milner J, Brown LA, Hsu F, Gilks B et al (2003) EMSY links the BRCA2 pathway to sporadic breast and ovarian cancer. Cell 115:523–535
79. Wilkerson PM, Dedes KJ, Wetterskog D, MacKay A, Lambros MB, Mansour M, Frankum J, Lord CJ, Natrajan R, Ashworth A et al (2011) Functional characterization of EMSY gene amplification in human cancers. J Pathol 225:29–42
80. Foulkes WD (2004) BRCA1 functions as a breast stem cell regulator. J Med Genet 41:1–5
81. Turner N, Tutt A, Ashworth A (2004) Hallmarks of "BRCAness" in sporadic cancers. Nat Rev Cancer 4:814–819
82. Foulkes WD, Smith IE, Reis-Filho JS (2010) Triple-negative breast cancer. N Engl J Med 363:1938–1948
83. Proia TA, Keller PJ, Gupta PB, Klebba I, Jones AD, Sedic M, Gilmore H, Tung N, Naber SP, Schnitt S et al (2011) Genetic predisposition directs breast cancer phenotype by dictating progenitor cell fate. Cell Stem Cell 8:149–163
84. Lim E, Vaillant F, Wu D, Forrest NC, Pal B, Hart AH, Asselin-Labat M-L, Gyorki DE, Ward T, Partanen A et al (2009) Aberrant luminal progenitors as the candidate target population for basal tumor development in BRCA1 mutation carriers. Nat Med 15:907–913
85. Molyneux G, Geyer FC, Magnay F-A, McCarthy A, Kendrick H, Natrajan R, MacKay A, Grigoriadis A, Tutt A, Ashworth A et al (2010) BRCA1 basal-like breast cancers originate from luminal epithelial progenitors and not from basal stem cells. Cell Stem Cell 7:403–417
86. Silver DP, Richardson AL, Eklund AC, Wang ZC, Szallasi Z, Li Q, Juul N, Leong CO, Calogrias D, Buraimoh A et al (2010) Efficacy of neoadjuvant cisplatin in triple-negative breast cancer. J Clin Oncol 28:1145–1153
87. O'Shaughnessy J, Osborne C, Pippen JE, Yoffe M, Patt D, Rocha C, Koo IC, Sherman BM, Bradley C (2011) Iniparib plus chemotherapy in metastatic triple-negative breast cancer. N Engl J Med 364:205–214
88. Konstantinopoulos PA, Spentzos D, Karlan BY, Taniguchi T, Fountzilas E, Francoeur N, Levine DA, Cannistra SA (2010) Gene expression profile of BRCAness that correlates with responsiveness to chemotherapy and with outcome in patients with epithelial ovarian cancer. J Clin Oncol 28:3555–3561
89. Graeser M, McCarthy A, Lord CJ, Savage K, Hills M, Salter J, Orr N, Parton M, Smith IE, Reis-Filho JS et al (2010) A marker of homologous recombination predicts pathologic complete response to neoadjuvant chemotherapy in primary breast cancer. Clin Cancer Res 16:6159–6168
90. Cancer Genome Atlas Network (2012) Comprehensive molecular portraits of human breast tumours. Nature 490(7418):61–70

DNA-PK in CLL Chemotherapy

L. Amrein, D. Davidson, R. Aloyz, and L. Panasci

1 Introduction

DNA is the principal target of many conventional anticancer agents, and inhibition of DNA repair is one of the most promising strategies in novel cancer therapy. Many studies demonstrated that nonhomologous end-joining (NHEJ) repair pathway proteins, especially DNA-dependent protein kinase (DNA-PK), is an attractive and effective target for the sensitization of cancer cells, including the most common type of leukemia in western countries, chronic lymphocytic leukemia (CLL), to DNA double-strand break (DSB)-inducing agents used in conventional cancer therapy. Nevertheless, promising results obtained *in vitro* cannot be translated to the clinic yet due to the nature of the DNA-PK inhibitors which are either nonspecific, for the first class of inhibitors, or degraded/eliminated from the human body before reaching the tumor site for the newer specific DNA-PK inhibitors.

2 CLL and Conventional Therapeutic Treatments

B-cell CLL is a complex disease characterized by actively dividing B-lymphocyte in the lymph nodes and bone marrow [1, 2] as well as the accumulation of quiescent lymphocytes in the peripheral blood of affected patients [3]. Although CLL has been described for a long time the cell of origin is unknown. This disease is the most common leukemia in western countries with approximately 15,500 new diagnoses and over than 4,000 deaths estimated per year in the United States only [4]. CLL cells express B-cell immunophenotypic markers, such as CD19, CD20, and CD23, along

L. Amrein • D. Davidson • R. Aloyz (✉) • L. Panasci
Lady Davis Institute, 3755 Cote ste Catherine, Montreal, QC, Canada
e-mail: raquel.aloyz@mcgill.ca

L. Panasci et al. (eds.), *Advances in DNA Repair in Cancer Therapy*, Cancer Drug Discovery and Development 72, DOI 10.1007/978-1-4614-4741-2_7,
© Springer Science+Business Media, LLC 2013

with the T-cell marker CD5. CLL lymphocytes are also characterized by the expression of very low amounts of surface immunoglobulin. The clinical course is very heterogeneous with overall survival ranging from several months to more than 15 years [5]. Several biological factors have been linked with the clinical heterogeneity of CLL. These include the Rai/Binet stage, parameters of cell division, β 2-microglobulin, somatic mutations of immunoglobulin heavy-chain variable region (IGHV) genes, cytogenetic aberrations (deletion 11, deletion 17, deletion 13, and trisomy 12), and expression of CD38 and ZAP70 (reviewed in 3, 6). Although the majority of patients are asymptomatic at diagnosis, the incessant accumulation of B-CLL lymphocytes leads to symptomatic disease requiring therapy. Conventional chemotherapeutic drugs used in the treatment of CLL include nitrogen mustard analogue (chlorambucil (CLB), cyclophosphamide, and bendamustine) or the nucleotides analogue fludarabine. Chemo-immunotherapy combines chemotherapeutic drugs with monoclonal antibodies (immunotherapy) such as combination therapy with fludarabine, cyclophosphamide, and the CD20 monoclonal antibody rituximab (FCR) which is now a standard of care, offering good overall response rates (ORR), complete remission (CR) rates, and increased median progression-free survival (PFS) [7, 8]. However, FCR is not suitable for all patients, has significant side effects, and appears too toxic for some elderly patients. Given that CLL predominates in the elderly community, the potential toxicity of therapeutic regimens is an important issue. In addition, comparative clinical trial of fludarabine and cyclophosphamide (FC) against fludarabine alone suggested a higher incidence of chemotherapy-related myeloid neoplasia (a long-term toxicity) after FC than after fludarabine treatment [9]. Results from another clinical trial in CLL patients after initial therapy with CLB compared with fludarabine in patients over 65 years of age demonstrate that despite higher ORR and CR rates, this did not translate into improved PFS or overall survival [10]. However, due to the fact that therapeutic regimens come with toxic side effects, some progress has been achieved within the last decade. Nevertheless, another significant problem in treating CLL is that although patients often initially respond to conventional treatment, they eventually become resistant to the drugs and even if new strategies comprising chemotherapy combinations or chemo-immunotherapy have been used, CLL is still considered as an incurable disease [11].

3 DNA Damage and DNA Repair Mechanism

Cells are continuously subjected to numerous exogenous (radiation and environmental genotoxic compounds) and endogenous (intermediate products from normal metabolism and errors during replication process) sources of DNA damages. To overcome these threats, cells developed robust, complex, and highly conserved DNA-damage surveillance network, beginning with rapid and efficient detection of the lesions followed by the induction of complex protein signaling cascades leading to DNA repair mechanisms to ensure genomic integrity and stability [12]. Defects in signaling and repair of DNA damage are causally linked with the development of genomic instability and human cancer. One of the most deleterious forms of DNA damage, the DNA double strand breaks (DSB) is repaired by two major DNA repair systems in eukaryotic cells, the homologous recombination (HR) and the NHEJ repair pathways [13, 14]. HR is error-free, depends on the presence of sister chromatids to provide a DNA template identical to the damaged one, and thus is active in late S and G2 phases of the cell cycle. NHEJ does not require a template, thereby; it is active throughout the cell cycle and is the predominant mechanism in higher eukaryotes [15, 16]. DNA-PK is a key component of the NHEJ pathway which plays an important role in V(D)J recombination and in the repair of DNA DSBs [17–20]. The carboxy-terminal region of DNA-PKcs contains a catalytic domain similar to the phosphatidylinositol 3-kinase (PI3K) superfamily involved in cell cycle control, DNA repair, and DNA damage responses [21]. DNA-PK acts as a sensor of DSB during NHEJ since it is activated to bind to the ends of DNA and targets other factors to the site of damage [22]. DNA-PK is a nuclear serine/threonine protein kinase comprising a DNA-binding subunit, the Ku autoantigen, and a large catalytic subunit (460 kDa), DNA-PKcs. The Ku autoantigen is a heterodimer of the Ku70 and Ku80 proteins that binds to DNA double-strand ends and recruits DNA-PKcs [23–25]. This active DNA-PK complex then acquires the capacity to phosphorylate many DNA-bound proteins containing Ser/Thr-Gln motif including c-jun, p53, Ku70, Ku80, X-ray cross-complementing group 4 (XRCC4), and DNA-PKcs itself [20, 26–30]. Mutations in either DNA-PKcs or in the Ku80 result in DSB repair defects that manifest themselves as X-ray sensitivity and impaired V(D)J recombination [31, 32]. In addition, previous reports showed that mutant cells deficient either in DNA-PKcs or in the Ku DNA-end binding activity also exhibit significant hypersensitivity to DSB-inducing agents [33, 34]. DNA-PKcs plays a central role in regulation of NHEJ since it remains quiescent until activation by DNA ends [24]. Many *in vitro* and *in vivo* phosphorylation sites of DNA-PKcs have been identified. The importance of DNA-PKcs autophosphorylation in the PQR cluster (Ser 2023–Ser 2056), the ABCDE cluster (Thr 2609–Thr 2647), Thr 3950, and Ser 3205 during the NHEJ process has been well defined [35–38].

4 DNA-PK Inhibitors

Wortmannin, vanillin, and quercetin are natural product classes inhibiting PI3K family members including DNA-PKcs [39, 40]. Wortmannin forms covalent adduct in a conserved lysine residue in the kinase domain of DNA-PKcs [41], while quercetine targets the ATP-binding site of the kinase resulting in irreversible inhibition of DNA-PK activity [42]. A more potent synthetic derivative of quercetine, LY294002 developed by Lilly Research Laboratories, also inhibits enzymatic phosphorylation of lipids and proteins [42]. These compounds were used *in vitro* to assess DNA-PK inhibition but due to their nonspecificity for this kinase a number of more specific DNA-PK inhibitors have been developed. As expected for specific DNA-PK inhibitors, compounds developed by ICOS Corporation (IC86621, IC486154, IC87102,

IC87261) directly inhibit the repair of DNA DSBs [43]. Research performed by KuDOS Pharmaceuticals Ltd led to the development of synthetic and specific DNA-PK inhibitors. They utilized LY294002 as a template and have identified several molecules including NU7026 and NU7441 with good selectivity for DNA-PK over other PI3K members. These inhibitors have demonstrated *in vitro* radio- and chemo-sensitization in several human tumor (including leukemia) cell lines [44–46]. Contrarily to wortmannin, the ICOS and KuDOS compounds target the DNA-PKcs ATP-binding pocket improving potency and selectivity for DNA-PK over other PI3K family enzymes. Also, wortmannin is an irreversible DNA-PKcs inhibitor while the inhibition by ICOS and KuDOS compounds is reversible [46].

5 Importance of DNA Repair in CLL

As stated above, chemotherapeutic drugs used for clinical treatment of CLL patients are DNA-damaging agent. The primary response of cells with excessive DNA damage is to repair the lesions. Maintenance of the switching mechanisms that shift cells from DNA repair to apoptosis is of central importance for avoiding progression to malignancy. It has been proposed that enzyme-mediated repair of DSBs is a major mechanism of resistance to both ionizing radiation (IR) and drugs that cause DSBs as intermediates in repair processes [12]. *In vitro* experiments demonstrating cross resistance between nitrogen mustards and mitomycin C in B-CLL lymphocytes support the concept that cross resistance to different DNA-damaging agents involves accelerated DNA repair [47]. Also, B-CLL cells resistant to γ-radiation-induced apoptosis are completely resistant to apoptosis induced by neocarzinostatin and etoposide, compounds that specifically cause DNA DSBs [48]. Because DSBs are repaired by HR and NHEJ, inhibitors of key component of these two pathways have been investigated in combination with conventional drugs in B-CLL lymphocytes. For example, inhibition of c-abl (this non-receptor protein kinase phosphorylates and activates Rad51, a key component of HR) sensitizes B-CLL lymphocytes to CLB and fludarabine *in vitro* [49–51]. One of these investigations led to a phase I clinical trial in CLL patients where the combination of CLB and imatinib resulted in a 45 % response rate in a heavily pretreated population with minimal toxicity [52]. NHEJ, the other major DNA repair pathway, is also an attractive target to overcome resistance in B-CLL.

6 Role of DNA-PK for CLL Treatment

Despite many studies with various human cell lines, the first study of regulation of DNA-PK activity and DNA-PKcs protein expression in freshly isolated primary B-lymphocytes was done in 1997. It was demonstrated for the first time that DNA-PK activity could be measured in primary quiescent human B-CLL lymphocytes and that the level of DNA-PK activity varied considerably amongst CLL samples with higher expression in previously clinically treated patient samples [53–55]. These results were concordant with our previous report demonstrating that lymphocytes from treated-resistant patients have an enhanced capacity to remove cross-links compared with those from untreated patients [56]. Similarly, changes in DNA-PK activity correlated with CLB resistance while sensitivity to topoisomerase II inhibitors (doxorubicin and etoposide) correlated with DNA-PKcs protein expression suggesting that DNA-PK plays an important role in regulating CLL response to DNA-damaging agents [54, 55, 57]. Also, inhibition of CLB-induced HR repair in CLL lymphocytes resulted in an increased DNA-PKcs autophosphorylation [51]. Major determinants of therapeutic resistance in B-CLL are deletion of p53 (chromosome 17), ATM (chromosome 11) gene, and/or mutation in p53 resulting in a dysfunctional p53-dependent DNA damage response pathway. B lymphocytes isolated from these CLL patients expressed higher DNA-PK activity than patient without these genetic abnormalities [58]. In accordance with the concept that regulation of DNA-PK activity occurs partially at the Ku level, the mechanism of regulation of DNA-PK activity in B-CLL lymphocytes proceeds initially through a variation in the Ku DNA end-binding activity and probably the expression of an altered form of the heterodimer. Furthermore, Ku expression and function in B-CLL cells play a pivotal role during the acquisition of resistance [53, 54]. These findings open the field for the investigation of NHEJ repair pathway inhibition to improve treatment and/or overcome the resistance to treatment in B-CLL patients.

7 DNA-PK Inhibitors to Improve CLL Treatment

Inhibition of DNA-PK and the consequent inhibition of DSB repair were speculated to be the mechanisms whereby wortmannin potentiates the cytotoxicity of ionizing radiation in a Chinese Hamster Ovary cell line [59]. In primary B-CLL lymphocytes, wortmannin enhanced CLB cytotoxicity and γ-radiation-induced apoptosis in cells sensitive and most importantly in lymphocytes resistant to DSB-inducing agent. Sensitivity to these DNA-damaging agents was associated with inhibition of DNA repair and in resistant lymphocytes, the increase in CLB sensitivity correlated with the ability of wortmannin to inhibit DNA-PK activity [48, 55]. Vanillin, another natural but nonspecific DNA-PK inhibitor, sensitizes B-CLL cells from drug-sensitive and -resistant lymphocytes to fludarabine but the authors did not find any correlation between either DNA-PKcs expression and fludarabine sensitivity or DNA-PKcs expression and inhibitor sensitization [60]. Nevertheless, wortmannin and vanillin inhibit all the PI3K family members rendering it difficult to determine the exact role of DNA-PK and the drug sensitization induced by these agents in B-CLL lymphocytes. Synthesis of specific DNA-PK inhibitors made possible studies of the real impact of DNA-PK inhibition on drug resistance and its potential advantage in CLL therapy. Although NU7026, a specific DNA-PK inhibitor, was not toxic by itself in primary B-CLL lymphocytes and a B-CLL cell line, when combined with γ-irradiations or CLB treatment, NU7026 inhibited NHEJ-mediated DNA repair and DNA-PKcs phosphorylation leading both sensitive and resistant cells to undergo apoptosis after DNA damage [48, 61]. These data confirmed results obtained with wortmannin suggesting that DSB end-ligation activity was dependent on DNA-PK activity in these cells. Importantly in primary B-CLL cells, NU7026 inhibits CLB-induced DNA-PKcs autophosphorylation but did not affect CLB-induced ATM (another PI3K family member implicated in DSB repair pathway) phosphorylation, suggesting that at the doses used, NU7026 is a specific DNA-PK inhibitor in these cells [61]. NU7441, another DNA-PK inhibitor developed from LY294002, increased CLB and fludarabine-induced DNA damage and apoptosis resulting in B-CLL cell sensitization to these conventional drugs [58, 60]. Furthermore, simultaneous inhibition of both the HR and the NHEJ (by specific inhibition of DNA-PK) pathways potentiated the synergistic effect of either inhibitor alone on CLB cytotoxicity in CLL lymphocytes and was associated with an increase in CLB-induced DNA damage and decreased DNA repair [51].

8 Limitation for DNA-PK Therapy

All the studies stated above demonstrated that DNA-PK inhibition enhances the effects of DNA-damaging compounds by preventing repair through the NHEJ pathway in primary B-CLL lymphocytes *in vitro*. All these results have clinical interest and can potentially increase therapeutic treatment for CLL patients. Unfortunately, natural compounds such as wortmannin and vanillin are not specific enough and current specific DNA-PK inhibitors such as NU7026 have poor *in vivo* bioavailability, largely due to rapid oxidative metabolism in the liver [62].

9 Conclusion

The primary response of cells to DNA damage is to repair the lesions. The balance between DNA repair and apoptosis is of central importance for avoiding the occurrence of cancer. The various mechanisms of DNA repair, which are important to maintain healthy cells, ironically can become the front line of resistance for malignant cells. Indeed, there is a dynamic interaction between the two major DNA repair pathways, HR and NHEJ, in CLL lymphocytes in response to drug-induced DNA damage and overactive NHEJ DSB repair allows human B-CLL cells to escape apoptosis in the presence of chemotherapy-induced DNA damage. The development of specific inhibitors of key proteins of DNA repair pathway, especially DNA-PK inhibitors, has helped circumvent the problem of resistance to drugs treatment at least *in vitro* and has important clinical implications. However, the problem which faces us is now to translate these discoveries from the bench to the bed side. The current step is to be able to optimize the structure of existing DNA-PK inhibitors to improve their *in vivo* properties for clinical administration.

References

1. Schmid C, Isaacson PG (1994) Proliferation centres in B-cell malignant lymphoma, lymphocytic (B-CLL): an immunophenotypic study. Histopathology 24(5):445–451
2. Messmer BT, Messmer D, Allen SL, Kolitz JE, Kudalkar P, Cesar D et al (2005) In vivo measurements document the dynamic cellular kinetics of chronic lymphocytic leukemia B cells. J Clin Invest 115(3):755–764
3. Hamblin TJ, Oscier DG (1997) Chronic lymphocytic leukaemia: the nature of the leukaemic cell. Blood Rev 11(3):119–128
4. Siegel R, Ward E, Brawley O, Jemal A (2011) Cancer statistics, 2011: The impact of eliminating socioeconomic and racial disparities on premature cancer deaths. CA Cancer J Clin 61(4):212–236
5. Chiorazzi N, Rai KR, Ferrarini M (2005) Chronic lymphocytic leukemia. N Engl J Med 352(8):804–815
6. Vroblova V, Smolej L, Vrbacky F, Jankovicova K, Hrudkova M, Maly J et al (2009) Biological prognostic markers in chronic lymphocytic leukemia. Acta Medica (Hradec Kralove) 52(1):3–8
7. Keating MJ, O'Brien S, Albitar M, Lerner S, Plunkett W, Giles F et al (2005) Early results of a chemoimmunotherapy regimen of fludarabine, cyclophosphamide, and rituximab as initial therapy for chronic lymphocytic leukemia. J Clin Oncol 23(18):4079–4088
8. Hallek M, Fischer K, Fingerle-Rowson G, Fink AM, Busch R, Mayer J et al (2010) Addition of rituximab to fludarabine and cyclophosphamide in patients with chronic lymphocytic leukaemia: a randomised, open-label, phase 3 trial. Lancet 376(9747):1164–1174
9. Smith MR, Neuberg D, Flinn IW, Grever MR, Lazarus HM, Rowe JM et al (2011) Incidence of Therapy-related Myeloid Neoplasia after Initial Therapy for CLL with Fludarabine-Cyclophosphamide versus Fludarabine: Long-Term Follow-up of US Intergroup Study E2997. Blood 118(13):3525–3527
10. Eichhorst BF, Busch R, Stilgenbauer S, Stauch M, Bergmann MA, Ritgen M et al (2009) First-line therapy with fludarabine compared with chlorambucil does not result in a major benefit for elderly patients with advanced chronic lymphocytic leukemia. Blood 114(16):3382–3391
11. Byrd JC, Stilgenbauer S, Flinn IW (2004) Chronic lymphocytic leukemia. Hematology (Am Soc Hematol Educ Prog) 1:163–183
12. Jackson SP (2002) Sensing and repairing DNA double-strand breaks. Carcinogenesis 23(5):687–696
13. De Silva IU, McHugh PJ, Clingen PH, Hartley JA (2000) Defining the roles of nucleotide excision repair and recombination in the repair of DNA interstrand cross-links in mammalian cells. Mol Cell Biol 20(21):7980–7990
14. McHugh PJ, Spanswick VJ, Hartley JA (2001) Repair of DNA interstrand crosslinks: molecular mechanisms and clinical relevance. Lancet Oncol 2(8):483–490
15. Takata M, Sasaki MS, Sonoda E, Morrison C, Hashimoto M, Utsumi H et al (1998) Homologous recombination and non-homologous end-joining pathways of DNA double-strand break repair have overlapping roles in the maintenance of chromosomal integrity in vertebrate cells. EMBO J 17(18):5497–5508
16. Rothkamm K, Kruger I, Thompson LH, Lobrich M (2003) Pathways of DNA double-strand break repair during the mammalian cell cycle. Mol Cell Biol 23(16):5706–5715
17. Finnie NJ, Gottlieb TM, Blunt T, Jeggo PA, Jackson SP (1996) DNA-dependent protein kinase defects are linked to deficiencies in DNA repair and V(D)J recombination. Philos Trans R Soc Lond B Biol Sci 351(1336):173–179
18. Jeggo PA, Taccioli GE, Jackson SP (1995) Menage a trois: double strand break repair, V(D)J recombination and DNA-PK. Bioessays 17(11):949–957
19. Weaver DT (1995) What to do at an end: DNA double-strand-break repair. Trends Genet 11(10):388–392

20. Smith GC, Jackson SP (1999) The DNA-dependent protein kinase. Genes Dev 13(8):916–934
21. Hartley KO, Gell D, Smith GC, Zhang H, Divecha N, Connelly MA et al (1995) DNA-dependent protein kinase catalytic subunit: a relative of phosphatidylinositol 3-kinase and the ataxia telangiectasia gene product. Cell 82(5):849–856
22. Durocher D, Jackson SP (2001) DNA-PK, ATM and ATR as sensors of DNA damage: variations on a theme? Curr Opin Cell Biol 13(2):225–231
23. Mimori T, Hardin JA (1986) Mechanism of interaction between Ku protein and DNA. J Biol Chem 261(22):10375–10379
24. Gottlieb TM, Jackson SP (1993) The DNA-dependent protein kinase: requirement for DNA ends and association with Ku antigen. Cell 72(1):131–142
25. Blier PR, Griffith AJ, Craft J, Hardin JA (1993) Binding of Ku protein to DNA. Measurement of affinity for ends and demonstration of binding to nicks. J Biol Chem 268(10):7594–7601
26. Lees-Miller SP, Chen YR, Anderson CW (1990) Human cells contain a DNA-activated protein kinase that phosphorylates simian virus 40 T antigen, mouse p53, and the human Ku autoantigen. Mol Cell Biol 10(12):6472–6481
27. Bannister AJ, Gottlieb TM, Kouzarides T, Jackson SP (1993) c-Jun is phosphorylated by the DNA-dependent protein kinase in vitro; definition of the minimal kinase recognition motif. Nucleic Acids Res 21(5):1289–1295
28. Anderson CW, Connelley MA, Zhang H, Sipley JD, Lees-Miller SP, Sakaguchi K et al (1994) The human DNA-activated protein kinase, DNA-PK, is activated by DNA breaks and phosphorylates nuclear DNA-binding substrates on serines and threonines following glutamine. J Prot Chem 13:500–501
29. Calsou P, Delteil C, Frit P, Drouet J, Salles B (2003) Coordinated assembly of Ku and p460 subunits of the DNA-dependent protein kinase on DNA ends is necessary for XRCC4-ligase IV recruitment. J Mol Biol 326(1):93–103
30. Mahaney BL, Meek K, Lees-Miller SP (2009) Repair of ionizing radiation-induced DNA double-strand breaks by non-homologous end-joining. Biochem J 417(3):639–650
31. Peterson SR, Kurimasa A, Oshimura M, Dynan WS, Bradbury EM, Chen DJ (1995) Loss of the catalytic subunit of the DNA-dependent protein kinase in DNA double-strand-break-repair mutant mammalian cells. Proc Natl Acad Sci USA 92(8):3171–3174
32. Taccioli GE, Rathbun G, Oltz E, Stamato T, Jeggo PA, Alt FW (1993) Impairment of V(D)J recombination in double-strand break repair mutants. Science 260(5105):207–210
33. Caldecott K, Jeggo P (1991) Cross-sensitivity of gamma-ray-sensitive hamster mutants to cross-linking agents. Mutat Res 255(2):111–121
34. Tanaka T, Yamagami T, Oka Y, Nomura T, Sugiyama H (1993) The scid mutation in mice causes defects in the repair system for both double-strand DNA breaks and DNA cross-links. Mutat Res 288(2):277–280
35. Chan DW, Chen BP, Prithivirajsingh S, Kurimasa A, Story MD, Qin J et al (2002) Autophosphorylation of the DNA-dependent protein kinase catalytic subunit is required for rejoining of DNA double-strand breaks. Genes Dev 16(18):2333–2338
36. Meek K, Douglas P, Cui X, Ding Q, Lees-Miller SP (2007) Trans Autophosphorylation at DNA-dependent protein kinase's two major autophosphorylation site clusters facilitates end processing but not end joining. Mol Cell Biol 27(10):3881–3890
37. Douglas P, Cui X, Block WD, Yu Y, Gupta S, Ding Q et al (2007) The DNA-dependent protein kinase catalytic subunit is phosphorylated in vivo on threonine 3950, a highly conserved amino acid in the protein kinase domain. Mol Cell Biol 27(5):1581–1591
38. Hammel M, Yu Y, Mahaney BL, Cai B, Ye R, Phipps BM et al (2010) Ku and DNA-dependent protein kinase dynamic conformations and assembly regulate DNA binding and the initial non-homologous end joining complex. J Biol Chem 285(2):1414–1423
39. Powis G, Bonjouklian R, Berggren MM, Gallegos A, Abraham R, Ashendel C et al (1994) Wortmannin, a potent and selective inhibitor of phosphatidylinositol-3-kinase. Cancer Res 54(9):2419–2423

40. Durant S, Karran P (2003) Vanillins–a novel family of DNA-PK inhibitors. Nucleic Acids Res 31(19):5501–5512
41. Wymann MP, Bulgarelli-Leva G, Zvelebil MJ, Pirola L, Vanhaesebroeck B, Waterfield MD et al (1996) Wortmannin inactivates phosphoinositide 3-kinase by covalent modification of Lys-802, a residue involved in the phosphate transfer reaction. Mol Cell Biol 16(4):1722–1733
42. Vlahos CJ, Matter WF, Hui KY, Brown RF (1994) A specific inhibitor of phosphatidylinositol 3-kinase, 2-(4-morpholinyl)-8-phenyl-4 H-1-benzopyran-4-one (LY294002). J Biol Chem 269(7):5241–5248
43. Kashishian A, Douangpanya H, Clark D, Schlachter ST, Eary CT, Schiro JG et al (2003) DNA-dependent protein kinase inhibitors as drug candidates for the treatment of cancer. Mol Cancer Ther 2(12):1257–1264
44. Hollick JJ, Golding BT, Hardcastle IR, Martin N, Richardson C, Rigoreau LJ et al (2003) 2,6-disubstituted pyran-4-one and thiopyran-4-one inhibitors of DNA-Dependent protein kinase (DNA-PK). Bioorg Med Chem Lett 13(18):3083–3086
45. Willmore E, de Caux S, Sunter NJ, Tilby MJ, Jackson GH, Austin CA et al (2004) A novel DNA-dependent protein kinase inhibitor, NU7026, potentiates the cytotoxicity of topoisomerase II poisons used in the treatment of leukemia. Blood 103(12):4659–4665
46. Leahy JJ, Golding BT, Griffin RJ, Hardcastle IR, Richardson C, Rigoreau L et al (2004) Identification of a highly potent and selective DNA-dependent protein kinase (DNA-PK) inhibitor (NU7441) by screening of chromenone libraries. Bioorg Med Chem Lett 14(24):6083–6087
47. Bramson J, McQuillan A, Aubin R, Alaoui-Jamali M, Batist G, Christodoulopoulos G et al (1995) Nitrogen mustard drug resistant B-cell chronic lymphocytic leukemia as an in vivo model for crosslinking agent resistance. Mutat Res 336(3):269–278
48. Deriano L, Guipaud O, Merle-Beral H, Binet JL, Ricoul M, Potocki-Veronese G et al (2005) Human chronic lymphocytic leukemia B cells can escape DNA damage-induced apoptosis through the nonhomologous end-joining DNA repair pathway. Blood 105(12):4776–4783
49. Aloyz R, Grzywacz K, Xu ZY, Loignon M, Alaoui-Jamali MA, Panasci L (2004) Imatinib sensitizes CLL lymphocytes to chlorambucil. Leukemia 18(3):409–414
50. Amrein L, Hernandez TA, Ferrario C, Johnston J, Gibson SB, Panasci L et al (2008) Dasatinib sensitizes primary chronic lymphocytic leukaemia lymphocytes to chlorambucil and fludarabine in vitro. Br J Haematol 143(5):698–706
51. Amrein L, Rachid Z, Jean-Claude B, Soulieres D, Aloyz R, Panasci L (2011) ZRF4, a combi-molecule with increased efficacy as compared with the individual components in chronic lymphocytic leukemia lymphocytes in vitro. Leukemia 25(9):1512–1516
52. Hebb J, Assouline S, Rousseau C, Desjardins P, Caplan S, Egorin MJ et al (2011) A phase I study of imatinib mesylate in combination with chlorambucil in previously treated chronic lymphocytic leukemia patients. Cancer Chemother Pharmacol 68(3):643–651
53. Muller C, Salles B (1997) Regulation of DNA-dependent protein kinase activity in leukemic cells. Oncogene 15(19):2343–2348
54. Muller C, Christodoulopoulos G, Salles B, Panasci L (1998) DNA-Dependent protein kinase activity correlates with clinical and in vitro sensitivity of chronic lymphocytic leukemia lymphocytes to nitrogen mustards. Blood 92(7):2213–2219
55. Christodoulopoulos G, Muller C, Salles B, Kazmi R, Panasci L (1998) Potentiation of chlorambucil cytotoxicity in B-cell chronic lymphocytic leukemia by inhibition of DNA-dependent protein kinase activity using wortmannin. Cancer Res 58(9):1789–1792
56. Torres-Garcia SJ, Cousineau L, Caplan S, Panasci L (1989) Correlation of resistance to nitrogen mustards in chronic lymphocytic leukemia with enhanced removal of melphalan-induced DNA cross-links. Biochem Pharmacol 38(18):3122–3123
57. Eriksson A, Lewensoh R, Larsson R, Nilsson A (2002) DNA-dependent protein kinase in leukaemia cells and correlation with drug sensitivity. Anticancer Res 22(3):1787–1793

58. Willmore E, Elliott SL, Mainou-Fowler T, Summerfield GP, Jackson GH, O'Neill F et al (2008) DNA-dependent protein kinase is a therapeutic target and an indicator of poor prognosis in B-cell chronic lymphocytic leukemia. Clin Cancer Res 14(12):3984–3992
59. Boulton S, Kyle S, Yalcintepe L, Durkacz BW (1996) Wortmannin is a potent inhibitor of DNA double strand break but not single strand break repair in Chinese hamster ovary cells. Carcinogenesis 17(11):2285–2290
60. Svirnovski AI, Serhiyenka TF, Kustanovich AM, Khlebko PV, Fedosenko VV, Taras IB et al (2010) DNA-PK, ATM and MDR proteins inhibitors in overcoming fludarabine resistance in CLL cells. Exp Oncol 32(4):258–262
61. Amrein L, Loignon M, Goulet AC, Dunn M, Jean-Claude B, Aloyz R et al (2007) Chlorambucil cytotoxicity in malignant B lymphocytes is synergistically increased by 2-(morpholin-4-yl)-benzo[h]chomen-4-one (NU7026)-mediated inhibition of DNA double-strand break repair via inhibition of DNA-dependent protein kinase. J Pharmacol Exp Ther 321(3):848–855
62. Nutley BP, Smith NF, Hayes A, Kelland LR, Brunton L, Golding BT et al (2005) Preclinical pharmacokinetics and metabolism of a novel prototype DNA-PK inhibitor NU7026. Br J Cancer 93(9):1011–1018

Poly(ADP) Ribose Polymerase at the Interface of DNA Damage Signaling and DNA Repair

Jana Krietsch, Michèle Rouleau, Michel Lebel, Guy Poirier, and Jean-Yves Masson

Chromosomal double-strand breaks (DSBs) are extremely hazardous to a cell as they do not leave an intact complementary strand to be used for repair. If not repaired accurately, the broken chromosomes undergo a wide variety of rearrangements such as translocations, mutations and deletions that may lead to cell death [1]. Genomic instability can promote cancer, developmental defects, tissue neurodegeneration, immunodeficiency, aging, as well as hypersensitivity to radiation. Each day a cell encounters approximately up to 50 DSBs, generated intrinsically such as during DNA synthesis when the processing replication fork encounters a damaged template [2]. DSBs can also be created during metabolic processes such as V(D)J recombination and class-switch recombination in vertebrate lymphocytes, meiotic recombination in germ cell lines, and mating type switching in yeast. Exogenous sources such as X-rays, gamma rays, UV light, topoisomerase I + II inhibitors can produce DSBs amongst other types of DNA damage. The cellular response to DNA

J. Krietsch
Cancer Research Unit, Laval University Medical Research Center, CHUQ-CRCHUL, Québec, QC, Canada G1V 4G2

Genome Stability Laboratory, Laval University Cancer Research Center, Hôtel-Dieu de Québec, Québec, Canada G1R 2J6

M. Rouleau • G. Poirier (✉)
Cancer Research Unit, Laval University Medical Research Center, CHUQ-CRCHUL, Québec, QC, Canada G1V 4G2
e-mail: guy.poirier@crchul.ulaval.ca

M. Lebel
Laval University Cancer Research Center, Hôtel-Dieu de Québec, Québec, Canada G1R 2J6

J.-Y Masson (✉)
Genome Stability Laboratory, Laval University Cancer Research Center, Hôtel-Dieu de Québec, Québec, Canada G1R 2J6
e-mail: Jean-Yves.Masson@crhdq.ulaval.ca

L. Panasci et al. (eds.), *Advances in DNA Repair in Cancer Therapy*, Cancer Drug Discovery and Development 72, DOI 10.1007/978-1-4614-4741-2_8,

© Springer Science+Business Media, LLC 2013

damage consists of multiple regulatory layers starting with sensing the damage, recruitment of repair proteins to the site of damage, and execution of DNA repair with possible outcomes concerning the cell's fate (such as apoptosis, entering terminal differentiation through senescence in order to prevent from inheriting damaged DNA) [3]. Interestingly, some members of the poly (ADP-ribose) polymerase (PARP) family have been implicated in DNA damage sensing as well as the repair of single strand breaks (SSBs) and DSBs, giving them a universal as well as a unique role in a cell's response to DNA damage [4]. Three PARPs that have been shown to be activated by DNA damage (PARP-1 as well as PARP-2 and possibly PARP-3) [5, 27] are therefore discussed in the following review with a focus on the two major pathways which have evolved to repair DNA DSBs: nonhomologous end joining (NHEJ) and homologous recombination (HR).

1 PARPs and Their Implications in Sensing and Repairing DNA Damage

The family of poly(ADP-ribose) polymerases (PARPs) also known as ADP-ribosyltransferases (ARTDs) consists of approximately 17 proteins in humans, estimated by the number of genes encoding proteins that possess an ADP-ribosyl-transferase catalytic domain [6]. PARP-1, PARP-2, PARP-3, and Tankyrases have been well described for their phylogenetically ancient, reversible posttranslational modification mechanism called poly(ADP-ribosyl)ation, which can modulate the function of their target proteins by regulating either enzymatic activities or molecular interactions between proteins, DNA, or RNA [7]. Responding to a large variety of cellular stresses, poly(ADP-ribosyl)ation is implicated in the maintenance of genomic stability, transcriptional regulation [8], energy metabolism, DNA methylation [9], and cell death [4, 10]. Upon activation, PARPs catalyze a reaction in which NAD^+ molecules are used to generate poly(ADP-ribose) molecules (pADPr) of varying length and complexity attached onto a number of acceptor proteins including PARPs themselves (automodification). As the first PARP discovered by Chambon and colleagues in 1963, the PARP-1 enzyme mediates the synthesis of an adenine-containing RNA-like polymer [11]. PARP-1 is one of the most abundant nuclear protein after histones.

The first function of PARPs in vitro was identified in response to DNA damage: Besides PARP-1, PARP-2, and PARP-3 have been shown to be enzymatically activated by encountering DNA strand breaks in vitro [5, 12, 27] with PARP-1 carrying out ~90% of the overall polymer synthesis and, notably, attaching the bulk of pADPr to itself [4]. The generation of knockout mice for PARP-1 further strengthened the hypothesis for a role for PARP-1 in DNA repair. The knock-out of PARP-1 or PARP-2 genes in mice is not lethal, suggesting that there is some redundancy between the function of these two PARPs. Importantly, PARP-1 knock-out mice led to the discovery of PARP-2. Notably, the double knock-out of PARP-1 and PARP-2 is not viable, indicating that poly(ADP-ribosyl)ation is

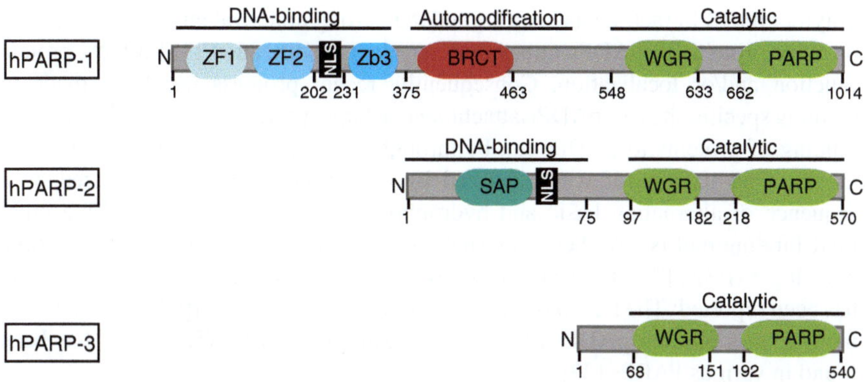

Fig. 1 Schematic comparison of the domain architecture of human PARP-1, PARP-2, and PARP-3. The following most significant domains are indicated: zinc finger (ZF); zinc binding (Zb); carboxyterminal domain (BRCT); the WGR domain, named after a conserved central motif (W-G-R); the PARP signature, representing the catalytic core needed for basal activity; nuclear localization signal (NLS); SAF/Acinus/PIAS-DNA-binding domain (SAP) (adapted from [27])

essential for early embryogenesis [13–15]. The modular structure of the PARP-1 protein is composed of at least six independent domains, containing two homologous zinc fingers (Zn1 and Zn2) at the extreme N-terminus that form the DNA binding module (Fig. 1). Recently, a third zinc binding domain (Zb3) has been identified [16, 17] which can bind DNA and seems not only to be critical for the DNA-dependent catalytic activity of PARP-1, but also involved in modulating chromatin structure. Indeed, Zb3 mutations in PARP-1 gene revealed a defect in the ability of PARP to compact chromatin. An internal automodification domain contains a BRCA1 C-terminal domain (BRCT) (shared by many DNA damage repair and cell cycle checkpoint proteins—essential for mediating protein–protein interactions) and three lysines that can be targeted for automodification. A catalytic domain is located at the C-terminus of PARP-1 and contains a region named PARP "signature," a highly conserved region in the PARP superfamily responsible for NAD^+ binding. In addition, the C-terminus also bears a WGR domain named after the highly conserved amino acid sequence in the motif (Trp, Gly, Arg) with an unknown function, which is also found in a variety of polyA polymerases. PARP-1, PARP-2, and PARP-3 share conserved WGR and catalytic domains. Interestingly, differing from PARP-1, the other two PARPs that can be activated by DNA damage do not contain the same DNA-binding module: Whereas PARP-2 contains a SAF/Acinus/PIAS (SAP) DNA binding domain, the DNA-binding domain of PARP-3 has not been characterized [6].

PARP-1 and PARP-2 are recognized as molecular sensors of SSBs and DSBs in vivo. The synthesis of pADPr chains is considered one of the earliest events of the DNA damage response as it occurs within seconds [3]. Besides the direct covalent modification on glutamate, aspartate, or lysine residues of various target proteins, some proteins have been elegantly shown to have a high affinity for the free polymer itself. In fact, it has been argued that strong noncovalent binding of PARP or other proteins to pADPr rather than covalent modification [18] affects protein function and/or localization. Consequently, recent progress has been made in defining specific sites for pADPr-attachment on target proteins [19–21]. Noncovalent binding of proteins to pADPr can be through at least four different PAR-binding motifs. One such motif was identified by our group and is characterized by a sequence of alternating basic and hydrophobic amino acids [22, 23]. Two other PAR-binding motifs have been described—the macrodomain and the PAR-binding zinc finger (PBZ) [24]. Only very recently a fourth type of polymer binding domain has been reported: The E3-ubiquitin ligase RNF146 contains a Trp-Trp-Glu (WWE) motif that is binding pADPr [25, 26]. Interestingly, this WWE domain has been found in various PARPs [27].

As mentioned above, PARP-1 is a molecular sensor of DNA strand breaks and the large size and negative charge of the polymer (which exceeds the charge density of DNA about two times) generated upon activation is playing a key role in the spatial and temporal organization of the DNA damage response. The in vivo half-life of the polymer generated upon PARP activation is rather short (seconds to minutes) and tightly regulated by the catalytic reactions of poly(ADP-ribose) glycohydrolase (PARG) and possibly ADP-ribose hydrolase (ARH) 3, which are so far the only glycohydrolases known to degrade the polymer [28, 29]. The fact that PARG and ARH3 antagonize PARP activity and thereby detach the polymer from PARP-1 itself re-enables the latter protein to bind DNA and start a new round of DNA damage signaling. Although the half-life of the polymer is extremely short, its impact on the cellular energy level can be dramatic as PARP hyperactivation following severe DNA damage consumes substantial amounts of the cytosolic and nuclear NAD^+ (and ATP) pool and thereby can result in cell death [30].

Interestingly, the ability of PARP-1 to disrupt and open chromatin structure by PARsylating histones (such as H1 and H2B) and destabilizing nucleosomes has been one of the earliest functions described for the proteins [31–33]. By disrupting the chromatin structure, DNA repair factors can gain access to a DNA damage site. Recent publications demonstrated that a variety of proteins implicated in DNA repair are recruited in a pADPr-dependent manner to DNA single or double strand breaks [34]. For instance, the Ataxia telangiectasia-mutated (ATM) protein is recruited to DNA DSBs in a way that is depending on polymer synthesis [34].

2 Roles of PARP-1 in Base Excision Repair

The role of PARP-1 in the repair of single-strand DNA breaks by base excision repair (BER) became already evident 30 years ago [35] and has since then been well examined by several investigators [36, 37]. Two Nature publications in 2005, from the Helleday and Ashworth groups, have revolutionized the understanding of PARP inhibitors in the context of DNA repair [38, 39]: The observation of antitumor effects of PARP inhibitors in a HR-deficient background has been explained as result from the disability of PARP-1 to respond to endogenous DNA damage through

BER [38]. However, the question whether SSBs increase after PARP inhibition is still matter of ongoing debates [40, 41]. Moreover, a lack of XRCC1 (another BER protein) in BRCA2 deficient cells (and thus deficient in HR) does not show the same effect as PARP inhibition, questioning the original explanation for increased sensitivity of HR-deficient cells by PARP inhibition. Even though it is well accepted that PARP-1 is implicated in BER, its exact role remains controversial: pADPr itself or automodified PARP-1 is said to be necessary for the recruitment of XRCC1, which further leads to the recruitment of polymerase β and DNA ligase III [42–44]. Although PARP-1 seems to attract SSB repair proteins, it seems not to be essential for SSB repair itself as PARP-1$^{-/-}$ knockout mice for example do not show any early onset of tumor formation [14]. It has been recently suggested that PARP inhibitors inhibit rather than trap PARP on the SSB intermediate which is formed during BER, thereby preventing accurate repair [40]. It is also well accepted that poly(ADP-ribosyl)ation of PARP-1 and histones due to the negative charge of the polymer leads to their dissociation from the DNA which further promotes local chromatin relaxation [45]. Consequently, one could argue that this alone can facilitate the assembly of repair proteins at the break site emphasizing a passive role for PARP-1 in BER. In association with PARP-1, PARP-2 has been implicated in BER through its ability to interact with XRCC1, DNA polymerase β and DNA ligase III. Whereas PARP-1 seems to affect early steps of BER, PARP-2 seems to be involved later in the process [46].

3 Double-Strand Break Repair by Homologous Recombination

Several lines of evidence have accumulated in the past years for a role of PARP-1 in the cellular response to DNA DSB repair. PARP-1 deficient cells are hypersensitive to DSB-inducing agents but most notably to camptothecin [47]. This phenotype is also observed in PARP-1(–/–) chicken DT40 mutants [48]. Camptothecin blocks topoisomerase-I in a state where it is covalently linked to nicked DNA. The resulting protein-DNA cross links are DNA replication and transcription blocks. Replication forks stalling at these lesions result in the formation of DNA DSBs that are repaired by HR [49]. HR can occur due to an availability of long sequence homologies in the sister chromatid after DNA replication. As the donor sequence used for HR is usually the sister chromatid, one of its key features is the preservation of the genetic material. However, the donor sequence might as well be another homologous region with consequences as deletions, inversions, or loss of heterozygosity [50]. Whereas NHEJ functions throughout the cell cycle, HR takes mainly place in S/G2 phase due to its necessity for a homolog template [51, 52].

HR is suggested to be initiated by MRE11-RAD50-NBS1 (MRN), CtIP, Exo1, DNA2, and BLM [53] in mammals, with 5′-3′ end resection to yield a 3′ single-stranded (ss) DNA overhang which is capable of invading duplex DNA containing a homologous sequence [54, 55] (Fig. 2). Interestingly, PARP-1 has been put in the context of MRN recruitment as it has been clearly demonstrated that PARP-1 can

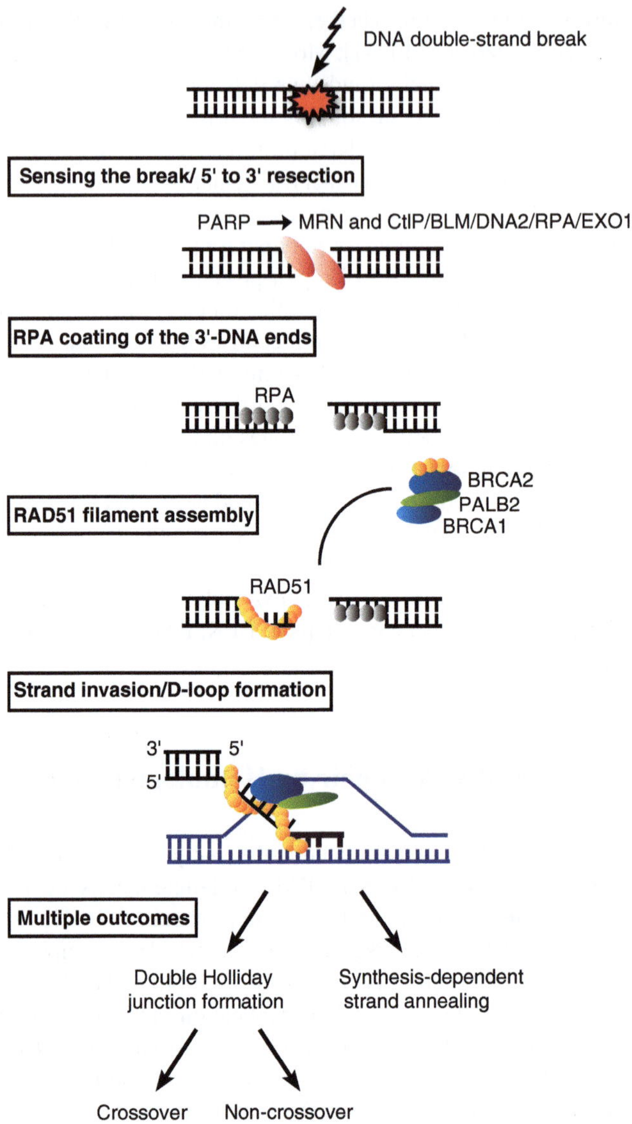

Fig. 2 Simplified overview of the homologous recombination (HR) repair pathway. Subsequent to DNA damage, the MRN complex (and associated resection machineries) binds and resects free DNA ends to create 3′ overhangs which are then bound by RPA. A complex of BRCA1, PALB2 and BRCA2 mediates the replacement of RPA by RAD51, which leads to the formation of the RAD51 filament coating the 3′ overhang. BRCA1/PALB2/BRCA2 then activates RAD51 to promote the invasion of an undamaged template in a step called strand invasion/ D-loop formation. Resolving of the D-loop structure can occur through synthesis-dependent strand annealing or double Holiday junction formation, generating either cross-over or non-cross-over products in the latter case

mediate the initial accumulation of the MRN complex to DSBs independent of γ-H2AX and MDC-1 [34]. This might have an implication in HR but also on a backup pathway of NHEJ (as discussed later in the text).

The replication protein A (RPA) has a high affinity for 3′-ssDNA tails and therefore binds to the newly generated 3′-ssDNA-overhang, a process that normally inhibits RAD51 loading and HR. HR mediators such as BRCA2 [56] and PALB2 [57] are helping to overcome that inhibition and lead to a displacement of RPA by RAD51 [58]. RAD51 itself, a DNA-dependent ATPase which is homolog to the bacterial RecA protein, is forming nucleoprotein filaments with DNA in a presynaptic step. RAD51 is recruited to DSBs in mammalian cells through BRCA2. Both, BRCA1 and 2 have been elegantly shown to be absolutely necessary for the HR reaction [59, 60] and there are several studies putting PALB2 (also known as FANCN) in the center of the BRCA1-BRCA2 complex [61, 62]. DSS1, a 70 amino acid protein, has been shown to be crucial for Rad51 foci formation as well and presumably for HR in mammalian cells [63]. A role for PARP-1 in that step of HR has been suggested to be rather of a regulatory nature than through a direct involvement in the actual mechanism: RAD51 foci are not only still forming in response to hydroxyurea in PARP-$1^{-/-}$ cells, but their number is also increasing in a PARP-1 deficient background [64]. In line with the latter finding it has been shown that in a PARP-1 deficient background (PARP-1 null MEFs) the spontaneous frequency of RAD51 foci is clearly enhanced [65]. Interestingly, pADPr, the product of catalytically active PARP, has been detected at HU-induced RPA foci raising the possibility that PARP-1 might for example prevent RAD51 from loading [66].

The following synaptic step is characterized by invasion of a homologous sequence to generate a D-loop structure (Fig. 2). Therewith the Rad51-ssDNA complex is binding to a complementary ssDNA region within the homologous duplex. Once formed, the D-loop structure has multiple fates: In the double-strand break repair (DSBR) model, the 3′ invading end from the broken chromosome is used to prime DNA synthesis templated by the donor duplex, whereas the other end of the break is presumably captured by the displaced strand from the donor duplex (D-loop) and is used to prime a second round of leading strand DNA synthesis. Therewith a so-called double Holliday Junction (dHJ) intermediate is formed that can, after branch migration and fill-in of the ssDNA, be resolved to form cross-over or non-cross-over products. In a second model called synthesis dependent strand annealing (SDSA), the invading strand that has been extended by DNA synthesis is displaced and anneals to complementary sequences exposed by 5′-3′ resection of the other side of the break. The remaining gaps can subsequently be filled in by newly synthesized DNA or by ligating the nicks [67]. SDSA will result only in non-cross-over products.

Collectively, there are several lines of evidence that PARP-1 regulates HR. PARP$1^{-/-}$ DT40 mutants showed more than threefold reduction in gene conversion [48]. Interestingly, the deletion of KU in PARP$1^{-/-}$ DT40 mutants completely reversed this phenotype suggesting that KU has a suppressive effect on HR. On the other hand, PARP-1 has been suggested to rather prevent HR, as the absence of

PARP-1 results in an increase of spontaneous somatic HR events in vivo [65]. PARP-1 also affects replication fork progression on damaged DNA. Indeed, fork progression is not slowed down in PARP1$^{-/-}$ DT40 cells treated with camptothecin. As fork slowing is correlated with the proficiency of HR, it implicates PARP-1 in the regulation of HR during DNA replication [68]. Additionally, by using the DNA fiber assay, Thomas Helleday and colleagues were able to show that PARP-1 is important for replication fork restart after blocking after HU treatment [66].

4 DNA Double-Strand Break Repair Through Nonhomologous End Joining

The repair of DSBs by HR has been demonstrated in practically all organisms examined from bacteria, yeast to human and seems to be conserved throughout evolution. Being described as a rather "error-free" pathway that is faithfully restoring genetic information it came as a big surprise to the DNA damage field that the major DSB repair pathway in higher eukaryotes is of a kind that does not rely on a homologous template but restores molecular integrity irrespective of the DNA sequence information. In nondividing haploid organisms or in diploid organisms that are not in the S-phase, a homologous template is not available for homology directed repair, setting the stage for a repair mechanism not relying on template homology, called NHEJ. The latter DSB repair pathway is effective throughout the cell cycle, but of particular importance during G0-, G1 and the early S-phase of cells. DNA DSB ends are often the result of damage to the sugar-phosphate backbone and/or the bases of the terminal nucleotides that have to be removed or processed prior to the religation step, explaining the fact that NHEJ is often mutagenic.

The most striking characteristic of the NHEJ pathway might be its high flexibility in terms of its templates, proteins involved and possible outcomes. The enzymes of the NHEJ pathway exhibit a remarkable tolerance concerning the DNA end substrate configurations they can act on. Different from other more distinct repair pathways, NHEJ enzymes act iteratively. Most of them can function independent of one another. As other repair pathways, NHEJ requires proteins that bring the ends in close proximity, nucleases/polymerases to process unligatable DNA ends and a ligase to restore integrity of the DNA strands [69]. From studies in which researchers investigated the status of Ku and DNA-PK$_{cs}$ in cell lines that are sensitive to ionizing radiation it became evident by their absence that these two proteins are implicated in NHEJ [70].

The generally accepted model of the "classical" NHEJ pathway is initiated with the heterodimeric complex of Ku70/Ku80 that binds to both ends of a broken DNA molecule (Fig. 3). This Ku-DNA complex acts presumably as a scaffold needed for the recruitment of DNA-PK$_{cs}$, which then functions as a molecular "bridge" between the two broken ends [71, 72]. Other than the Ku70/Ku80 complex, the association of Ku70/80 to the DNA-PK$_{cs}$ is transient and most likely stimulated by

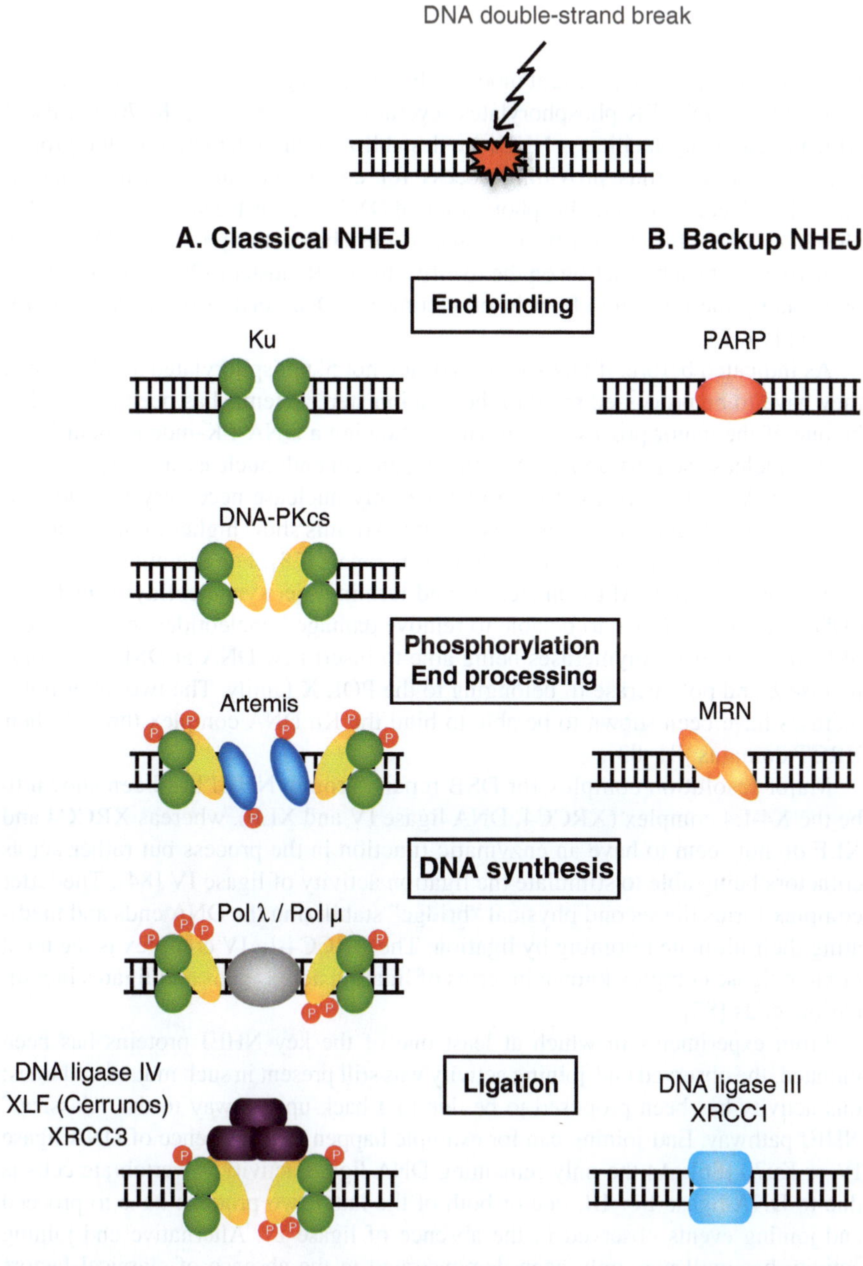

Fig. 3 DNA double-strand break repair through nonhomologous end joining (NHEJ). (**a**). The classical NHEJ pathway is initiated with Ku70/80 binding to the free DNA ends. The subsequent recruitment of the catalytic subunit of DNA-PK leads to the assembly of the end-bridging DNA-PK complex. DNA-PK then phosphorylates many proteins including Ku70 and itself. This loosens the DNA-PK DNA-binding which gives access to end processing proteins (such as Artemis/ PNK/ APLF/ TdT). After a fill-in of missing nucleotides by polymerase λ and μ the ends are joined by DNA ligase IV in a complex with its accessory factors (XRCC4 and XLF). (**b**) In the absence of or in competition to Ku70 it has been shown that PARP-1 can bind free DNA ends. Ends might further be processed by the MRN complex prior to a ligation by DNA ligase III/XRCC1

free DNA ends [73]. In a current model, it has been suggested that upon recruitment in a manner, DNA-PK phosphorylates several proteins including Ku70 and itself, which presumably facilitates NHEJ by destabilizing the interaction of the protein itself with DNA, thus providing access for end processing enzymes such as Artemis. Whereas the autophosphorylation of DNA-PK$_{cs}$ on the six-residue ABCDE cluster (T2609 cluster) has been shown to destabilize the protein DNA-binding properties, a phosphorylation on the five-residue PQR cluster (S2056), in return has presumably the opposite effect in protecting the DNA ends from excessive processing [74, 75].

As indicated before, if DNA DSB ends are not 5′ phosphorylated and ligatable, they have to be processed prior to the ligation step. Artemis has been revealed to be one of the major processing enzymes, showing a DNA-PK-independent 5′- to 3′- exonuclease activity and a DNA-PK-dependent endonuclease activity [76, 77]. However Artemis does not seem to be the only nuclease necessary for end-processing in DNA DSB repair, as cells lacking Artemis show higher radiosensitivity but do not have major defects in DNA DSB repair [78]. For example polynucleotide kinase (PNK), APLF nucleases and terminal deoxynucleotidyl transferase (TdT) have been shown to be able to remove damaged nucleotides in the context of NHEJ [79, 80]. Polymerases being able to insert new DNA at DSBs are polymerase λ and polymerase μ, belonging to the POL X family. The two latter polymerases have been shown to be able to bind the Ku:DNA complex through their BRCT domains [81–83].

Major resolution complex for DSB repair through NHEJ has been shown to be the X4-L4 complex (XRCC4, DNA ligase IV and XLF), whereas XRCC4 and XLF do not seem to have an enzymatic function in the process but rather act as cofactors being able to stimulate the ligation activity of ligase IV [84]. The latter complex forms the second physical "bridge" stabilizing the DNA ends and mediating their ultimate rejoining by ligation. The XRCC4-ligIV complex is the most flexible ligase complex known in terms of ligating across gaps and ligates incompatible ends [85].

From experiments in which at least one of the key NHEJ proteins has been mutated, the observed end-joining activity was still present in such mutant cell lines; this activity has been proposed to be due to a back-up pathway to the "classical" NHEJ pathway. End-joining can for example happen in the absence of DNA ligase IV or Ku70 [86]. As the only remaining DNA ligase activity in vertebrate cells is due to DNA ligase I or III, one or both of the latter two proteins have to proceed end-joining events observed in the absence of ligase IV. Alternative end-joining activity has until now only been demonstrated in the absence of classical factors therewith in the absence of the "classical" NHEJ, indicating an actual backup rather than a coexisting alternative pathway [87]. However the possibility that the NHEJ happening in the absence of Ku70 and ligase IV, can act alternatively to the classical pathway has not yet been disproven. From in vivo experiments in *S. cerevisiae* and mammals it has been elegantly shown that the variation of the ligation product is diminished as terminal microhomology occurs [88].

Besides the key factors described above, there have been other proteins shown to have an impact on the NHEJ reaction. Interestingly, the MRN complex which is known to coordinate DNA DSB repair by HR has recently been shown to promote efficient NHEJ in a $XRCC4^{+/+}$ and $XRCC4^{-/-}$ background in mice embryonic stem cells [89]. As accessory factors for the ligase reaction through its ability to interact with XRCC4, Polynucleotide kinase (PNK), aprataxin (APTX) and aprataxin- and PNK-like factor (APLF) have been identified [90]. Interestingly, PARP-3 has been suggested very recently to accelerate DNA ligation during NHEJ in the context of APLF [12].

The affinity of PARP-1 for a blunt ended and 3′ single-base overhang DSBs has been shown to be greater than the one of DNA-PK, with a fourfold lower affinity of PARP-1 for SSBs compared to blunt-ended DSBs [91]. Also PARP-1 has been demonstrated to directly interact with Ku proteins in vitro and in vivo, whereas Ku70, Ku80 and DNA-PKcs are able to bind pADPr [23]. PARP-1's PARylation of Ku leads to a decreased binding to DSBs [92]. Moreover, several studies implicated PARP-1 functionally in NHEJ: PARP-1 and Ku80, both being highly abundant in the cell, have been shown to compete for free DNA ends in vitro presumably through two distinct NHEJ pathways. Whereas the Ku complex is one of the key factors for the classical NHEJ pathway, PARP-1 seems to also interact with ligase III in the backup pathway [93–95].

5 Regulation of the DNA DSB Repair Pathway Choice (I Suggest to Rephrase this Title): Collaboration or Competition?

Several factors are channeling the DSB repair pathway choice between NHEJ and HR. It is generally accepted that the cell-cycle phase is one of them. Early studies in vertebrates showed that NHEJ-deficient *scid* (carrying a loss-of-function mutation in DNA-PKcs) cells and *Ku70*$^{-/-}$ chicken DT40 cells were hypersensitive to IR only in G1 and early S-phase whereas HR-defective $Rad54^{-/-}$ cells were IR sensitive in late S/G2 phase [96]. The Cdk1 kinase has recently been shown to have control over the key recombination steps giving an elegant explanation for the fluctuating HR efficiency throughout the cell cycle [97]. Being at the same time one of the main engines for the cell cycle, Cdk1 would be an excellent tool to control the DSB repair pathway choice. Indeed a recent publication suggests that HR and NHEJ are oppositely affected by Cdk1 activity: Whereas HR is activated, NHEJ seems to be repressed [98]. Moreover the level of several critical HR proteins (BRCA1, Rad51/52) has been shown to increase from S to G2 phase and that steps of HR are activated by CDKs [99] suggesting another potential for regulating the pathway choice through the level of proteins expressed for the corresponding pathway. A similar observation has been made for the protein level of DNA-PK [100].

The nature of the DNA lesion plays an additional role to the choice of DSB repair pathway: RAG-mediated DSBs during V(D)J-recombination are certainly repaired through NHEJ [101] whereas Spo11-mediated DSBs generated during meiosis for instance will be repaired by HR [102]. Besides the key players in HR and NHEJ it has recently been shown that ~ 15–20% of ionizing irradiation induced foci (IRIF) require additional proteins, such as ATM, Artemis, the MRN-complex, γ-H2AX, 53BP1, MDC1 and RNF8, RNF168 for repair, some of them being implicated in both DSB repair pathways [103]. As an example, 53BP1 has been implicated in NHEJ [104] whereas 53BP1 deficiency rescues HR in a BRCA1 deficient background by a mechanism dependent on ATM-mediated resection. Interestingly, loss of 53BP1 does not complement the loss of BRCA2, which might be explained by genetic studies that put BRCA2 more downstream in HR in a process following end-resection [105, 106].

Moreover, the complexity of chromatin may influence repair pathway choice as it has recently been shown that X-ray induced DSBs located in close proximity to heterochromatin predominantly use HR for repair [107]. Especially the distance of ionizing radiation-induced foci to heterochromatin and the ATM-dependent phosphorylation of Kap-1 which promotes chromatin relaxation seem to somehow affect repair [108].

An important regulatory step involved in pathway choice is the process of DSB resection, comprising the 5′- to-3′ nucleolytic processing of DNA ends by the MRN complex in conjunction with auxiliary factors including CtIP, RECQ helicases, Exo1 and DNA2, being necessary for HR but not for NHEJ. An observation suggesting that competition exists between the two major DSB repair pathways is given by the fact that NHEJ mutants (e.g. Ku70 deficient cells) that have enhanced end resection show increased HR whereas mutants with decreased end resection (e.g., Sae2/CtIP) have increased NHEJ. Possibly, since Ku70 binds DNA ends, it thereby prevents the initial step of HR, the end resection. Surprisingly, Ku depletion in chicken cells actually leads to an overall increased resistance to ionizing irradiation during late S/G2 phase which can be interpreted as Ku interfering with HR under normal conditions in the latter cell cycle phases [109]. Additionally, impairing DNA-PK from binding to a DSB end dramatically promotes the initiation step of HR [108]. Interestingly, from double mutant analysis for NHEJ and HR components it is suggested that the concomitant loss of a protein involved in HR and a protein involved in NHEJ results in a more severe phenotype than one would expect from loss of either single pathway [110], promoting rather collaboration of the two pathways.

Interestingly, in a study that highlighted rather competition than collaboration between the major DSB repair pathways it has been elegantly shown that PARP-1 is hyperactivated in BRCA2 deficient cells but this hyperactivation cannot be explained by an accumulation of DNA damage, which normally triggers PARP activity [111]. A new model has been suggested only very recently proposing that in a BRCA2 deficient background PARP-1 might prevent DSB repair through NHEJ, possibly by blocking DNA-PK and Artemis. By adding PARP inhibitors to HR deficient cells, error-prone NHEJ is promoted and the unrestricted NHEJ could then induce genomic instability and eventual lethality [112].

Notably, the opposite effect to PARP inhibition has been described for 53BP1 in a BRCA1 negative background: By depletion of 53BP1 ATM-dependent processing of DNA ends is restored which can generate single-stranded DNA which is competent for HR. Thus, the loss of 53BP1 in a BRCA1 negative cell can overcome PARP inhibitor sensitivity [106, 113].

6 Conclusions

To summarize, more than 40 years of research in the PARP- and pADPr fields have uncovered implications in various layers of the DNA damage response to DNA DSBs: The initial processes starting with sensing the DSB and signaling of the latter in order to recruit other repair proteins to the damage site implies PARP-1 and the polymer generated at the damage site. Furthermore, an automodification of the protein leads to its detachment from the DNA which guaranties access for other proteins but also enables another round of damage signaling [114]. Interestingly, the polymer generated at the damage site has an important impact on the local chromatin structure due to its largely negative charge. By disrupting the chromatin structure surrounding the damage site, access to the DNA is facilitated [4].

Besides PARPs implication in sensing and signaling of DNA damage and a role in BER, first lines of evidence have been given that even the choice for the DSB repair pathway is influenced by PARP-1, as the protein seems to block DNA-PK_{cs} and therewith classical NHEJ [112]. At the same time PARP-1 itself has been shown to be involved in the backup-pathway of NHEJ [95] as well as suppressing HR, indicated by an increase of RAD51 foci in a PARP-1 deficient background [64]. PARP-3 on the other hand seems to interact with APLF in NHEJ [12].

Taken together, PARPs are multifunctional regulators of the DNA damage response, expanding the current model of action for PARP inhibition in HR-deficient cancer cells. A mechanism called synthetic lethality explains the original model, meaning that two genetic lesions together lead to cell death whereas a defect in only one of these genes does not. In BRCA1- or BRCA2-deficient cancer cells for example where HR is hampered, the cytotoxic effect of PARP inhibitors has been originally suggested to be due to the cells inability to overcome SSBs by BER, which can further degenerate during replication to form DSBs. These DSBs can in healthy cells but not in HR-deficient cancer cells be repaired by HR [115] (Fig. 4a). This view was recently challenged, mostly because it was very difficult to detect increased SSBs after PARP inhibition [111]. The current view involves the aberrant activation of NHEJ, rather than inhibition of BER by PARP inhibitors in HR-deficient cells, leading to genomic instability and cell death [112] (Fig. 4b). Hence, even though PARP inhibitors have been put with widespread enthusiasm into clinical trials, the exact molecular effects are still debated and under investigation at the cellular level.

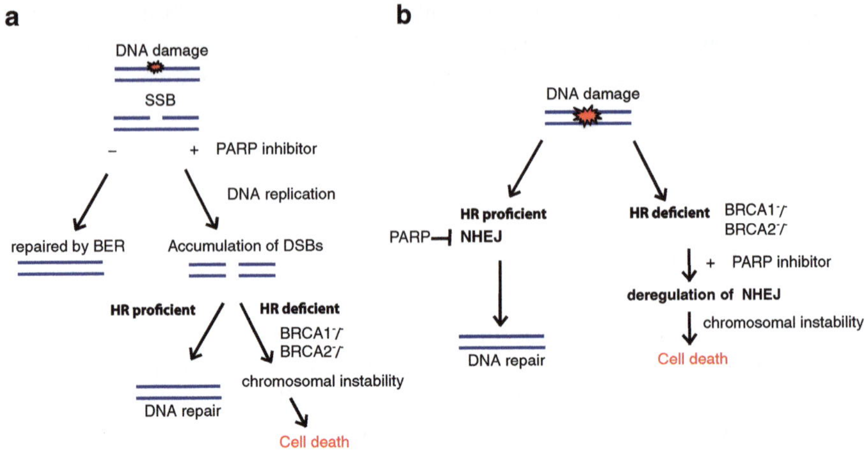

Fig. 4 Models explaining the lethality of HR-deficient cells with PARP inhibitors. (**a**) The synthetic lethality pathway model based on a deficiency in single-strand break repair. (**b**) Model based on error-prone NHEJ. Details are given in the text

How these inhibitors work in the appropriate clinical context still remains elusive. Hence, the PARP field awaits many scientific surprises with fundamental and clinical relevance.

Acknowledgments Research from the authors has been supported by the CIHR and the Cancer Research Society. G.P. is a Canada Research Chair in Proteomics and M.L. and J.-Y.M. are FRSQ Senior Investigators JK is a FQRNT scholar.

References

1. Moynahan ME, Jasin M (2010) Mitotic homologous recombination maintains genomic stability and suppresses tumorigenesis. Nat Rev Mol Cell Biol 11(3):196–207
2. Vilenchik MM, Knudson AG (2003) Endogenous DNA double-strand breaks: production, fidelity of repair, and induction of cancer. Proc Natl Acad Sci USA 100(22):12871–12876
3. Ciccia A, Elledge SJ (2010) The DNA damage response: making it safe to play with knives. Mol Cell 40(2):179–204
4. Krishnakumar R, Kraus WL (2010) The PARP side of the nucleus: molecular actions, physiological outcomes, and clinical targets. Mol Cell 39(1):8–24
5. Benjamin RC, Gill DM (1980) ADP-ribosylation in mammalian cell ghosts. Dependence of poly(ADP-ribose) synthesis on strand breakage in DNA. J Biol Chem 255(21):10493–10501
6. Hottiger MO, Hassa PO, Luscher B, Schuler H, Koch-Nolte F (2010) Toward a unified nomenclature for mammalian ADP-ribosyltransferases. Trends Biochem Sci 35(4):208–219
7. D'Amours D, Desnoyers S, D'Silva I, Poirier GG (1999) Poly(ADP-ribosyl)ation reactions in the regulation of nuclear functions. Biochem J 342(Pt 2):249–268
8. Krishnakumar R, Gamble MJ, Frizzell KM, Berrocal JG, Kininis M, Kraus WL (2008) Reciprocal binding of PARP-1 and histone H1 at promoters specifies transcriptional outcomes. Science 319(5864):819–821

9. Caiafa P, Guastafierro T, Zampieri M (2009) Epigenetics: poly(ADP-ribosyl)ation of PARP-1 regulates genomic methylation patterns. FASEB J 23(3):672–678
10. Heeres JT, Hergenrother PJ (2007) Poly(ADP-ribose) makes a date with death. Curr Opin Chem Biol 11(6):644–653
11. Chambon P, Weill JD, Mandel P (1963) Nicotinamide mononucleotide activation of new DNA-dependent polyadenylic acid synthesizing nuclear enzyme. Biochem Biophys Res Commun 11:39–43
12. Rulten SL, Fisher AE, Robert I, Zuma MC, Rouleau M, Ju L, Poirier G, Reina-San-Martin B, Caldecott KW (2011) PARP-3 and APLF function together to accelerate nonhomologous end-joining. Mol Cell 41(1):33–45
13. Menissier de Murcia J, Ricoul M, Tartier L, Niedergang C, Huber A, Dantzer F, Schreiber V, Ame JC, Dierich A, LeMeur M et al (2003) Functional interaction between PARP-1 and PARP-2 in chromosome stability and embryonic development in mouse. EMBO J 22(9):2255–2263
14. de Murcia JM, Niedergang C, Trucco C, Ricoul M, Dutrillaux B, Mark M, Oliver FJ, Masson M, Dierich A, LeMeur M et al (1997) Requirement of poly(ADP-ribose) polymerase in recovery from DNA damage in mice and in cells. Proc Natl Acad Sci USA 94(14):7303–7307
15. Masutani M, Nozaki T, Nishiyama E, Shimokawa T, Tachi Y, Suzuki H, Nakagama H, Wakabayashi K, Sugimura T (1999) Function of poly(ADP-ribose) polymerase in response to DNA damage: gene-disruption study in mice. Mol Cell Biochem 193(1–2):149–152
16. Langelier MF, Servent KM, Rogers EE, Pascal JM (2008) A third zinc-binding domain of human poly(ADP-ribose) polymerase-1 coordinates DNA-dependent enzyme activation. J Biol Chem 283(7):4105–4114
17. Langelier MF, Ruhl DD, Planck JL, Kraus WL, Pascal JM (2010) The Zn3 domain of human poly(ADP-ribose) polymerase-1 (PARP-1) functions in both DNA-dependent poly(ADP-ribose) synthesis activity and chromatin compaction. J Biol Chem 285(24):18877–18887
18. Hassa PO, Hottiger MO (2008) The diverse biological roles of mammalian PARPS, a small but powerful family of poly-ADP-ribose polymerases. Front Biosci 13:3046–3082
19. Altmeyer M, Messner S, Hassa PO, Fey M, Hottiger MO (2009) Molecular mechanism of poly(ADP-ribosyl)ation by PARP1 and identification of lysine residues as ADP-ribose acceptor sites. Nucleic Acids Res 37(11):3723–3738
20. Haenni SS, Hassa PO, Altmeyer M, Fey M, Imhof R, Hottiger MO (2008) Identification of lysines 36 and 37 of PARP-2 as targets for acetylation and auto-ADP-ribosylation. Int J Biochem Cell Biol 40(10):2274–2283
21. Kanai M, Hanashiro K, Kim SH, Hanai S, Boulares AH, Miwa M, Fukasawa K (2007) Inhibition of Crm1-p53 interaction and nuclear export of p53 by poly(ADP-ribosyl)ation. Nat Cell Biol 9(10):1175–1183
22. Pleschke JM, Kleczkowska HE, Strohm M, Althaus FR (2000) Poly(ADP-ribose) binds to specific domains in DNA damage checkpoint proteins. J Biol Chem 275(52):40974–40980
23. Gagne JP, Isabelle M, Lo KS, Bourassa S, Hendzel MJ, Dawson VL, Dawson TM, Poirier GG (2008) Proteome-wide identification of poly(ADP-ribose) binding proteins and poly(ADP-ribose)-associated protein complexes. Nucleic Acids Res 36(22):6959–6976
24. Kleine H, Luscher B (2009) Learning how to read ADP-ribosylation. Cell 139(1):17–19
25. Zhang Y, Liu S, Mickanin C, Feng Y, Charlat O, Michaud GA, Schirle M, Shi X, Hild M, Al B et al (2011) RNF146 is a poly(ADP-ribose)-directed E3 ligase that regulates axin degradation and Wnt signalling. Nat Cell Biol 13(5):623–629
26. Andrabi SA, Kang HC, Haince JF, Lee YI, Zhang J, Chi Z, West AB, Koehler RC, Poirier GG, Dawson TM et al (2011) Iduna protects the brain from glutamate excitotoxicity and stroke by interfering with poly(ADP-ribose) polymer-induced cell death. Nat Med 17(6):692–699
27. Schreiber V, Dantzer F, Ame JC, de Murcia G (2006) Poly(ADP-ribose): novel functions for an old molecule. Nat Rev Mol Cell Biol 7(7):517–528
28. Meyer-Ficca ML, Meyer RG, Coyle DL, Jacobson EL, Jacobson MK (2004) Human poly(ADP-ribose) glycohydrolase is expressed in alternative splice variants yielding isoforms that localize to different cell compartments. Exp Cell Res 297(2):521–532

29. Oka S, Kato J, Moss J (2006) Identification and characterization of a mammalian 39-kDa poly(ADP-ribose) glycohydrolase. J Biol Chem 281(2):705–713
30. David KK, Andrabi SA, Dawson TM, Dawson VL (2009) Parthanatos, a messenger of death. Front Biosci 14:1116–1128
31. Poirier GG, de Murcia G, Jongstra-Bilen J, Niedergang C, Mandel P (1982) Poly(ADP-ribosyl)ation of polynucleosomes causes relaxation of chromatin structure. Proc Natl Acad Sci USA 79(11):3423–3427
32. Mathis G, Althaus FR (1987) Release of core DNA from nucleosomal core particles following (ADP-ribose)n-modification in vitro. Biochem Biophys Res Commun 143(3):1049–1054
33. Huletsky A, de Murcia G, Muller S, Hengartner M, Menard L, Lamarre D, Poirier GG (1989) The effect of poly(ADP-ribosyl)ation on native and H1-depleted chromatin. A role of poly(ADP-ribosyl)ation on core nucleosome structure. J Biol Chem 264(15):8878–8886
34. Haince JF, McDonald D, Rodrigue A, Dery U, Masson JY, Hendzel MJ, Poirier GG (2008) PARP1-dependent kinetics of recruitment of MRE11 and NBS1 proteins to multiple DNA damage sites. J Biol Chem 283(2):1197–1208
35. Durkacz BW, Omidiji O, Gray DA, Shall S (1980) (ADP-ribose)n participates in DNA excision repair. Nature 283(5747):593–596
36. Allinson SL, Dianova II, Dianov GL (2003) Poly(ADP-ribose) polymerase in base excision repair: always engaged, but not essential for DNA damage processing. Acta Biochim Pol 50(1):169–179
37. Petermann E, Keil C, Oei SL (2005) Importance of poly(ADP-ribose) polymerases in the regulation of DNA-dependent processes. Cell Mol Life Sci 62(7–8):731–738
38. Bryant HE, Schultz N, Thomas HD, Parker KM, Flower D, Lopez E, Kyle S, Meuth M, Curtin NJ, Helleday T (2005) Specific killing of BRCA2-deficient tumours with inhibitors of poly(ADP-ribose) polymerase. Nature 434(7035):913–917
39. Farmer H, McCabe N, Lord CJ, Tutt AN, Johnson DA, Richardson TB, Santarosa M, Dillon KJ, Hickson I, Knights C et al (2005) Targeting the DNA repair defect in BRCA mutant cells as a therapeutic strategy. Nature 434(7035):917–21
40. Strom CE, Johansson F, Uhlen M, Al-Khalili Szigyarto C, Erixon K, Helleday T (2011) Poly (ADP-ribose) polymerase (PARP) is not involved in base excision repair but PARP inhibition traps a single-strand intermediate. Nucleic Acids Res 39(8):3166–3175
41. Pachkowski BF, Tano K, Afonin V, Elder RH, Takeda S, Watanabe M, Swenberg JA, Nakamura J (2009) Cells deficient in PARP-1 show an accelerated accumulation of DNA single strand breaks, but not AP sites, over the PARP-1-proficient cells exposed to MMS. Mutat Res 671(1–2):93–99
42. Leppard JB, Dong Z, Mackey ZB, Tomkinson AE (2003) Physical and functional interaction between DNA ligase IIIalpha and poly(ADP-Ribose) polymerase 1 in DNA single-strand break repair. Mol Cell Biol 23(16):5919–5927
43. El-Khamisy SF, Masutani M, Suzuki H, Caldecott KW (2003) A requirement for PARP-1 for the assembly or stability of XRCC1 nuclear foci at sites of oxidative DNA damage. Nucleic Acids Res 31(19):5526–5533
44. Prasad R, Lavrik OI, Kim SJ, Kedar P, Yang XP, Vande Berg BJ, Wilson SH (2001) DNA polymerase beta -mediated long patch base excision repair. Poly(ADP-ribose)polymerase-1 stimulates strand displacement DNA synthesis. J Biol Chem 276(35):32411–32414
45. Rouleau M, Aubin RA, Poirier GG (2004) Poly(ADP-ribosyl)ated chromatin domains: access granted. J Cell Sci 117(Pt 6):815–825
46. Schreiber V, Ame JC, Dolle P, Schultz I, Rinaldi B, Fraulob V, Menissier-de Murcia J, de Murcia G (2002) Poly(ADP-ribose) polymerase-2 (PARP-2) is required for efficient base excision DNA repair in association with PARP-1 and XRCC1. J Biol Chem 277(25): 23028–23036
47. Bowman KJ, Newell DR, Calvert AH, Curtin NJ (2001) Differential effects of the poly (ADP-ribose) polymerase (PARP) inhibitor NU1025 on topoisomerase I and II inhibitor cytotoxicity in L1210 cells in vitro. Br J Cancer 84(1):106–112

48. Hochegger H, Dejsuphong D, Fukushima T, Morrison C, Sonoda E, Schreiber V, Zhao GY, Saberi A, Masutani M, Adachi N et al (2006) Parp-1 protects homologous recombination from interference by Ku and Ligase IV in vertebrate cells. EMBO J 25(6):1305–1314
49. Pommier Y, Redon C, Rao VA, Seiler JA, Sordet O, Takemura H, Antony S, Meng L, Liao Z, Kohlhagen G et al (2003) Repair of and checkpoint response to topoisomerase I-mediated DNA damage. Mutat Res 532(1–2):173–203
50. Aguilera A, Gomez-Gonzalez B (2008) Genome instability: a mechanistic view of its causes and consequences. Nat Rev Genet 9(3):204–217
51. Takata M, Sasaki MS, Sonoda E, Morrison C, Hashimoto M, Utsumi H, Yamaguchi-Iwai Y, Shinohara A, Takeda S (1998) Homologous recombination and non-homologous end-joining pathways of DNA double-strand break repair have overlapping roles in the maintenance of chromosomal integrity in vertebrate cells. EMBO J 17:5497–5508
52. Rodrigue A, Lafrance M, Gauthier MC, McDonald D, Hendzel M, West SC, Jasin M, Masson JY (2006) Interplay between human DNA repair proteins at a unique double-strand break in vivo. EMBO J 25(1):222–231
53. Nimonkar AV, Genschel J, Kinoshita E, Polaczek P, Campbell JL, Wyman C, Modrich P, Kowalczykowski SC (2011) BLM-DNA2-RPA-MRN and EXO1-BLM-RPA-MRN constitute two DNA end resection machineries for human DNA break repair. Genes Dev 25(4):350–362
54. Sartori AA, Lukas C, Coates J, Mistrik M, Fu S, Bartek J, Baer R, Lukas J, Jackson SP (2007) Human CtIP promotes DNA end resection. Nature 450(7169):509–514
55. Nicolette ML, Lee K, Guo Z, Rani M, Chow JM, Lee SE, Paull TT (2010) Mre11-Rad50-Xrs2 and Sae2 promote 5′ strand resection of DNA double-strand breaks. Nat Struct Mol Biol 17(12):1478–1485
56. Sung P (2005) Mediating repair. Nat Struct Mol Biol 12(3):213–214
57. Buisson R, Dion-Cote AM, Coulombe Y, Launay H, Cai H, Stasiak AZ, Stasiak A, Xia B, Masson JY (2010) Cooperation of breast cancer proteins PALB2 and piccolo BRCA2 in stimulating homologous recombination. Nat Struct Mol Biol 17(10):1247–1254
58. Liu J, Doty T, Gibson B, Heyer WD (2010) Human BRCA2 protein promotes RAD51 filament formation on RPA-covered single-stranded DNA. Nat Struct Mol Biol 17(10):1260–1262
59. Moynahan ME, Chiu JW, Koller BH, Jasin M (1999) Brca1 controls homology-directed DNA repair. Mol Cell 4(4):511–518
60. Moynahan ME, Pierce AJ, Jasin M (2001) BRCA2 is required for homology-directed repair of chromosomal breaks. Mol Cell 7(2):263–272
61. Sy SM, Huen MS, Chen J (2009) PALB2 is an integral component of the BRCA complex required for homologous recombination repair. Proc Natl Acad Sci USA 106(17):7155–7160
62. Zhang F, Ma J, Wu J, Ye L, Cai H, Xia B, Yu X (2009) PALB2 links BRCA1 and BRCA2 in the DNA-damage response. Curr Biol 19(6):524–529
63. Yang H, Jeffrey PD, Miller J, Kinnucan E, Sun Y, Thoma NH, Zheng N, Chen PL, Lee WH, Pavletich NP (2002) BRCA2 function in DNA binding and recombination from a BRCA2-DSS1-ssDNA structure. Science 297(5588):1837–1848
64. Schultz N, Lopez E, Saleh-Gohari N, Helleday T (2003) Poly(ADP-ribose) polymerase (PARP-1) has a controlling role in homologous recombination. Nucleic Acids Res 31(17): 4959–4964
65. Claybon A, Karia B, Bruce C, Bishop AJ (2011) PARP1 suppresses homologous recombination events in mice in vivo. Nucleic Acids Res 38(21):7538–7545
66. Bryant HE, Petermann E, Schultz N, Jemth AS, Loseva O, Issaeva N, Johansson F, Fernandez S, McGlynn P, Helleday T (2009) PARP is activated at stalled forks to mediate Mre11-dependent replication restart and recombination. EMBO J 28(17):2601–2615
67. Mimitou EP, Symington LS (2009) Nucleases and helicases take center stage in homologous recombination. Trends Biochem Sci 34(5):264–272
68. Sugimura K, Takebayashi S, Taguchi H, Takeda S, Okumura K (2008) PARP-1 ensures regulation of replication fork progression by homologous recombination on damaged DNA. J Cell Biol 183(7):1203–1212

69. Lieber MR (2010) The mechanism of double-strand DNA break repair by the nonhomologous DNA end-joining pathway. Annu Rev Biochem 79:181–211
70. Ferguson DO, Alt FW (2001) DNA double strand break repair and chromosomal translocation: lessons from animal models. Oncogene 20(40):5572–5579
71. Ochi T, Sibanda BL, Wu Q, Chirgadze DY, Bolanos-Garcia VM, Blundell TL (2010) Structural biology of DNA repair: spatial organisation of the multicomponent complexes of nonhomologous end joining. J Nucleic Acids 2010:1–19. (doi:pii: 621695.)
72. Gottlieb TM, Jackson SP (1994) Protein kinases and DNA damage. Trends Biochem Sci 19(11):500–503
73. Yaneva M, Kowalewski T, Lieber MR (1997) Interaction of DNA-dependent protein kinase with DNA and with Ku: biochemical and atomic-force microscopy studies. EMBO J 16(16): 5098–5112
74. Meek K, Dang V, Lees-Miller SP (2008) DNA-PK: the means to justify the ends? Adv Immunol 99:33–58
75. Uematsu N, Weterings E, Yano K, Morotomi-Yano K, Jakob B, Taucher-Scholz G, Mari PO, van Gent DC, Chen BP, Chen DJ (2007) Autophosphorylation of DNA-PKCS regulates its dynamics at DNA double-strand breaks. J Cell Biol 177(2):219–229
76. Pawelczak KS, Turchi JJ (2010) Purification and characterization of exonuclease-free Artemis: Implications for DNA-PK-dependent processing of DNA termini in NHEJ-catalyzed DSB repair. DNA Repair (Amst) 9(6):670–677
77. Gu J, Li S, Zhang X, Wang LC, Niewolik D, Schwarz K, Legerski RJ, Zandi E, Lieber MR (2010) DNA-PKcs regulates a single-stranded DNA endonuclease activity of Artemis. DNA Repair (Amst) 9(4):429–437
78. Wang J, Pluth JM, Cooper PK, Cowan MJ, Chen DJ, Yannone SM (2005) Artemis deficiency confers a DNA double-strand break repair defect and Artemis phosphorylation status is altered by DNA damage and cell cycle progression. DNA Repair (Amst) 4(5):556–570
79. Chappell C, Hanakahi LA, Karimi-Busheri F, Weinfeld M, West SC (2002) Involvement of human polynucleotide kinase in double-strand break repair by non-homologous end joining. EMBO J 21(11):2827–2832
80. Mahaney BL, Meek K, Lees-Miller SP (2009) Repair of ionizing radiation-induced DNA double-strand breaks by non-homologous end-joining. Biochem J 417(3):639–650
81. Capp JP, Boudsocq F, Bertrand P, Laroche-Clary A, Pourquier P, Lopez BS, Cazaux C, Hoffmann JS, Canitrot Y (2006) The DNA polymerase lambda is required for the repair of non-compatible DNA double strand breaks by NHEJ in mammalian cells. Nucleic Acids Res 34(10):2998–3007
82. Capp JP, Boudsocq F, Besnard AG, Lopez BS, Cazaux C, Hoffmann JS, Canitrot Y (2007) Involvement of DNA polymerase mu in the repair of a specific subset of DNA double-strand breaks in mammalian cells. Nucleic Acids Res 35(11):3551–3560
83. Covo S, Blanco L, Livneh Z (2004) Lesion bypass by human DNA polymerase mu reveals a template-dependent, sequence-independent nucleotidyl transferase activity. J Biol Chem 279(2):859–865
84. Ahnesorg P, Smith P, Jackson SP (2006) XLF interacts with the XRCC4-DNA ligase IV complex to promote DNA nonhomologous end-joining. Cell 124(2):301–313
85. Deshpande RA, Wilson TE (2007) Modes of interaction among yeast Nej1, Lif1 and Dnl4 proteins and comparison to human XLF, XRCC4 and Lig4. DNA Repair (Amst) 6(10):1507–1516
86. Bennardo N, Cheng A, Huang N, Stark JM (2008) Alternative-NHEJ is a mechanistically distinct pathway of mammalian chromosome break repair. PLoS Genet 4(6):e1000110
87. Simsek D, Jasin M (2010) Alternative end-joining is suppressed by the canonical NHEJ component Xrcc4-ligase IV during chromosomal translocation formation. Nat Struct Mol Biol 17(4):410–416
88. Daley JM, Palmbos PL, Wu D, Wilson TE (2005) Nonhomologous end joining in yeast. Annu Rev Genet 39:431–451
89. Xie A, Kwok A, Scully R (2009) Role of mammalian Mre11 in classical and alternative nonhomologous end joining. Nat Struct Mol Biol 16(8):814–818

90. Koch CA, Agyei R, Galicia S, Metalnikov P, O'Donnell P, Starostine A, Weinfeld M, Durocher D (2004) Xrcc4 physically links DNA end processing by polynucleotide kinase to DNA ligation by DNA ligase IV. EMBO J 23(19):3874–3885
91. D'Silva I, Pelletier JD, Lagueux J, D'Amours D, Chaudhry MA, Weinfeld M, Lees-Miller SP, Poirier GG (1999) Relative affinities of poly(ADP-ribose) polymerase and DNA-dependent protein kinase for DNA strand interruptions. Biochim Biophys Acta 1430(1):119–126
92. Li B, Navarro S, Kasahara N, Comai L (2004) Identification and biochemical characterization of a Werner's syndrome protein complex with Ku70/80 and poly(ADP-ribose) polymerase-1. J Biol Chem 279(14):13659–13667
93. Wang M, Wu W, Rosidi B, Zhang L, Wang H, Iliakis G (2006) PARP-1 and Ku compete for repair of DNA double strand breaks by distinct NHEJ pathways. Nucleic Acids Res 34(21):6170–6182
94. Audebert M, Salles B, Calsou P (2004) Involvement of poly(ADP-ribose) polymerase-1 and XRCC1/DNA ligase III in an alternative route for DNA double-strand breaks rejoining. J Biol Chem 279(53):55117–55126
95. Mansour WY, Rhein T, Dahm-Daphi J (2010) The alternative end-joining pathway for repair of DNA double-strand breaks requires PARP1 but is not dependent upon microhomologies. Nucleic Acids Res 38(18):6065–6077
96. Lee SE, Mitchell RA, Cheng A, Hendrickson EA (1997) Evidence for DNA-PK-dependent and -independent DNA double-strand break repair pathways in mammalian cells as a function of the cell cycle. Mol Cell Biol 17(3):1425–1433
97. Ira G, Pellicioli A, Balijja A, Wang X, Fiorani S, Carotenuto W, Liberi G, Bressan D, Wan L, Hollingsworth NM et al (2004) DNA end resection, homologous recombination and DNA damage checkpoint activation require CDK1. Nature 431(7011):1011–1017
98. Zhang Y, Shim EY, Davis M, Lee SE (2009) Regulation of repair choice: Cdk1 suppresses recruitment of end joining factors at DNA breaks. DNA Repair (Amst) 8(10):1235–1241
99. Shrivastav M, De Haro LP, Nickoloff JA (2008) Regulation of DNA double-strand break repair pathway choice. Cell Res 18(1):134–147
100. Koike M, Awaji T, Kataoka M, Tsujimoto G, Kartasova T, Koike A, Shiomi T (1999) Differential subcellular localization of DNA-dependent protein kinase components Ku and DNA-PKcs during mitosis. J Cell Sci 112(Pt 22):4031–4039
101. Soulas-Sprauel P, Rivera-Munoz P, Malivert L, Le Guyader G, Abramowski V, Revy P, de Villartay JP (2007) V(D)J and immunoglobulin class switch recombinations: a paradigm to study the regulation of DNA end-joining. Oncogene 26(56):7780–7791
102. Cole F, Keeney S, Jasin M (2010) Evolutionary conservation of meiotic DSB proteins: more than just Spo11. Genes Dev 24(12):1201–1207
103. Riballo E, Kuhne M, Rief N, Doherty A, Smith GC, Recio MJ, Reis C, Dahm K, Fricke A, Krempler A et al (2004) A pathway of double-strand break rejoining dependent upon ATM, Artemis, and proteins locating to gamma-H2AX foci. Mol Cell 16(5):715–724
104. Nakamura K, Sakai W, Kawamoto T, Bree RT, Lowndes NF, Takeda S, Taniguchi Y (2006) Genetic dissection of vertebrate 53BP1: a major role in non-homologous end joining of DNA double strand breaks. DNA Repair (Amst) 5(6):741–749
105. Bouwman P, Aly A, Escandell JM, Pieterse M, Bartkova J, van der Gulden H, Hiddingh S, Thanasoula M, Kulkarni A, Yang Q et al (2010) 53BP1 loss rescues BRCA1 deficiency and is associated with triple-negative and BRCA-mutated breast cancers. Nat Struct Mol Biol 17(6):688–695
106. Bunting SF, Callen E, Wong N, Chen HT, Polato F, Gunn A, Bothmer A, Feldhahn N, Fernandez-Capetillo O, Cao L et al (2010) 53BP1 inhibits homologous recombination in Brca1-deficient cells by blocking resection of DNA breaks. Cell 141(2):243–254
107. Beucher A, Birraux J, Tchouandong L, Barton O, Shibata A, Conrad S, Goodarzi AA, Krempler A, Jeggo PA, Lobrich M (2009) ATM and Artemis promote homologous recombination of radiation-induced DNA double-strand breaks in G2. EMBO J 28(21):3413–3427
108. Shibata A, Conrad S, Birraux J, Geuting V, Barton O, Ismail A, Kakarougkas A, Meek K, Taucher-Scholz G, Lobrich M et al (2011) Factors determining DNA double-strand break repair pathway choice in G2 phase. EMBO J 30(6):1079–1092

109. Fukushima T, Takata M, Morrison C, Araki R, Fujimori A, Abe M, Tatsumi K, Jasin M, Dhar PK, Sonoda E et al (2001) Genetic analysis of the DNA-dependent protein kinase reveals an inhibitory role of Ku in late S-G2 phase DNA double-strand break repair. J Biol Chem 276(48):44413–44418
110. Mills KD, Ferguson DO, Essers J, Eckersdorff M, Kanaar R, Alt FW (2004) Rad54 and DNA Ligase IV cooperate to maintain mammalian chromatid stability. Genes Dev 18(11):1283–1292
111. Gottipati P, Vischioni B, Schultz N, Solomons J, Bryant HE, Djureinovic T, Issaeva N, Sleeth K, Sharma RA, Helleday T (2010) Poly(ADP-ribose) polymerase is hyperactivated in homologous recombination-defective cells. Cancer Res 70(13):5389–5398
112. Patel AG, Sarkaria JN, Kaufmann SH (2011) Nonhomologous end joining drives poly(ADP-ribose) polymerase (PARP) inhibitor lethality in homologous recombination-deficient cells. Proc Natl Acad Sci USA 108(8):3406–3411
113. Kass EM, Moynahan ME, Jasin M (2010) Loss of 53BP1 is a gain for BRCA1 mutant cells. Cancer Cell 17(5):423–425
114. Mortusewicz O, Ame JC, Schreiber V, Leonhardt H (2007) Feedback-regulated poly(ADP-ribosyl)ation by PARP-1 is required for rapid response to DNA damage in living cells. Nucleic Acids Res 35(22):7665–7675
115. Aly A, Ganesan S (2011) BRCA1, PARP, and 53BP1: conditional synthetic lethality and synthetic viability. J Mol Cell Biol 3(1):66–74

Cellular Protection Against the Antitumor Drug Bleomycin

Dindial Ramotar and Mustapha Aouida

1 Introduction

Bleomycin is a basic hydrophilic antibiotic isolated as a metal complex from the culture medium of the microorganism *Streptomyces verticillus* [1, 2]. Bleomycin comprises of several species differing only in the terminal amine moiety (Fig. 1), with bleomycin-A_2 being the most abundant form [2–5]. By the late 60s, substantial evidence had accumulated showing that bleomycin can diminish the growth of experimentally induced tumors in animal models, and dramatically decrease the size of human tumors [6–10]. It has been postulated that bleomycin mediates the cell killing by directly attacking the DNA [11, 12]. This notion rapidly gained support from subsequent independent studies showing that bleomycin triggers the induction of lysogenic phage in bacteria, a result of DNA damage, and induces mitotic recombination and mutations in many model systems including the budding yeast *Saccharomyces cerevisiae*, *Aspergillus*, and *Drosophila* [13–18]. Later studies also showed that bleomycin can induce micronuclei formation and chromosome aberrations in human lymphocytes [19]. The accumulated findings strongly suggest that bleomycin mediates its effect as a chemotherapeutic agent primarily by damaging the DNA [20–23]. However, additional studies showed that RNA is also damaged by bleomycin, raising the possibility that, besides DNA, RNA could be a major target [24].

D. Ramotar, Ph.D. (✉)
Maisonneuve-Rosemont Hospital Research Center, University of Montreal,
5415, Boul. de l'Assomption, Montreal, QC, Canada H1T 2M4
e-mail: dramotar.hmr@ssss.gouv.qc.ca

M. Aouida, Ph.D.
Biological and Environmental Sciences and Engineering Division,
Center for Desert Agriculture, 4700 King Abdullah University of Science and Technology,
Building 2 3231-WS15, P.Box: 2863, Thuwal 23955-6900, Saudi Arabia
e-mail: mustapha.aouida@kaust.edu.sa

L. Panasci et al. (eds.), *Advances in DNA Repair in Cancer Therapy*,
Cancer Drug Discovery and Development 72, DOI 10.1007/978-1-4614-4741-2_9,

© Springer Science+Business Media, LLC 2013

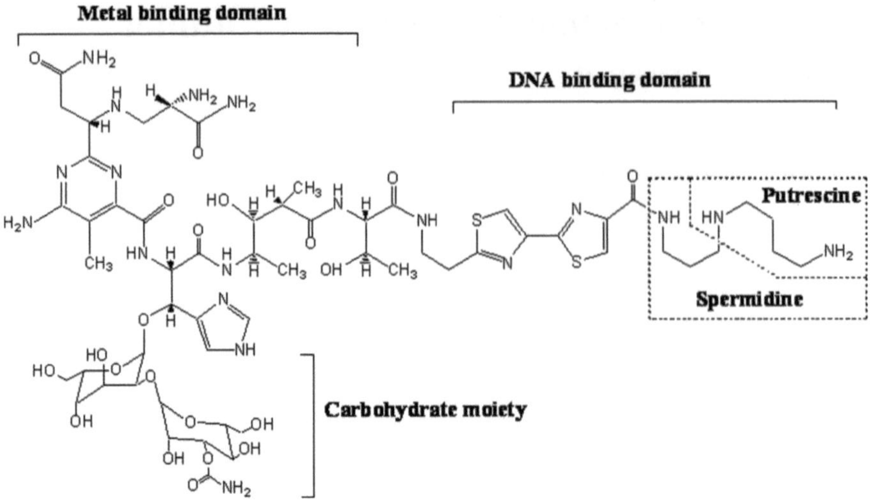

Fig. 1 Structure of the antitumor drug bleomycin depicting the three domains

Bleomycin is widely used in the clinic as a mixture (blenoxane), consisting primarily of the isomers bleomycin-A_2 and bleomycin-B_2, as well as several additional minor species including bleomycin-A_5 [4]. It is used only in combination therapy with a number of other antineoplastic agents such as etoposide [4, 25, 26]. Bleomycin is most effective against lymphomas, testicular carcinomas, and squamous cell carcinomas of the cervix, head, and neck [27, 28]. One useful property of bleomycin is that it does not appear to cause myelosuppression, a phenomenon that leads, for example, to the decrease of cells responsible for immune response, as compared to other cytotoxic antineoplastic drugs [28, 29]. Moreover, bleomycin is eliminated rapidly from the circulatory system by renal excretion. At least half of the drug is cleared from the blood within 2–4 h, except for patients with impaired renal function [29]. Like many other antitumor drugs, bleomycin also manifests clinical limitations. For example, at high doses (i.e., >400 U or ~235 mg), bleomycin can induce pulmonary fibrosis, a condition characterized as a diffuse disease of the lung parenchyma that can cause respiratory insufficiency leading to fatal hypoxemia [30, 31]. The exact mechanisms by which bleomycin induces pulmonary fibrosis is not known, but findings from several experimental animal models suggest that the onset of the disease is triggered by lipid peroxidation [32, 33]. Another common factor that limits the clinical application of bleomycin is tumor resistance [28]. So far, a clear mechanism has not yet emanated to account for tumor resistance towards bleomycin. While several possible mechanisms are likely to involve (i) decreased drug uptake, (ii) increased drug extrusion, (iii) enhanced repair of bleomycin-induced DNA lesions, and (iv) increased inactivation of bleomycin [34–38], recent studies provide strong evidence to support the former possibility [34–38]. This has been aided by the advent of fluorescently labeled bleomycin (F-BLM) and the identification of a transporter that mediates uptake of F-BLM. It seems logical that the existence of a transporter would serve as a key regulatory step to control drug entry and that defects in this process is likely to engender striking resistance towards this chemotherapeutic agent, which are elaborated in detail below.

2 Structure of Bleomycin

The structure of bleomycin consists of three functional domains, including a metal binding domain, a DNA binding domain, and the carbohydrate moiety (Fig. 1). The metal domain also binds to molecular oxygen, in addition to the minor groove of DNA. This domain is thus largely responsible for the anti-tumor activity of bleomycin. It has a flexible requirement for metal ions, as it is capable of binding to both redox-active transition metal ions such as iron and copper and non-redox active metal ions such as zinc, cobalt, and cadmium [39–44]. The metal ion plays two roles in bleomycin-induced genotoxicity, i.e., one is to facilitate contact between bleomycin and DNA, and the other is to activate oxygen such that a reactive radical species is generated [20, 40, 41, 45–47]. Among the metal ions, cobalt forms the most stable complex with bleomycin. Despite this, iron is the metal ion predominantly used in clinical preparations of bleomycin, as it enhances the production of DNA lesions [47, 48].

The function of the two other domains of bleomycin is not clearly established. The DNA binding domain bears a bithiazole group required for DNA binding and for sequence-selective DNA cleavage [22, 49, 50]. In some species of bleomycin, such as bleomycin-A_5, the DNA binding domain also contains the chemical composition of polyamines, and thus, this form of the drug is considered to be a polyamine analogue [51]. The role of the carbohydrate moiety of bleomycin is far more elusive. Removal of the carbohydrate moiety from bleomycin does not alter the resulting deglycobleomycin ability to cleave DNA, excluding a role for this region in incising DNA [52]. No additional studies have been conducted with deglycobleomycin to examine whether it is capable of entering cells, or causing cell killing.

To date several chemical modifications have been introduced to alter the structure of bleomycin, but so far none has led to a more potent antitumor activity without the ability to cause pulmonary fibrosis. Thus, enhancing the antitumor effect of bleomycin will likely rely on modulation of cellular molecules that would improve, for example, uptake of bleomycin or its interaction with nucleic acids in cancer cells.

3 Bleomycin Induced DNA Lesions

Bleomycin can enter into mammalian cells, through an active transport pathway (see below), where a fraction of the drug reaches the nucleus to inflict a narrow set of DNA lesions through a multistep process [53]. In the earliest events, bleomycin

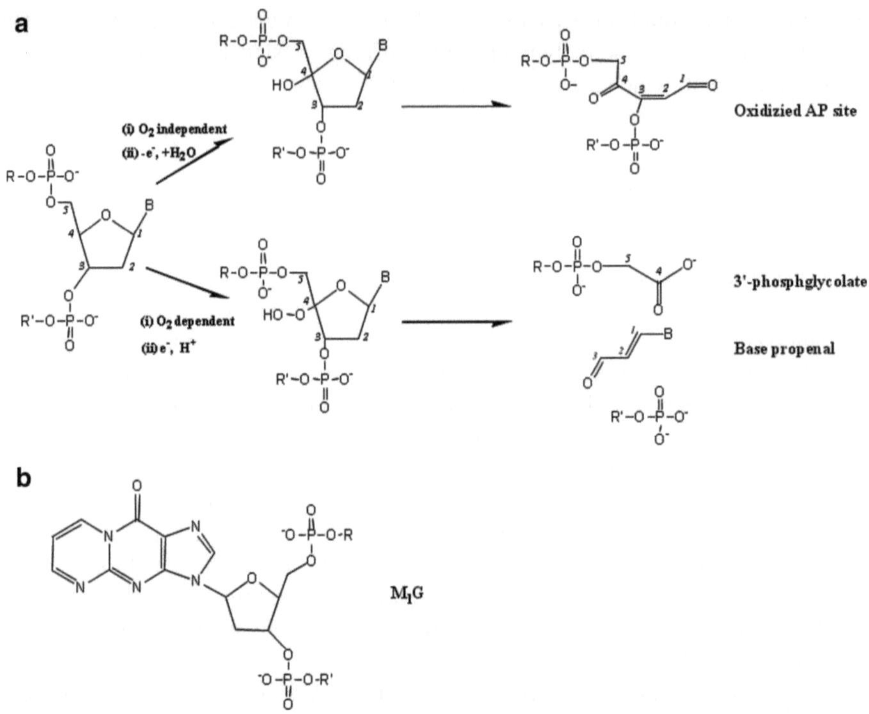

Fig. 2 Structure of bleomycin-induced DNA lesions. Production of the various types of bleomycin-induced lesions is dependent on oxygenation conditions. (**a**) In the absence of oxygen, bleomycin produces primarily oxidized apurinic/apyrimidinic (AP) site, while in the presence of oxygen it generates mostly DNA strand breaks, such as 3′-phosphoglycolate. (**b**) The M_1G lesion, 3-(2′-deoxy-β-D-erythro-pentofuranosyl)-pyrimido[1,2-a]purin-10(3H)-one, is produced by reaction of the base propenal with deoxyguanosine

binds to reduce iron (Fe II) and molecular oxygen followed by its conversion into an activated form [54]. The activated bleomycin (Blm-Fe(II)-O_2) complex then acts as an oxidant, abstracting a hydrogen atom from the 4′-carbon of deoxyribose to produce an unstable sugar carbon-radical and a single electron reduced form of activated bleomycin (Blm-Fe(III)-OH•), which can carry out multiple attacks on DNA [55–58]. The unstable sugar generated by activated bleomycin can be rearranged to generate at least four types of oxidative DNA lesions (Fig. 2). These lesions are structurally and chemically related to some of the lesions produced by ionizing radiation, and include the following: (i) *Oxidized* (*ketoaldehyde*) *apurinic/apyrimidinic* (*AP*) *sites*, where the entire base is lost, resulting in no template information for DNA polymerase [53, 59], (ii) *DNA single strand breaks* where the 3′-ends are terminated with a portion of the deoxyribose ring to form 3′-phosphoglycolate (3′-PG) which effectively blocks DNA synthesis [59, 60]. The remaining portion of the fragmented sugar, left attached to the base, exists in the free base propenal form, and exhibits a high propensity to undergo secondary reactions to form a variety of base adducts [59, 60], (iii) *Pyrimidopurinone of deoxyguanosine* (M_1G) is the most abundant base adduct produced when the malondialdehyde moiety of the propenal base reacts with guanine [61]. The M_1G lesion is also generated by aerobic metabolism and it is detected at levels of ~5,000 adducts/cell in normal human liver [62–64]. This lesion is mutagenic in bacterial test systems [62–64], and (iv) *Bi-stranded DNA lesions*, which are produced at certain sequences, such as CGCC, when the Fe.bleomycin complex induces an AP site on one strand, and directly opposes strand break on the complementary strand [44, 65–67]. This lesion requires a single activated bleomycin molecule, which binds to both strands of the duplex DNA [44]. The bi-stranded lesions can be converted to double strand breaks following spontaneous cleavage of the AP site by primary amines (e.g., histone amine) in vivo [65–67].

The extent of formation of the various bleomycin-induced lesions depends on the redox status of the cells [68–71]. In the presence of oxygen, bleomycin produces primarily DNA strand breaks, but under low oxygen tension it forms largely AP sites in the DNA [53, 59, 65, 72]. Thus, the redox state of the cells is likely to dictate the types of DNA lesions that are generated by bleomycin. These DNA lesions are also influenced by bleomycin concentrations. At high concentrations, bleomycin releases all four bases from DNA in the order of preference thymine > cytosine > adenosine > guanine [53, 56, 73]. At lower concentrations, bleomycin exhibits significant base sequence specificity. Although bleomycin cuts mixed sequence DNAs with a disposition for GC=GT>GA>>GG, it efficiently cleaves regions of (AT)n•(TA)n and hardly at (ATT)n•(TTA)n, (ATT)n•(AAT)n, (AC) n•(GT)n, and (A)n•(T)n raising the possibility that AT rich regions of the genome are more susceptible to lesions formed by bleomycin [74–76]. The structure of DNA also plays a role in the outcome of bleomycin-induced DNA lesions [77]. DNA that is pre-exposed to other DNA damaging agents, such as cisplatin, alters the pattern of lesions produce by bleomycin [78–80]. Thus, the clinical application of bleomycin together with other DNA damaging agents is likely to produce irreparable DNA lesions.

Several studies clearly demonstrate that bleomycin-induced DNA lesions are mutagenic [81–85]. For example, introduction of bleomycin-treated vectors into mammalian cells, followed by recovery, revealed that the vectors contain high levels of base substitutions and single-base deletions [81, 82]. The base substitutions are likely to be misincorporation of nucleotides by DNA polymerase at unrepaired oxidized AP sites, while the one-base deletions may arise from incorrect repair of bi-stranded DNA lesions [81, 82]. Thus, the normal cells of a cancer patient exposed to bleomycin must rely on enzymes to efficiently repair bleomycin-induced DNA lesions to prevent the production of lethal mutations that can lead to toxic side effects and secondary tumors. Likewise, tumor cells are likely to employ even more efficient DNA repair mechanisms to evade the genotoxic effects of bleomycin.

4 Bleomycin-Induced RNA Cleavage

Several reports demonstrated that bleomycin can also cleave many different RNAs including HIV-1 reverse transcriptase mRNA, transfer RNAs, ribosomal RNA, and RNA present in RNA•DNA heteroduplex [86–91]. Incision of RNA also occurs via an oxidative pathway reminiscent of the cleavage mechanism of DNA [92, 93]. Furthermore, RNA cleavage occurs preferentially at 5′-GU-3′ sequences similar to the site-specific 5′-GT-3′ incision observed in DNA [86, 92]. Besides these similarities, there are distinct differences between RNA and DNA with respect to cleavage with bleomycin. A notable difference is that not all RNA molecules, e.g., *Escherichia coli* $tRNA^{Tyr}$ and yeast mitochondrial $tRNA^{Asp}$, are cleaved by bleomycin [86, 92]. This observation led to the suggestion that bleomycin-induced cleavage of RNA is structure specific. Another key difference is that double stranded RNA is not cleaved by bleomycin [93]. Moreover, the extent of RNA cleavage is significantly less than that of DNA [86]. Finally, the cleavage of RNA, but not DNA, is impeded with as low as 0.5 mM Mg^{2+} ions [87]. Since Mg^{2+} ions are required to maintain most RNA structure and function, it is postulated that the Mg^{2+} ions bind to RNA at the same exact site that coincides with binding of bleomycin [87]. The exceptional selectivity for destruction of certain RNAs, even with excess non-substrate RNAs, led to the suggestion that at least one unique RNA species could be targeted for destruction by bleomycin during chemotherapy. However, the following findings stand against RNA being a therapeutic target: (i) most RNA molecules exist in multiple copies and that destruction of a few molecules is unlikely to cause cell death, unless bleomycin is able to target a specific essential RNA species present in extremely low abundance, (ii) RNA cleavage by bleomycin is inhibited by the physiological concentration of Mg^{2+} (2 mM), and (iii) cleavage of RNA is structure specific and occurs much slower than DNA [86, 87, 93]. As such, it can be inferred that DNA is the most likely target during bleomycin chemotherapy.

5 Other Cellular Targets

In addition to DNA and RNA, bleomycin can also attack the integrity of the cell wall (a complex structure composed mainly of glucans, mannoproteins, and chitin) of microbes. At high doses, or under prolonged exposure, bleomycin can create small incisions in the cell wall thereby exposing the protoplast [94, 95]. The protoplast is osmotically fragile and this can lead to plasma membrane rupturing and cell death [94, 95]. Because the sugar constituents of the cell wall have a stereochemistry at the C-5 position that is similar to the C-4 position of the deoxyribose moiety of DNA, it is believed that bleomycin destroys the cell wall via oxidative damage to the sugar [95, 96]. Another relevant target affected by bleomycin is the plasma membrane, which is believed to undergo lipid peroxidation, and this may constitute the initiation process of bleomycin-induced pulmonary fibrosis [32, 97, 98].

6 Prokaryotic and Eukaryotic Defense Mechanisms Against Bleomycin Toxicity

6.1 Cell Wall and Membrane Barriers

The highly reactive nature of bleomycin towards various cellular components, particularly DNA, suggests that organisms must employ multiple defense mechanisms to combat the deleterious effects of this drug. Some of these defense mechanisms include the barrier afforded by the cell wall and plasma membrane, proteins that bind and sequester bleomycin, and proteins that repair bleomycin-induced DNA lesions [95, 99, 100]. The contribution of each mechanism towards the protection against bleomycin toxicity is often determined by measuring the sensitivity of mutants to the drug. This type of analysis revealed that in the budding yeast *S. cerevisiae*, the cell wall appears to play a minor passive role in the protection against bleomycin toxicity. This is supported by the fact that some, and not all, cell wall-defective mutants displayed only a modest sensitivity towards bleomycin (Leduc, A and Ramotar, D., unpublished).

Previous report claimed the presence of a receptor protein that exists on the plasma membrane of mammalian and yeast cells that may mediate bleomycin internalization [99, 101]. This putative receptor (~250-kDa in size) was initially identified by its specific interaction with labeled cobalt-bleomycin complex [99]. However, no further study was undertaken to identify and characterize this plasma membrane protein, although it could hold the key to provide a rational explanation for why certain tumor types, and not others, can be reduced by bleomycin chemotherapy (see below). Aside from the prediction of a bleomycin-receptor, it is equally plausible that the plasma membrane may harbor a specific efflux pump to limit bleomycin uptake. So far, there is no direct evidence that any of the known drug efflux pumps has a role in expelling bleomycin from cells as part of a detoxification process, although a genetic approach is currently being used in this laboratory to exploit the yeast system to search for such an efflux pump.

6.2 Bleomycin Binding Proteins

The transposon Tn5, commonly used for insertion mutagenesis in many Gram-negative bacteria, was serendipitously discovered to harbor a gene *ble* that renders cells resistant to bleomycin [102–104]. Two other genes *Sa ble* and *Sh ble*, have been characterized and shown to encode proteins that are homologous to *Tn5 ble* [102, 105–107]. The *Sh ble* gene from *Streptoalloteichus hindustanus* encodes a 14-kDa protein that confers resistance to bleomycin by sequestering the drug [108, 109]. The X-ray crystal structure of Sh ble revealed that it consists of two halves that are identically folded despite no sequence similarity [109]. The structure further revealed that the Sh ble dimer binds to two molecules of bleomycin [109].

In vitro assays demonstrated that this protein prevents the action of bleomycin on DNA. At concentrations as low as 1 μM, bleomycin can completely degrade 0.2 μg of chromosomal, linear, or covalently closed circular (CCC) DNA within few minutes at ambient temperature, a process that is completely inhibited in the presence of a fivefold molar excess of the Sh ble protein [110]. It is likely that the related ble members may also function to sequester bleomycin, and possess no direct role in DNA repair as previously suggested [111, 112]. A ble-related protein is also present in the bleomycin producing strain *S. verticillus*, raising the possibility that the ble-related protein could have yet another role by sequestering bleomycin in *S. verticillus* for efficient transport to the exterior [113, 114]. To date, database searches reveal that eukaryotes do not possess the ble-related protein, and suggest that higher organisms may have evolved other mechanisms to mount a defense against bleomycin.

6.3 Bleomycin Hydrolase

Earlier studies demonstrated that bleomycin can be metabolically inactivated in normal and tumor tissues by an enzyme called bleomycin hydrolase, and that such inactivation may play a role in bleomycin resistance [115–117]. This is supported by the correlation that tissues with low levels of bleomycin hydrolase are usually sensitive to bleomycin, and tumor cells that acquire resistance to bleomycin possess higher levels of activity [35, 36, 38, 118, 119]. To better understand the role of bleomycin hydrolase, the enzyme was characterized and shown to be a thiol protease that hydrolyzes the β-aminoalanine amide moiety at the carboxyl terminus of bleomycin to generate the inactive deamido metabolite [35, 37, 118]. Using a specific thiol protease inhibitor (E64) that blocks bleomycin hydrolase activity, it was further shown that cells become more sensitive to bleomycin [120]. This finding quickly led to the isolation of the bleomycin hydrolase corresponding gene from yeast and mammalian cells [121–124]. Expression of the yeast bleomycin hydrolase gene *BLH1* in mammalian cells conferred nearly eightfold increase resistance to bleomycin, and which was blocked by the E64 inhibitor [125]. One would expect that removal of the *BLH1* gene from yeast would cause a bleomycin-hypersensitive phenotype. However, two independent studies showed conflicting data regarding the role of Blh1 in the detoxification of bleomycin in the yeast model system [121, 122]. While one study showed that *blh1Δ* mutant is mildly sensitive to bleomycin, another clearly established that the mutant is not at all sensitive [121, 122]. Additional findings revealed that *blh1Δ* mutants are not sensitive to bleomycin [126]. Moreover, overexpression of the *BLH1* gene in yeast cells confers no additional resistance to bleomycin [126]. Thus, the role of bleomycin hydrolase in producing tumor resistance is controversial. The situation is further complicated by the fact that the Blh1 protein, also called Gal6, is under the control of the Gal4 transcriptional activator [127]. Blh1/Gal6 binds specifically to the Gal4 transcription factor DNA binding site and acts as a repressor to negatively control the galactose metabolism pathway [127–129]. On the basis of the foregoing studies, it would appear that bleomycin hydrolase has a more general role in the cells to degrade proteins, or perhaps to degrade transcription factors to regulate gene expression [130–132]. In support of this, a more recent study documented that Blh1 is one of the proteases required to process Huntington protein to generate the N-terminal cleave form thought to be involved in the pathogenesis of the disease [133]. In fact, both yeast and human BLH1 play a more direct physiological role in protecting cells against homocysteine toxicity, a risk factor for Alzheimer's disease, by hydrolyzing intracellular homocysteine-thiolactone [134, 135].

6.4 DNA Repair Pathways

Repairing of bleomycin-induced DNA lesions is likely the most crucial mechanism employed by cells to avert bleomycin-induced genotoxicity. Thus, organisms exposed to bleomycin must recruit a variety of enzymes and/or proteins to repair the diverse types of bleomycin-induced DNA lesions. While such enzymes are still being characterized in eukaryotic cells, the bacterium *E. coli* has two well documented enzymes, i.e., endonuclease IV and exonuclease III, that repair bleomycin-induced DNA lesions [100, 136, 137]. Both enzymes possess (i) a 3′-diesterase that removes 3′-blocking groups (such as 3′-phosphoglycolate) at strand breaks, and (ii) an AP endonuclease that cleaves AP sites. These enzymatic activities regenerate 3′-hydroxyl groups that allow DNA repair synthesis by DNA polymerase [137–139]. *E. coli* mutants lacking both endonuclease IV and exonuclease III are severely impaired in the removal of bleomycin-induced DNA lesions, and, as a consequence, display extreme hypersensitivity to bleomycin [137]. Between the two enzymes, endonuclease IV plays a more predominant role in repairing bleomycin-damaged DNA [140]. This is supported by two independent studies, the first showing that mutants deficient in endonuclease IV are substantially more sensitive to bleomycin than exonuclease III-deficient mutants [137, 141]. The second study demonstrated that purified endonuclease IV is more active at processing bleomycin-induced DNA lesions in vitro, as compared to purified exonuclease III [140].

While the *E. coli* studies were in progress, the first eukaryotic homologue of endonuclease IV, called Apn1, was discovered in *S. cerevisiae* [142, 143]. Surprisingly, yeast mutants lacking Apn1 are not sensitive to bleomycin, leading to the prediction that yeast may use alternative enzyme(s) to combat the genotoxic effects of bleomycin [143, 144]. Consequently, a rigorous search was initiated for a possible auxiliary enzyme(s) in yeast that might repair bleomycin-induced DNA lesions. One approach exploited the power of biochemistry to detect enzymatic activities that would process lesions along defined DNA substrates. In one case, a highly sensitive assay was developed consisting of a double stranded DNA substrate where one strand (*[^{32}P]-labeled) bears a single-strand break terminated with 3′-phosphoglycolate (Fig. 3). This biochemical assay identified an extremely weak 3′-diesterase in total extracts derived from an *apn1Δ* mutant (i.e., lacking the major

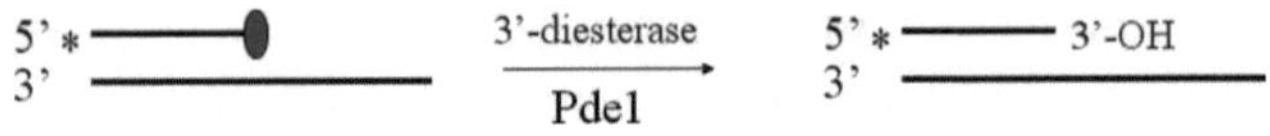

Fig. 3 Depiction of an oligonucleotide DNA substrate bearing a 3′-phophoglycolate terminus. The 3′-phosphoglycolate (*oval shape*) is produced by bleomycin and requires processing by a 3′-repair diesterase in order to regenerate a 3′-hydroxyl group for DNA polymerase activity. Labeling (*asterisk*) the 5′-end with ^{32}P allows detection of the processed product by polyarcylamide gels

3′-diesterase/AP endonuclease activity of Apn1). The weak activity removed the 3′-phosphoglycolate (PG) from the labeled DNA strand to produce 3′-OH (Fig. 3) [145]. The enzyme, called Pde1, was partially purified and also found to have an AP endonuclease in addition to the 3′-diesterase activity [146]. Immediately following this report, the gene (*APN2/ETH1*) encoding Pde1 was isolated by two independent laboratories, and the deduced amino acid sequence was found to share 19 % identity with the *E. coli* exonuclease III [146, 147]. Thus, Pde1/Apn2/Eth1 is the yeast homologue of *E. coli* exonuclease III. The most surprising finding is that yeast mutants lacking both Apn1 and Pde1 (Apn2/Eth1), if at all, showed very mild sensitivity to bleomycin (D.R., unpublished). However, the *apn1Δ pde1Δ* double mutants are exquisitely sensitive to the alkylating agent methyl methane sulfonate, which produces natural AP sites, as opposed to oxidized AP sites generated by bleomycin [53]. It is therefore possible that the 3′-phosphoglycolate and the oxidized AP site lesions produced by bleomycin are inaccessible or refractory to cleavage by the 3′-diesterase/AP endonuclease activities of either Apn1 or Pde1 in vivo. If so, yeast may possess yet other "backup" enzymes to initiate the repair of bleomycin-induced DNA lesions. This possibility is supported by the discovery of the hPNKP gene encoding the human polynucleotide kinase, which possesses two enzymatic activities, a kinase that phosphorylates the 5′-hydroxyl group of DNA and a strong 3′-diesterase activity that repairs oxidative DNA lesions in *E. coli* [148]. hPNPK is unrelated to any of the known 3′-diesterase/AP endonuclease belonging to the endonuclease IV or exonuclease III family, but it may share a related active site [100, 149]. A gene (*TPP1*) encoding a yeast homolog of the human hPNKP has been subsequently isolated, but gene knock out of *TPP1* alone has no effect on bleomycin sensitivity, unless TPP1 is also deleted in a background lacking both Apn1 and Apn2 [148, 150].

The repair of bleomycin-induced DNA lesions is not restricted to enzymes with the ability to cleave AP sites or remove 3′-blocking groups, as other DNA repair pathways also participate in the repair process. In yeast, the recombination and the post-replication DNA repair pathways, respectively, represented by the Rad52 and Rad6 proteins are involved in the repair of bleomycin-induced DNA lesions [151, 152]. These two pathways also repair a wide spectrum of other DNA lesions including those generated by the alkylating agent methyl methane sulfonate, 4-nitroquinoline-1-oxide (which forms bulky DNA adducts), and γ-rays. The *rad52Δ* and *rad6Δ* mutants are hypersensitive to a large number of DNA damaging agents including bleomycin [153–156]. On the basis of cell killing and growth kinetic analyses, both Rad52 and Rad6 showed different contribution to the repair of bleomycin-induced DNA lesions [152]. At bleomycin concentrations ranging from 10 to 15 μg/ml in culture media the Rad52 pathway is required to repair bleomycin-induced DNA damage, while at higher concentrations (15–30 μg/ml culture media), the Rad6 pathway plays a more prominent role [152]. This disparate response can be explained if distinct lesions are generated in yeast cells in a manner that depends on the bleomycin dose.

A few studies also implicated the involvement of other proteins in the repair of bleomycin damaged DNA. For example, the Ku proteins, a heterodimer composed of a 70-kDa subunit and a 80-kDa subunit that is involved in nonhomologous end joining of DNA, is implicated in the repair of bleomycin-induced DNA lesions [157, 158]. However, a number of laboratories could not convincingly confirm the earlier findings that Ku heterodimer-deficient yeast mutants (*hdf1Δ hdf2Δ*) are sensitive to bleomycin [157, 158] (Masson, J-Y and Ramotar, D., unpublished). This discrepancy may be related to the yeast strain background used in the initial studies [157, 158]. Proteins that remodel the chromatin structure are also involved in protecting the genome from the genotoxic effects of bleomycin. It has been shown that either bleomycin or methyl methane sulfonate can activate the Mec1 kinase in yeast, leading to direct phosphorylation of serine 129 of histone H2A [159]. A mutation (S129A) that prevented the phosphorylation of H2A causes cells to be hypersensitive to both bleomycin and MMS [159]. The investigators proposed that phosphorylation of H2A is required to relax the chromatin to either allow gene expression to facilitate repair, or to permit access of repair proteins and other factors directly to the DNA lesions [159].

While it is clear that DNA repair plays an important role in the protection against bleomycin-induced DNA lesions, there is no direct evidence that the overproduction of DNA repair proteins can contribute to enhance bleomycin resistance in cells. At least in yeast, overproduction of some of the DNA repair proteins described above does not confer bleomycin resistance to parental cells [144, 156]. This is in discord with one of the earlier predictions that tumor resistance to bleomycin may be attributed to elevated DNA repair activities [38]. Irrespective of whether DNA repair activities are subsequently discovered to be elevated in bleomycin resistant tumors, any attempts to promote the antitumor potential of bleomycin should take into consideration the possibility of diminishing the DNA repair capacity of tumor cells.

7 Transport of Bleomycin into Yeast and Mammalian Cells

The first evidence for a possible transporter for the uptake of bleomycin into cells came from a study showing that the plasma membrane of yeast and mammalian cells contained a protein that binds to bleomycin carrying labeled cobalt [^{57}Co] [99]. Further characterization of this plasma membrane protein was hindered as it requires the use of [^{57}Co]-bleomycin. As such, an alternative approach was devised that employs coupling the dye fluorescein to bleomycin-A_5 followed by the purification of the conjugated drug (F-BLM) using high performance liquid chromatography [160]. The resulting purified F-BLM retained the ability to inflict damage to the DNA in a manner analogous to the native bleomycin-A_5 [160]. Moreover, yeast mutants that are defective in the repair of bleomycin-A_5-induced lesions exhibit sensitivity to F-BLM, indicating that this modified drug has the capability of entering the cell and damaged the DNA. In fact, F-BLM enters parent yeast cells in a concentration- and time-dependent manner and raises a distinct possibility that a plasma membrane transporter exists to allow bleomycin entry into the cell [160]. If this is indeed the case, yeast mutants lacking the transporter function are expected to be resistant to bleomycin. This notion prompted the search for a possible transporter of bleomycin starting with yeast as a model system because of the ease, for example, of identifying genes belonging to the same functional pathways by various high throughput analyses [161]. To find the bleomycin transporter, a collection consisting of ~ 4,000 yeast haploid mutant each lacking a nonessential gene was screened for those showing sensitivity or resistance to bleomycin [162]. This large scale approach revealed over 200 mutants displaying remarkable sensitivity to bleomycin and are deleted for genes encoding proteins belonging to several functional groups including DNA repair and chromatin structure, transcription, and cell cycle [163]. A large group of genes belong to the vacuolar pathway, highlighting the importance of the vacuoles in detoxifying bleomycin [160, 164]. Amongst these sensitive genes none has been characterized in any details with respect to potential target for therapy.

Of importance, the large scale screen revealed five mutants displaying sharp resistance to bleomycin as compare to the parent [163]. Amongst these mutants one lacked the *AGP2* gene and exhibited the greatest resistance (~3,000-fold more) to bleomycin, but not to other chemotherapeutic drugs such as cisplatin, camptothecin, and etoposide [163]. Of all the hypersensitive and resistant mutants, at least 76 are deleted for genes encoding proteins that share significant level of identity with a human protein [163]. Thus, it appears that both yeast and human cells may conserve the same biological processes to regulate the toxicity of bleomycin. Since one of the objectives of the large scale screen is to identify a plasma membrane transporter that when deleted causes resistance to bleomycin, most of the ensuing analyses have been directed toward the molecular characterization of the phenotypes associated with deletion of the *AGP2* gene.

7.1 Yeast Agp2 Is a Transporter of Bleomycin

AGP2 encodes a 67.2-kDa plasma membrane protein that shares significant homology with the amino acid transporter family. This transporter is involved in the uptake of L-carnitine, which serves as a carrier for acetyl-CoA, from the peroxisome to the mitochondria for complete oxidation [165]. Cells lacking Agp2 are defective in mediating the uptake of F-BLM [163]. This has been supported by epifluorescence analysis revealing that there was no detectable staining of F-BLM in the *agp2Δ* mutants as compared to the parent [163]. F-BLM uptake could be effectively restored in the *agp2Δ* mutants by the reintroduction of the *AGP2* gene, strongly indicating that Agp2 is the main channel to allow entry of bleomycin into the cell [163]. It is no surprise then that coincubation of parent cells with L-carnitine sharply reduced the uptake of F-BLM into the cells. More importantly, overproduction of the transporter stimulated the uptake of F-BLM and selectively sensitized the cells to killing by bleomycin, a consequence of substantially elevated damage to the chromosomal DNA [163].

7.2 Human hCT2 Is a Transporter of Bleomycin

On the basis of the above studies in yeast, it seems reasonable to postulate that mammalian permeases with the ability to transport L-carnitine could be a candidate to mediate uptake of bleomycin. At least, two high affinity L-carnitine transporters hCT2 and OCTN2 are present in humans [166, 167]. hCT2 is expressed mainly in the bone marrow and testis, while OCTN2 is expressed in multiple tissues [166]. As testis expressed the utmost levels of hCT2 and that testicular cancer is highly responsive to bleomycin therapy raise the possibility that there might be a correlation between hCT2 expression and bleomycin response [168]. Indeed, RT-PCR analysis confirmed that the testicular cancer cell line NT2/D1expressed high levels of hCT2, while its expression is undetectable in the colon carcinoma cell line HCT116, and only weakly detected in the breast cancer MCF-7 and the lung fibroblast LL47 cells [169]. In fact, the testicular cancer cell line NT2/D1 displayed more than 300-fold more sensitivity to bleomycin than the colon cancer cell line HCT116, suggesting that the higher sensitivity of NT2/D1 is related to the higher expression level of hCT2 [169]. This differential response between the two cell lines was not observed if the cells were challenged with other genotoxic chemotherapeutic agents such as cisplatin [169].

Since hCT2 has been shown to be a high affinity transporter of L-carnitine [166], it seems no surprise that L-carnitine acts as a competitive inhibitor and block the uptake of bleomycin if both gain entry into the cells via the same transporter [169]. In fact, L-carnitine protected the testicular cancer cell line from the genotoxic effects of bleomycin [169]. These observations led to the prediction that if hCT2 expression is downregulated, it would block entry of F-BLM and cause the cells to become resistant to bleomycin. Indeed, using siRNA technology that specifically depleted the expression of hCT2 sharply reduced the uptake of labeled L-carnitine into NT2/D1 cells; highlighting the effectiveness of the siRNA. Furthermore, the diminished expression of hCT2 conferred upon these cells resistance to bleomycin, but not to other anticancer agents such as cisplatin [169]. It seems logical that the complete loss of hCT2 expression, such as *hct2*$^{-/-}$ homozygous null cells, would make cells even more resistant to bleomycin. However, these latter cells are not available to test this possibility [169].

A further prediction from the above findings is that cell lines devoid of hCT2 expression, such as the colon cancer cell line HCT116, when forced to express the transporter should be sensitized to bleomycin. The transient transfection of a plasmid designed to drive hCT2 expression as a hCT2-GFP fusion protein from the CMV promoter directed the production of an ~93 kDa protein that localized to the plasma membrane [169]. This fusion protein stimulated the uptake of F-BLM into the HCT116 cells [169]. These HCT116 cells overexpressing hCT2-GFP showed enhance sensitivity to bleomycin, and not to cisplatin, as compared to cells carrying only the GFP portion [169]. These findings are in agreement with the notion that hCT2 has a specific role in transporting bleomycin into the cell [164].

Currently, Hodgkin's lymphoma patients are being treated with bleomycin, although a fraction of these patients remain unresponsive to the drug [170]. To check if this is correlated with a reduced level of the transporter, hCT2 expression level was determined in a panel of established Hodgkin's lymphoma cell lines that include Namalwa, Raji, Daudi, H2, DHL16, RL, and SR. Interestingly, amongst these cell lines only H2 expressed the highest level of hCT2 and displayed significant sensitivity to bleomycin, as compared to the other cell lines with low levels of hCT2 [169]. Thus, lymphoma patients expressing high levels of hCT2 are likely to show favorable clinical respond towards bleomycin.

8 Summary and Perspectives

So far, much of our understanding of the various mechanisms leading to bleomycin resistance emanate from the yeast *S. cerevisiae*, mainly because of the availability of ready-to-use experimental tools. Based on current information, it would appear that the principal defense mechanism against bleomycin involves entry of the drug into cells (Fig. 4). The observations that the levels of the hCT2 transporter correlate with the sensitivity of cells towards bleomycin, strongly suggests that it could play a key role in specifically regulating cellular resistance to bleomycin. Thus, it can be inferred that high hCT2 activity levels in tumor cell samples would be indicative of responsiveness towards bleomycin-therapy, while low hCT2 activity would correlate with drug resistance. As such, hCT2 could be a determining factor for patients' response to treatment regimens consisting of bleomycin. To date, no studies have been performed to closely examine if hCT2 is regulated and to explore ways to stimulate its expression, for example, with small molecules or hormones [171]. Such efforts would have broad implications by enhancing the uptake of bleomycin in tumors such as breast, colon, and ovarian which are generally refractory to the drug therapy and may be due to the poor expression of hCT2.

In short, it is reasonable to propose that the major mechanism leading to bleomycin resistance occurs at the level of drug uptake and that mutations impairing the transporter activity could exist and easily explain why the remaining fraction of testicular cancer patients are resistant to bleomycin therapy.

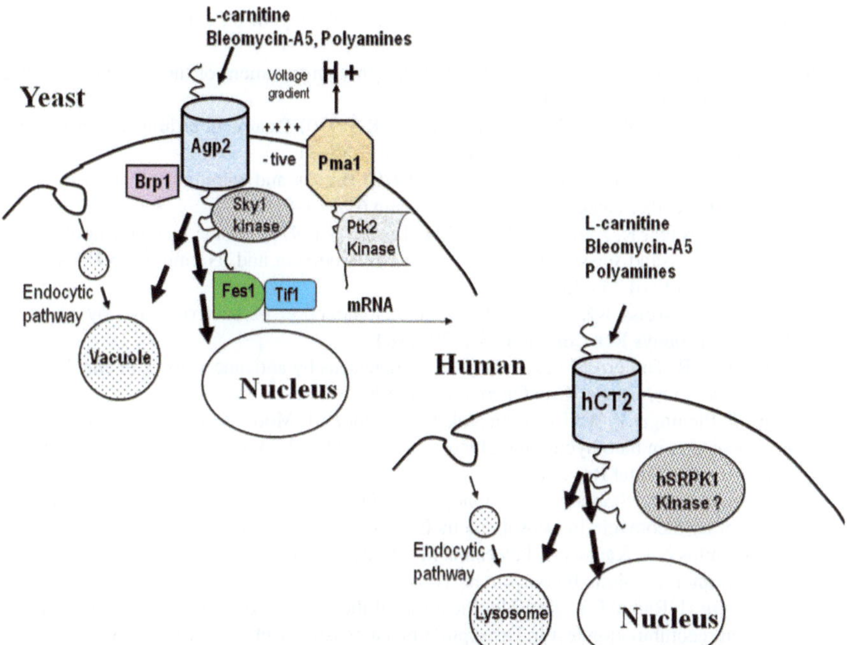

Fig. 4 A model illustrating the transport and detoxification pathway of BLM-A5 in yeast and human cells. In yeast, the drug enters the cell via the transporter Agp2, and its activity might be influenced by the kinases Ptk2 and Sky1. Following uptake, BLM-A5 is channeled to the vacuole for detoxification, as well as to the nucleus to destroy the DNA. Interruption of the endocytic pathway to the vacuoles leads to hypersensitivity towards BLM-A5. In humans, hCT2 is responsible for transporting BLM-A5, into the cells. hSRPK1 may regulate hCT2

References

1. Umezawa H (1965) Bleomycin and other antitumor antibiotics of high molecular weight. Antimicrob Agents Chemother 5:1079–1085
2. Umezawa H, Maeda K, Takeuchi T, Okami Y (1966) New antibiotics, bleomycin A and B. J Antibiot (Tokyo) 19:200–209
3. Umezawa H, Ishizuka M, Maeda K, Takeuchi T (1967) Studies on bleomycin. Cancer 20:891–895
4. Umezawa H (1971) Natural and artificial bleomycins: chemistry and antitumor activities. Pure Appl Chem 28:665–680
5. Fisher LM, Kuroda R, Sakai TT (1985) Interaction of bleomycin A2 with deoxyribonucleic acid: DNA unwinding and inhibition of bleomycin-induced DNA breakage by cationic thiazole amides related to bleomycin A2. Biochemistry 24:3199–3207
6. Suzuki H, Nagai K, Yamaki H, Tanaka N, Umezawa H (1968) Mechanism of action of bleomycin. Studies with the growing culture of bacterial and tumor cells. J Antibiot (Tokyo) 21:379–386
7. Kanno T, Nakazawa T, Sugimoto T (1969) Study of bleomycin on brain tumors. 1. Inhibitory effect of bleomycin on cultured brain tumor cells. Seishin Igaku Kenkyusho Gyosekishu 16:23–31

8. Terasima T, Umezawa H (1970) Lethal effect of bleomycin on cultured mammalian cells. J Antibiot (Tokyo) 23:300–304
9. Ichikawa T, Nakano I, Hirokawa I (1969) Bleomycin treatment of the tumors of penis and scrotum. J Urol 102:699–707
10. Oka S, Sato K, Nakai Y, Kurita K, Hashimoto K (1970) Treatment of lung cancer with bleomycin. II. Sci Rep Res Inst Tohoku Univ Med 17:77–91
11. Terasima T, Yasukawa M, Umezawa H (1970) Breaks and rejoining of DNA in cultured mammalian cells treated with bleomycin. Gann 61:513–516
12. Suzuki H, Nagai K, Akutsu E, Yamaki H, Tanaka N (1970) On the mechanism of action of bleomycin. Strand scission of DNA caused by bleomycin and its binding to DNA in vitro. J Antibiot (Tokyo) 23:473–480
13. Haidle CW, Weiss KK, Mace ML Jr (1972) Induction of bacteriophage by bleomycin. Biochem Biophys Res Commun 48:1179–1184
14. Ferguson LR, Turner PM (1988) 'Petite' mutagenesis by anticancer drugs in *Saccharomyces cerevisiae*. Eur J Cancer Clin Oncol 24:591–596
15. Koy JF, Pleninger P, Wall L, Pramanik A, Martinez M, Moore CW (1995) Genetic changes and bioassays in bleomycin- and phleomycin-treated cells, and their relationship to chromosomal breaks. Mutat Res 336:19–27
16. Demopoulos NA, Stamatis ND, Yannopoulos G (1980) Induction of somatic and male crossing-over by bleomycin in Drosophila melanogaster. Mutat Res 78:347–351
17. Demopoulos NA, Kappas A, Pelecanos M (1982) Recombinogenic and mutagenic effects of the antitumor antibiotic bleomycin in *Aspergillus nidulans*. Mutat Res 102:51–57
18. Cederberg H, Ramel C (1989) Modifications of the effect of bleomycin in the somatic mutation and recombination test in Drosophila melanogaster. Mutat Res 214:69–80
19. Hoffmann GR, Colyer SP, Littlefield LG (1993) Induction of micronuclei by bleomycin in G0 human lymphocytes: I. Dose–response and distribution. Environ Mol Mutagen 21:130–135
20. Burger RM, Peisach J, Horwitz SB (1981) Activated bleomycin. A transient complex of drug, iron, and oxygen that degrades DNA. J Biol Chem 256:11636–11644
21. Burger RM, Peisach J, Horwitz SB (1982) Stoichiometry of DNA strand scission and aldehyde formation by bleomycin. J Biol Chem 257:8612–8614
22. Hecht SM (1986) DNA strand scission by activated bleomycin group antibiotics. Fed Proc 45:2784–2791
23. Kane SA, Hecht SM (1994) Polynucleotide recognition and degradation by bleomycin. Prog Nucleic Acid Res Mol Biol 49:313–352
24. Hecht SM (2000) Bleomycin: new perspectives on the mechanism of action. J Nat Prod 63:158–168
25. Wharam MD, Phillips TL, Kane L, Utley JF (1973) Response of a murine solid tumor to in vivo combined chemotherapy and irradiation. Radiology 109:451–455
26. Jani JP, Mistry JS, Morris G, Davies P, Lazo JS, Sebti SM (1992) In vivo circumvention of human colon carcinoma resistance to bleomycin. Cancer Res 52:2931–2937
27. Povirk LF, Austin MJ (1991) Genotoxicity of bleomycin. Mutat Res 257:127–143
28. Lazo JS, Sebti SM, Schellens JH (1996) Bleomycin. Cancer Chemother Biol Response Modif 16:39–47
29. Dorr RT (1992) Bleomycin pharmacology: mechanism of action and resistance, and clinical pharmacokinetics. Semin Oncol 19:3–8
30. Sikic BI (1986) Biochemical and cellular determinants of bleomycin cytotoxicity. Cancer Surv 5:81–91
31. Harrison JH Jr, Lazo JS (1987) High dose continuous infusion of bleomycin in mice: a new model for drug-induced pulmonary fibrosis. J Pharmacol Exp Ther 243:1185–1194
32. Ekimoto H, Takahashi K, Matsuda A, Takita T, Umezawa H (1985) Lipid peroxidation by bleomycin-iron complexes in vitro. J Antibiot (Tokyo) 38:1077–1082
33. Wang Q, Wang Y, Hyde DM, Gotwals PJ, Lobb RR, Ryan ST, Giri SN (2000) Effect of antibody against integrin alpha4 on bleomycin-induced pulmonary fibrosis in mice. Biochem Pharmacol 60:1949–1958

34. Miyaki M, Ono T, Hori S, Umezawa H (1975) Binding of bleomycin to DNA in bleomycin-sensitive and -resistant rat ascites hepatoma cells. Cancer Res 35:2015–2019
35. Akiyama S, Ikezaki K, Kuramochi H, Takahashi K, Kuwano M (1981) Bleomycin-resistant cells contain increased bleomycin-hydrolase activities. Biochem Biophys Res Commun 101:55–60
36. Morris G, Mistry JS, Jani JP, Mignano JE, Sebti SM, Lazo JS (1992) Neutralization of bleomycin hydrolase by an epitope-specific antibody. Mol Pharmacol 42:57–62
37. Sebti SM, Jani JP, Mistry JS, Gorelik E, Lazo JS (1991) Metabolic inactivation: a mechanism of human tumor resistance to bleomycin. Cancer Res 51:227–232
38. Urade M, Ogura T, Mima T, Matsuya T (1992) Establishment of human squamous carcinoma cell lines highly and minimally sensitive to bleomycin and analysis of factors involved in the sensitivity. Cancer 69:2589–2597
39. Oppenheimer NJ, Rodriguez LO, Hecht SM (1980) Metal binding to modified bleomycins. Zinc and ferrous complexes with an acetylated bleomycin. Biochemistry 19:4096–4103
40. Ehrenfeld GM, Rodriguez LO, Hecht SM, Chang C, Basus VJ, Oppenheimer NJ (1985) Copper(I)-bleomycin: structurally unique complex that mediates oxidative DNA strand scission. Biochemistry 24:81–92
41. Ehrenfeld GM, Shipley JB, Heimbrook DC, Sugiyama H, Long EC, van Boom JH, van der Marel GA, Oppenheimer NJ, Hecht SM (1987) Copper-dependent cleavage of DNA by bleomycin. Biochemistry 26:931–942
42. Levy MJ, Hecht SM (1988) Copper(II) facilitates bleomycin-mediated unwinding of plasmid DNA. Biochemistry 27:2647–2650
43. Petering DH, Mao Q, Li W, DeRose E, Antholine WE (1996) Metallobleomycin-DNA interactions: structures and reactions related to bleomycin-induced DNA damage. Met Ions Biol Syst 33:619–648
44. Hoehn ST, Junker HD, Bunt RC, Turner CJ, Stubbe J (2001) Solution structure of co(iii)-bleomycin-ooh bound to a phosphoglycolate lesion containing oligonucleotide: implications for bleomycin-induced double-strand dna cleavage. Biochemistry 40:5894–5905
45. Oppenheimer NJ, Chang C, Rodriguez LO, Hecht SM (1981) Copper(I). bleomycin. A structurally unique oxidation-reduction active complex. J Biol Chem 256:1514–1517
46. Sausville EA, Peisach J, Horwitz SB (1976) A role for ferrous ion and oxygen in the degradation of DNA by bleomycin. Biochem Biophys Res Commun 73:814–822
47. Sausville EA, Stein RW, Peisach J, Horwitz SB (1978) Properties and products of the degradation of DNA by bleomycin and iron(II). Biochemistry 17:2746–2754
48. Sausville EA, Peisach J, Horwitz SB (1978) Effect of chelating agents and metal ions on the degradation of DNA by bleomycin. Biochemistry 17:2740–2746
49. Kane SA, Natrajan A, Hecht SM (1994) On the role of the bithiazole moiety in sequence-selective DNA cleavage by Fe.bleomycin. J Biol Chem 269:10899–10904
50. Abraham AT, Zhou X, Hecht SM (1999) DNA cleavage by Fe(II). bleomycin conjugated to a solid support. J Am Chem Soc 121:1982–1983
51. Aouida M, Leduc A, Poulin R, Ramotar D (2005) AGP2 encodes the major permease for high affinity polyamine import in *Saccharomyces cerevisiae*. J Biol Chem 280:24267–24276
52. Leitheiser CJ, Rishel MJ, Wu X, Hecht SM (2000) Solid-phase synthesis of bleomycin group antibiotics. Elaboration of deglycobleomycin A(5). Org Lett 2:3397–3399
53. Burger RM (1998) Cleavage of nucleic acids by bleomycin. Chem Rev 98:1153–1169
54. Burger RM, Peisach J, Blumberg WE, Horwitz SB (1979) Iron-bleomycin interactions with oxygen and oxygen analogues. Effects on spectra and drug activity. J Biol Chem 254: 10906–10912
55. Povirk LF, Wubter W, Kohnlein W, Hutchinson F (1977) DNA double-strand breaks and alkali-labile bonds produced by bleomycin. Nucleic Acids Res 4:3573–3580
56. Ekimoto H, Kuramochi H, Takahashi K, Matsuda A, Umezawa H (1980) Kinetics of the reaction of bleomycin-Fe(II)-O_2 complex with DNA. J Antibiot (Tokyo) 33:426–434
57. Burger RM, Berkowitz AR, Peisach J, Horwitz SB (1980) Origin of malondialdehyde from DNA degraded by Fe(II) x bleomycin. J Biol Chem 255:11832–11838

58. Burger RM, Peisach J, Horwitz SB (1981) Mechanism of bleomycin action: in vitro studies. Life Sci 28:715–727
59. Worth L Jr, Frank BL, Christner DF, Absalon MJ, Stubbe J, Kozarich JW (1993) Isotope effects on the cleavage of DNA by bleomycin: mechanism and modulation. Biochemistry 32:2601–2609
60. Giloni L, Takeshita M, Johnson F, Iden C, Grollman AP (1981) Bleomycin-induced strand-scission of DNA. Mechanism of deoxyribose cleavage. J Biol Chem 256:8608–8615
61. Dedon PC, Plastaras JP, Rouzer CA, Marnett LJ (1998) Indirect mutagenesis by oxidative DNA damage: formation of the pyrimidopurinone adduct of deoxyguanosine by base propenal. Proc Natl Acad Sci USA 95:11113–11116
62. Chaudhary AK, Nokubo M, Reddy GR, Yeola SN, Morrow JD, Blair IA, Marnett LJ (1994) Detection of endogenous malondialdehyde-deoxyguanosine adducts in human liver. Science 265:1580–1582
63. Fink SP, Reddy GR, Marnett LJ (1997) Mutagenicity in *Escherichia coli* of the major DNA adduct derived from the endogenous mutagen malondialdehyde. Proc Natl Acad Sci USA 94:8652–8657
64. Mao H, Schnetz-Boutaud NC, Weisenseel JP, Marnett LJ, Stone MP (1999) Duplex DNA catalyzes the chemical rearrangement of a malondialdehyde deoxyguanosine adduct. Proc Natl Acad Sci USA 96:6615–6620
65. Absalon MJ, Wu W, Kozarich JW, Stubbe J (1995) Sequence-specific double-strand cleavage of DNA by Fe-bleomycin. 2. Mechanism and dynamics. Biochemistry 34:2076–2086
66. Dedon PC, Goldberg IH (1992) Free-radical mechanisms involved in the formation of sequence-dependent bistranded DNA lesions by the antitumor antibiotics bleomycin, neocarzinostatin, and calicheamicin. Chem Res Toxicol 5:311–332
67. Steighner RJ, Povirk LF (1990) Bleomycin-induced DNA lesions at mutational hot spots: implications for the mechanism of double-strand cleavage. Proc Natl Acad Sci USA 87:8350–8354
68. Onishi T, Iwata H, Takagi Y (1975) Effects of reducing and oxidizing agents on the action of bleomycin. J Biochem (Tokyo) 77:745–752
69. Burger RM, Peisach J, Horwitz SB (1982) Effects of O_2 on the reactions of activated bleomycin. J Biol Chem 257:3372–3375
70. Burger RM, Blanchard JS, Horwitz SB, Peisach J (1985) The redox state of activated bleomycin. J Biol Chem 260:15406–15409
71. Templin J, Berry L, Lyman S, Byrnes RW, Antholine WE, Petering DH (1992) Properties of redox-inactivated bleomycins. In vitro DNA damage and inhibition of Ehrlich cell proliferation. Biochem Pharmacol 43:615–623
72. Absalon MJ, Kozarich JW, Stubbe J (1995) Sequence-specific double-strand cleavage of DNA by Fe-bleomycin. 1. The detection of sequence-specific double-strand breaks using hairpin oligonucleotides. Biochemistry 34:2065–2075
73. Povirk LF (1979) Catalytic release of deoxyribonucleic acid bases by oxidation and reduction of an iron.bleomycin complex. Biochemistry 18:3989–3995
74. Kuroda R, Shinomiya M, Otsuka M (1992) Origin of sequence specific cleavage of DNA by bleomycins. Nucleic Acids Symp Ser 27:9–10
75. Nightingale KP, Fox KR (1993) DNA structure influences sequence specific cleavage by bleomycin. Nucleic Acids Res 21:2549–2555
76. Kuroda R, Satoh H, Shinomiya M, Watanabe T, Otsuka M (1995) Novel DNA photocleaving agents with high DNA sequence specificity related to the antibiotic bleomycin A2. Nucleic Acids Res 23:1524–1530
77. Kane SA, Hecht SM, Sun JS, Garestier T, Helene C (1995) Specific cleavage of a DNA triple helix by FeII.bleomycin. Biochemistry 34:16715–16724
78. Mascharak PK, Sugiura Y, Kuwahara J, Suzuki T, Lippard SJ (1983) Alteration and activation of sequence-specific cleavage of DNA by bleomycin in the presence of the antitumor drug cis-diamminedichloroplatinum(II). Proc Natl Acad Sci USA 80:6795–6798

79. Hertzberg RP, Caranfa MJ, Hecht SM (1985) DNA methylation diminishes bleomycin-mediated strand scission. Biochemistry 24:5286–5289
80. Hertzberg RP, Caranfa MJ, Hecht SM (1988) Degradation of structurally modified DNAs by bleomycin group antibiotics. Biochemistry 27:3164–3174
81. Bennett RA, Swerdlow PS, Povirk LF (1993) Spontaneous cleavage of bleomycin-induced abasic sites in chromatin and their mutagenicity in mammalian shuttle vectors. Biochemistry 32:3188–3195
82. Dar ME, Jorgensen TJ (1995) Deletions at short direct repeats and base substitutions are characteristic mutations for bleomycin-induced double- and single-strand breaks, respectively, in a human shuttle vector system. Nucleic Acids Res 23:3224–3230
83. Pavon V, Esteve I, Guerrero R, Villaverde A, Gaju N (1995) Induced mutagenesis by bleomycin in the purple sulfur bacterium Thiocapsa roseopersicina. Curr Microbiol 30: 117–120
84. Steighner RJ, Povirk LF (1990) Effect of in vitro cleavage of apurinic/apyrimidinic sites on bleomycin-induced mutagenesis of repackaged lambda phage. Mutat Res 240:93–100
85. Tates AD, van Dam FJ, Natarajan AT, Zwinderman AH, Osanto S (1994) Frequencies of HPRT mutants and micronuclei in lymphocytes of cancer patients under chemotherapy: a prospective study. Mutat Res 307:293–306
86. Carter BJ, de Vroom E, Long EC, van der Marel GA, van Boom JH, Hecht SM (1990) Site-specific cleavage of RNA by Fe(II).bleomycin. Proc Natl Acad Sci USA 87:9373–9377
87. Huttenhofer A, Hudson S, Noller HF, Mascharak PK (1992) Cleavage of tRNA by Fe(II)-bleomycin. J Biol Chem 267:24471–24475
88. Holmes CE, Hecht SM (1993) Fe.bleomycin cleaves a transfer RNA precursor and its "transfer DNA" analog at the same major site. J Biol Chem 268:25909–25913
89. Hecht SM (1994) RNA degradation by bleomycin, a naturally occurring bioconjugate. Bioconjug Chem 5:513–526
90. Keck MV, Hecht SM (1995) Sequence-specific hydrolysis of yeast tRNA(Phe) mediated by metal-free bleomycin. Biochemistry 34:12029–12037
91. Morgan MA, Hecht SM (1994) Iron(II) bleomycin-mediated degradation of a DNA-RNA heteroduplex. Biochemistry 33:10286–10293
92. Holmes CE, Carter BJ, Hecht SM (1993) Characterization of iron (II).bleomycin-mediated RNA strand scission. Biochemistry 32:4293–4307
93. Holmes CE, Duff RJ, van der Marel GA, van Boom J, Hecht SM (1997) On the chemistry of RNA degradation by Fe.bleomycin. Bioorg Med Chem 5:1235–1248
94. Moore CW, Del Valle R, McKoy J, Pramanik A, Gordon RE (1992) Lesions and preferential initial localization of [S-methyl-3 H]bleomycin A2 on *Saccharomyces cerevisiae* cell walls and membranes. Antimicrob Agents Chemother 36:2497–2505
95. Beaudouin R, Lim ST, Steide JA, Powell M, McKoy J, Pramanik AJ, Johnson E, Moore CW, Lipke PN (1993) Bleomycin affects cell wall anchorage of mannoproteins in *Saccharomyces cerevisiae*. Antimicrob Agents Chemother 37:1264–1269
96. Lim ST, Jue CK, Moore CW, Lipke PN (1995) Oxidative cell wall damage mediated by bleomycin-Fe(II) in *Saccharomyces cerevisiae*. J Bacteriol 177:3534–3539
97. Hay J, Shahzeidi S, Laurent G (1991) Mechanisms of bleomycin-induced lung damage. Arch Toxicol 65:81–94
98. Arndt D, Zeisig R, Bechtel D, Fichtner I (2001) Liposomal bleomycin: increased therapeutic activity and decreased pulmonary toxicity in mice. Drug Deliv 8:1–7
99. Pron G, Belehradek J Jr, Mir LM (1993) Identification of a plasma membrane protein that specifically binds bleomycin. Biochem Biophys Res Commun 194:333–337
100. Ramotar D (1997) The apurinic-apyrimidinic endonuclease IV family of DNA repair enzymes. Biochem Cell Biol 75:327–336
101. Pron G, Mahrour N, Orlowski S, Tounekti O, Poddevin B, Belehradek J Jr, Mir LM (1999) Internalisation of the bleomycin molecules responsible for bleomycin toxicity: a receptor-mediated endocytosis mechanism. Biochem Pharmacol 57:45–56

102. Genilloud O, Garrido MC, Moreno F (1984) The transposon Tn5 carries a bleomycin-resistance determinant. Gene 32:225–233
103. Collis CM, Hall RM (1985) Identification of a Tn5 determinant conferring resistance to phleomycins, bleomycins, and tallysomycins. Plasmid 14:143–151
104. Blot M, Meyer J, Arber W (1991) Bleomycin-resistance gene derived from the transposon Tn5 confers selective advantage to *Escherichia coli* K-12. Proc Natl Acad Sci USA 88:9112–9116
105. Semon D, Movva NR, Smith TF, el Alama M, Davies J (1987) Plasmid-determined bleomycin resistance in *Staphylococcus aureus*. Plasmid 17:46–53
106. Drocourt D, Calmels T, Reynes JP, Baron M, Tiraby G (1990) Cassettes of the *Streptoalloteichus hindustanus* ble gene for transformation of lower and higher eukaryotes to phleomycin resistance. Nucleic Acids Res 18:4009
107. Gennimata D, Davies J, Tsiftsoglou AS (1996) Bleomycin resistance in *Staphylococcus aureus* clinical isolates. J Antimicrob Chemother 37:65–75
108. Gatignol A, Baron M, Tiraby G (1987) Phleomycin resistance encoded by the ble gene from transposon Tn 5 as a dominant selectable marker in *Saccharomyces cerevisiae*. Mol Gen Genet 207:342–348
109. Dumas P, Bergdoll M, Cagnon C, Masson JM (1994) Crystal structure and site-directed mutagenesis of a bleomycin resistance protein and their significance for drug sequestering. EMBO J 13:2483–2492
110. Hayes JD, Wolf CR (1990) Molecular mechanisms of drug resistance. Biochem J 272:281–295
111. Maruyama M, Kumagai T, Matoba Y, Hayashida M, Fujii T, Hata Y, Sugiyama M (2001) Crystal structures of the transposon Tn5-carried bleomycin resistance determinant uncomplexed and complexed with bleomycin. J Biol Chem 276:9992–9999
112. Blot M, Heitman J, Arber W (1993) Tn5-mediated bleomycin resistance in *Escherichia coli* requires the expression of host genes. Mol Microbiol 8:1017–1024
113. Calcutt MJ, Schmidt FJ (1994) Bleomycin biosynthesis: structure of the resistance genes of the producer organism. Ann N Y Acad Sci 721:133–137
114. Sugiyama M, Kumagai T, Matsuo H, Bhuiyan MZ, Ueda K, Mochizuki H, Nakamura N, Davies JE (1995) Overproduction of the bleomycin-binding proteins from bleomycin-producing *Streptomyces verticillus* and a methicillin-resistant *Staphylococcus aureus* in *Escherichia coli* and their immunological characterisation. FEBS Lett 362:80–84
115. Umezawa H, Hori S, Sawa T, Yoshioka T, Takeuchi T (1974) A bleomycin-inactivating enzyme in mouse liver. J Antibiot (Tokyo) 27:419–424
116. Sebti SM, Lazo JS (1988) Metabolic inactivation of bleomycin analogs by bleomycin hydrolase. Pharmacol Ther 38:321–329
117. Nishimura C, Suzuki H, Tanaka N, Yamaguchi H (1989) Bleomycin hydrolase is a unique thiol aminopeptidase. Biochem Biophys Res Commun 163:788–796
118. Lazo JS, Humphreys CJ (1983) Lack of metabolism as the biochemical basis of bleomycin-induced pulmonary toxicity. Proc Natl Acad Sci USA 80:3064–3068
119. Ferrando AA, Velasco G, Campo E, Lopez-Otin C (1996) Cloning and expression analysis of human bleomycin hydrolase, a cysteine proteinase involved in chemotherapy resistance. Cancer Res 56:1746–1750
120. Jani JP, Mistry JS, Morris G, Lazo JS, Sebti SM (1992) In vivo sensitization of human lung carcinoma to bleomycin by the cysteine proteinase inhibitor E-64. Oncol Res 4:59–63
121. Enenkel C, Wolf DH (1993) BLH1 codes for a yeast thiol aminopeptidase, the equivalent of mammalian bleomycin hydrolase. J Biol Chem 268:7036–7043
122. Magdolen U, Muller G, Magdolen V, Bandlow W (1993) A yeast gene (BLH1) encodes a polypeptide with high homology to vertebrate bleomycin hydrolase, a family member of thiol proteinases. Biochim Biophys Acta 1171:299–303
123. Bromme D, Rossi AB, Smeekens SP, Anderson DC, Payan DG (1996) Human bleomycin hydrolase: molecular cloning, sequencing, functional expression, and enzymatic characterization. Biochemistry 35:6706–6714
124. O'Farrell PA, Gonzalez F, Zheng W, Johnston SA, Joshua-Tor L (1999) Crystal structure of human bleomycin hydrolase, a self-compartmentalizing cysteine protease. Structure 7:619–627

125. Pei Z, Calmels TP, Creutz CE, Sebti SM (1995) Yeast cysteine proteinase gene ycp1 induces resistance to bleomycin in mammalian cells. Mol Pharmacol 48:676–681
126. Wang H, Ramotar D (2002) Cellular resistance to bleomycin in *Saccharomyces cerevisiae* is not affected by changes in bleomycin hydrolase levels. Biochem Cell Biol 80:789–796
127. Zheng W, Xu HE, Johnston SA (1997) The cysteine-peptidase bleomycin hydrolase is a member of the galactose regulon in yeast. J Biol Chem 272:30350–30355
128. Xu HE, Johnston SA (1994) Yeast bleomycin hydrolase is a DNA-binding cysteine protease. Identification, purification, biochemical characterization. J Biol Chem 269:21177–21183
129. Joshua-Tor L, Xu HE, Johnston SA, Rees DC (1995) Crystal structure of a conserved protease that binds DNA: the bleomycin hydrolase, Gal6. Science 269:945–950
130. Zheng W, Johnston SA, Joshua-Tor L (1998) The unusual active site of Gal6/bleomycin hydrolase can act as a carboxypeptidase, aminopeptidase, and peptide ligase. Cell 93:103–109
131. Koldamova RP, Lefterov IM, DiSabella MT, Almonte C, Watkins SC, Lazo JS (1999) Human bleomycin hydrolase binds ribosomal proteins. Biochemistry 38:7111–7117
132. Schwartz DR, Homanics GE, Hoyt DG, Klein E, Abernethy J, Lazo JS (1999) The neutral cysteine protease bleomycin hydrolase is essential for epidermal integrity and bleomycin resistance. Proc Natl Acad Sci USA 96:4680–4685
133. Ratovitski T, Chighladze E, Waldron E, Hirschhorn RR, Ross CA (2011) Cysteine proteases bleomycin hydrolase and cathepsin Z mediate N-terminal proteolysis and toxicity of mutant huntingtin. J Biol Chem 286:12578–12589
134. Zimny J, Sikora M, Guranowski A, Jakubowski H (2006) Protective mechanisms against homocysteine toxicity: the role of bleomycin hydrolase. J Biol Chem 281:22485–22492
135. Suszynska J, Tisonczyk J, Lee HG, Smith MA, Jakubowski H (2010) Reduced homocysteine-thiolactonase activity in Alzheimer's disease. J Alzheimers Dis 19:1177–1183
136. Demple B, Harrison L (1994) Repair of oxidative damage to DNA: enzymology and biology. Annu Rev Biochem 63:915–948
137. Ramotar D, Popoff SC, Demple B (1991) Complementation of DNA repair-deficient *Escherichia coli* by the yeast Apn1 apurinic/apyrimidinic endonuclease gene. Mol Microbiol 5:149–155
138. Demple B, Johnson A, Fung D (1986) Exonuclease III and endonuclease IV remove 3′ blocks from DNA synthesis primers in H_2O_2-damaged *Escherichia coli*. Proc Natl Acad Sci USA 83:7731–7735
139. Levin JD, Johnson AW, Demple B (1988) Homogeneous *Escherichia coli* endonuclease IV. Characterization of an enzyme that recognizes oxidative damage in DNA. J Biol Chem 263:8066–8071
140. Levin JD, Demple B (1996) In vitro detection of endonuclease IV-specific DNA damage formed by bleomycin in vivo. Nucleic Acids Res 24:885–889
141. Cunningham RP, Saporito SM, Spitzer SG, Weiss B (1986) Endonuclease IV (nfo) mutant of *Escherichia coli*. J Bacteriol 168:1120–1127
142. Popoff SC, Spira AI, Johnson AW, Demple B (1990) Yeast structural gene (APN1) for the major apurinic endonuclease: homology to *Escherichia coli* endonuclease IV. Proc Natl Acad Sci USA 87:4193–4197
143. Ramotar D, Popoff SC, Gralla EB, Demple B (1991) Cellular role of yeast Apn1 apurinic endonuclease/3′-diesterase: repair of oxidative and alkylation DNA damage and control of spontaneous mutation. Mol Cell Biol 11:4537–4544
144. Masson JY, Ramotar D (1996) The *Saccharomyces cerevisiae* IMP2 gene encodes a transcriptional activator that mediates protection against DNA damage caused by bleomycin and other oxidants. Mol Cell Biol 16:2091–2100
145. Sander M, Ramotar D (1997) Partial purification of Pde1 from *Saccharomyces cerevisiae*: enzymatic redundancy for the repair of 3′-terminal DNA lesions and abasic sites in yeast. Biochemistry 36:6100–6106
146. Bennett RA (1999) The *Saccharomyces cerevisiae* ETH1 gene, an inducible homolog of exonuclease III that provides resistance to DNA-damaging agents and limits spontaneous mutagenesis. Mol Cell Biol 19:1800–1809

147. Johnson RE, Torres-Ramos CA, Izumi T, Mitra S, Prakash S, Prakash L (1998) Identification of APN2, the *Saccharomyces cerevisiae* homolog of the major human AP endonuclease HAP1, and its role in the repair of abasic sites. Genes Dev 12:3137–3143
148. Jilani A, Ramotar D, Slack C, Ong C, Yang XM, Scherer SW, Lasko DD (1999) Molecular cloning of the human gene, PNKP, encoding a polynucleotide kinase 3′-phosphatase and evidence for its role in repair of DNA strand breaks caused by oxidative damage. J Biol Chem 274:24176–24186
149. Yang X, Tellier P, Masson JY, Vu T, Ramotar D (1999) Characterization of amino acid substitutions that severely alter the DNA repair functions of *Escherichia coli* endonuclease IV. Biochemistry 38:3615–3623
150. Vance JR, Wilson TE (2001) Repair of DNA strand breaks by the overlapping functions of lesion-specific and non-lesion-specific DNA 3′ phosphatases. Mol Cell Biol 21:7191–7198
151. Abe H, Wada M, Kohno K, Kuwano M (1994) Altered drug sensitivities to anticancer agents in radiation-sensitive DNA repair deficient yeast mutants. Anticancer Res 14:1807–1810
152. Keszenman DJ, Salvo VA, Nunes E (1992) Effects of bleomycin on growth kinetics and survival of *Saccharomyces cerevisiae*: a model of repair pathways. J Bacteriol 174:3125–3132
153. Resnick MA, Martin P (1976) The repair of double-strand breaks in the nuclear DNA of *Saccharomyces cerevisiae* and its genetic control. Mol Gen Genet 143:119–129
154. Jentsch S, McGrath JP, Varshavsky A (1987) The yeast DNA repair gene RAD6 encodes a ubiquitin-conjugating enzyme. Nature 329:131–134
155. Madura K, Prakash S, Prakash L (1990) Expression of the *Saccharomyces cerevisiae* DNA repair gene RAD6 that encodes a ubiquitin conjugating enzyme, increases in response to DNA damage and in meiosis but remains constant during the mitotic cell cycle. Nucleic Acids Res 18:771–778
156. He CH, Masson JY, Ramotar D (1996) A *Saccharomyces cerevisiae* phleomycin-sensitive mutant, ph140, is defective in the RAD6 DNA repair gene. Can J Microbiol 42:1263–1266
157. Mages GJ, Feldmann HM, Winnacker EL (1996) Involvement of the *Saccharomyces cerevisiae* HDF1 gene in DNA double-strand break repair and recombination. J Biol Chem 271:7910–7915
158. Feldmann H, Driller L, Meier B, Mages G, Kellermann J, Winnacker EL (1996) HDF2, the second subunit of the Ku homologue from *Saccharomyces cerevisiae*. J Biol Chem 271: 27765–27769
159. Downs JA, Lowndes NF, Jackson SP (2000) A role for *Saccharomyces cerevisiae* histone H2A in DNA repair. Nature 408:1001–1004
160. Aouida M, Leduc A, Wang H, Ramotar D (2004) Characterization of a transport and detoxification pathway for the antitumour drug bleomycin in *Saccharomyces cerevisiae*. Biochem J 384:47–58
161. Steinmetz LM, Scharfe C, Deutschbauer AM, Mokranjac D, Herman ZS, Jones T, Chu AM, Giaever G, Prokisch H, Oefner PJ, Davis RW (2002) Systematic screen for human disease genes in yeast. Nat Genet 31:400–404
162. Chang M, Bellaoui M, Boone C, Brown GW (2002) A genome-wide screen for methyl methanesulfonate-sensitive mutants reveals genes required for S phase progression in the presence of DNA damage. Proc Natl Acad Sci USA 99:16934–16939
163. Aouida M, Page N, Leduc A, Peter M, Ramotar D (2004) A genome-wide screen in *Saccharomyces cerevisiae* reveals altered transport as a mechanism of resistance to the anticancer drug bleomycin. Cancer Res 64:1102–1109
164. Berg K, Dietze A, Kaalhus O, Hogset A (2005) Site-specific drug delivery by photochemical internalization enhances the antitumor effect of bleomycin. Clin Cancer Res 11:8476–8485
165. van Roermund CW, Hettema EH, van den Berg M, Tabak HF, Wanders RJ (1999) Molecular characterization of carnitine-dependent transport of acetyl-CoA from peroxisomes to mitochondria in *Saccharomyces cerevisiae* and identification of a plasma membrane carnitine transporter, Agp2p. EMBO J 18:5843–5852
166. Enomoto A, Wempe MF, Tsuchida H, Shin HJ, Cha SH, Anzai N, Goto A, Sakamoto A, Niwa T, Kanai Y, Anders MW, Endou H (2002) Molecular identification of a novel carnitine

transporter specific to human testis. Insights into the mechanism of carnitine recognition. J Biol Chem 277:36262–36271
167. Tamai I, Ohashi R, Nezu J, Yabuuchi H, Oku A, Shimane M, Sai Y, Tsuji A (1998) Molecular and functional identification of sodium ion-dependent, high affinity human carnitine transporter OCTN2. J Biol Chem 273:20378–20382
168. Einhorn LH (2002) Curing metastatic testicular cancer. Proc Natl Acad Sci USA 99: 4592–4595
169. Aouida M, Poulin R, Ramotar D (2010) The human carnitine transporter SLC22A16 mediates high affinity uptake of the anticancer polyamine analogue bleomycin-A5. J Biol Chem 285:6275–6284
170. Canellos GP (2004) Lymphoma: present and future challenges. Semin Hematol 41:26–31
171. Sato N, Ito K, Onogawa T, Akahira J, Unno M, Abe T, Niikura H, Yaegashi N (2007) Expression of organic cation transporter SLC22A16 in human endometria. Int J Gynecol Pathol 26:53–60

ATR as a Therapeutic Target

Fiona K. Middleton and Nicola J. Curtin

1 An Introduction to ATR

Ataxia Telangiectasia Mutated and Rad3-related (ATR) is a vital sensor of a variety of DNA lesions and is critical to cell cycle arrest at the S and G2 checkpoints as well as initiation of DNA repair via homologous recombination repair (HRR). ATR is a member of the PI-3K like family of kinases (PIKKs), which include Ataxia Telangiectasia Mutated (ATM) and DNA-PK_{CS}(DNA-dependent protein kinase catalytic subunit) [1]; protein kinases that are also involved in the complex network of DNA damage signalling and repair mechanisms known as the DNA damage response (DDR). The DDR comprises sensor proteins which detect the DNA damage and signal to transducer proteins, e.g. p53 and checkpoint kinases which then transmit this information to downstream effector proteins. These effectors activate the appropriate damage response, be it cell cycle arrest and DNA repair or apoptosis. Many of the phosphorylation substrates of ATR are also common to ATM, and the two are both involved in HRR in response to double strand breaks (DSBs). There is also crosstalk between the two PIKKs. ATM and ATR phosphorylate >900 sites on >700 proteins in response to DNA damage induced experimentally, highlighting the complexity of the network. The majority of phosphorylated proteins are involved in DNA replication, recombination and repair plus cell cycle regulation [2].

F.K. Middleton • N.J. Curtin (✉)
Northern Institute for Cancer Research, Newcastle University,
Paul O'Gorman Building, Framlington Place, Newcastle upon Tyne NE2 4HH, UK
e-mail: nicola.curtin@newcastle.ac.uk

L. Panasci et al. (eds.), *Advances in DNA Repair in Cancer Therapy*,
Cancer Drug Discovery and Development 72, DOI 10.1007/978-1-4614-4741-2_10,

© Springer Science+Business Media, LLC 2013

1.1 Investigating the Role of ATR

The first indications that ATR may be involved in cell cycle checkpoints came from the cloning of ATR using homology sequencing with Rad3 of *S. pombe* , which is known to be involved in cell cycle arrest [3]. The critical role of ATR for viability was established when it was discovered that $ATR^{-/-}$ mice die on embryonic day 7 [4]. This may explain why no humans with germ-line homozygous ATR deletions have been identified, which, for a long time, made characterising ATR challenging. Blastocysts from the $ATR^{-/-}$ mice could be harvested prior to death, cultured and analysed. This showed that this embryonic lethality was due to increased apoptosis and chromosomal fragmentation [4]. Such fragmentation is also observed in cells undergoing mitotic catastrophe; a characteristic of cells entering mitosis prematurely following incomplete DNA synthesis [5]. This was one of the first indications that ATR is critical to cell cycle arrest.

In humans, the only well characterised disorder associated with ATR is the autosomal-recessive disease Seckel syndrome [6]. Sufferers have a hypomorphic mutation in the ATR gene resulting in low levels of the protein and exhibit growth retardation and microcephaly; characteristics which are similar to those exhibited by sufferers of other disorders associated with impaired cell cycle arrest such as Nijmegen breakage syndrome [7]. Examination of lymphoblast cell lines from Seckel patients has shown that they exhibit a three- to seven-fold increase in chromosomal fragmentation following inhibition of DNA synthesis [8] which is similar to that seen in $ATR^{-/-}$ murine cells [4]. Interestingly, Seckel syndrome patients do not have an increased prevalence of cancer, which is surprising as in vivo studies showed $ATR^{+/-}$ mice have an increased incidence of tumour formation [4].

The role of ATR in cell cycle arrest was confirmed upon the development of cells expressing an inducible ATR kinase-dead (ATR-KD) cell line where the ATR-KD acts as a dominant negative inhibitor of the native protein [9]. When ATR was no longer active, these cells did not arrest following DNA damage and, in particular, lacked G2 arrest [9]. Further work in ATR-KD cells showed that ATR becomes active following different types of DNA damage compared to ATM [10, 11].

1.2 ATR: A Sensor of DNA Damage

ATR is activated by the single stranded–double stranded DNA (ssDNA–dsDNA) junctions, which arise principally at stalled replication forks, resected double strand breaks (DSBs) and nucleotide excision repair (NER) intermediates. Stalled replication forks occur when the dNTP pool is depleted preventing further DNA synthesis, when the number of origins of replication exceeds dNTP supply or when the replication machinery encounters a DNA lesion. This tends to occur when the advancing replication fork reaches lesions such as single-strand breaks (SSBs), bulky adducts and interstrand cross links (ICLs) [12, 13]. Under these circumstances the polymerase on the damaged strand may arrest while the opposing polymerase continues [14], thus creating the ssDNA–dsDNA structure. Many of these lesions occur endogenously; reactive oxygen species (ROS) are the major source of the 10^4–10^5 base lesions that are generated per cell each day [15] which can lead to SSBs and also the 50 DSB generated/cell/day [16]. Bulky adducts and ICLs can be created by environmental mutagens such as UV light and tobacco smoke [17].

The nucleotide excision repair (NER) pathway is used to remove bulky adducts such as UV-induced (6–4) photoproducts [11], which occur on a single strand of the DNA [18]. These adducts are removed by cleaving away 20–30 nucleotides of DNA around the damage, leaving a portion of ssDNA and thus generating the ssDNA–dsDNA structure.

Resected double strand breaks (DSBs) also create the ssDNA–dsDNA structures that activate ATR [19]. A major cause of DSBs is ionising radiation (IR). IR has been shown to primarily activate ATM rather than ATR. ATM is recruited to DSBs by the MRN complex (composed of Mre11, Rad50 and Nbs1) and upon its activation stimulates resection by phosphorylating the MRN complex and EXO [20]. The exonuclease activity then acts on one of the DNA strands causing the ssDNA–dsDNA structure to form, thus activating ATR [20]. In vitro work has shown that TopBP1, an important component in HRR, interacts with Nbs1 [21] and TopBP1 feeds back to activate ATR [22].

ATR activation is initiated by the binding of the single-stranded binding replication protein A (RPA) to the single-stranded portion of the ssDNA–dsDNA junction [23]. The role of RPA is twofold: to protect the single-stranded DNA overhang from exonuclease activity thus preventing formation of lethal DSBs, and to activate the downstream ATR pathway to initiate cell cycle arrest and DNA repair. RPA interacts with the ATR interacting protein ATRIP which is in complex with ATR, thus recruiting ATR to the DNA [24, 25]. The Rad17-RFC complex is simultaneously recruited to the DNA [26] mediating loading of the Rad9-Rad1-Hus1 (9-1-1) complex onto the DNA and thus recruiting TopBP1. TopBP1 is ultimately responsible for the activation of ATR kinase [22]. Active ATR kinase can then signal to downstream transducer proteins.

The mismatch repair system (MMR) is also important in ATR activation. MMR repairs any DNA base mismatches or insertion deletion loops that most often occur due to alkylating mutagens [18]. O^6-methylguanine is a common result of exposure to alkylating agents and is a major target of the MMR machinery. Key protein complexes within this repair system are the MutSα and MutLα complexes which are required for recognition and excision of mis-incorporated bases (including O^6-methylguanine). Evidence has shown that the ATR-ATRIP complex is recruited to O^6-methylguanine in a MutLα and MutSα-dependent manner and that this recruitment activates ATR [27]. The MSH2/MSH6 heterodimer binds the mismatched DNA, and immunoprecipitation experiments showed that ATR interacts with both of these component proteins. Furthermore, siRNA depletion of MSH2 has demonstrated that MSH2 is required for CHK1^{ser317} phosphorylation [28]. There is additional evidence of the interaction between the MMR machinery and

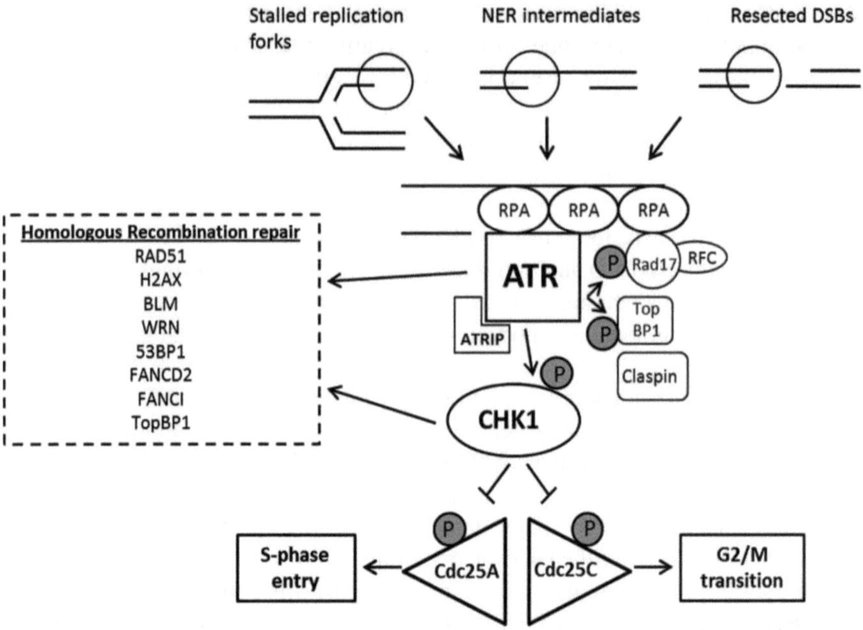

Fig. 1 Role of ATR in S/G2 checkpoint control and DNA DSB repair via homologous recombination. ATR-ATRIP is recruited by RPA to ssDNA–dsDNA junctions. The Rad17-RFC complex is also recruited to the DNA leading to TopBP1 recruitment and ultimate activation of ATR kinase. ATR phosphorylates a number of downstream targets, the best described of which is CHK1. CHK1 becomes active and phosphorylates downstream targets including the Cdc25 phosphatases. Cdc25A mediates S-phase entry by dephosphorylating CDK2/Cyclin A and Cdc25C controls G2/M transition via dephosphorylation of Cdc2/Cyclin B. Both ATR and CHK1 interact with and phosphorylate proteins involved in HRR repair

the ATR pathway as MutSα interacts with ATR, CHK1 and TopBP1 in a purified system [29]. MMR-proficient cells also form ATR foci following DNA damage with the alkylating agent temozolomide (TMZ) whereas their MMR-deficient counterparts do not [30].

1.3 Involvement of ATR in S/G2 Arrest

To prevent any damaged DNA being replicated, the cell must first arrest its cell cycle in order to repair this damage. ATR kinase is pivotal to arrest at the S and G2 checkpoints (Fig. 1). A key downstream target of ATR is CHK1. When Rad17 is recruited to the DNA, it interacts with claspin [31]; a protein responsible for the regulation of CHK1 phosphorylation by ATR. CHK1 is an essential kinase [32] that is phosphorylated by ATR on serine residues 317 and 345. Phosphorylation of CHK1^{ser345} by ATR is essential for CHK1 kinase activation [33] and this phosphorylation event is often used as a marker of ATR activity [34–36]. Upon phosphorylation at these residues CHK1 becomes active triggering autophosphorylation at serine 296 [36]. CHK1 then dissociates from the chromatin [37] and can phosphorylate a number of targets. Cell cycle arrest is achieved by phosphorylation of the cell cycle Cdc25 checkpoint phosphatase proteins rendering them inactive [38]. Inactive Cdc25A dephosphorylates the CDK2/Cyclin A or E complex that promotes S-phase entry [39]. Similarly, Cdc25C activates the Cdc2/Cyclin B complex by removing the inactivating phosphates on threonine 14 and tyrosine 15 of Cdc2 [39] to promote entry into mitosis. These two residues are phosphorylated by Wee1 [40]. The effect of CHK1 on cell cycle progression is therefore threefold: Cdc25A is phosphorylated and inactivated preventing S phase arrest, Wee1 is phosphorylated and stabilised [40] resulting in phosphorylation of Cdc2 at Thr14 and Tyr15, and Cdc25C is phosphorylated and inactivated allowing Cdc2 to remain phosphorylated thus preventing the G2/M transition.

1.4 ATR's Response to Endogenous Damage

While much of the research surrounding ATR concentrates on its role following exogenous DNA damage, it is important to remember that much of the damage incurred by the cell is from endogenous or environmental sources. As previously mentioned, ROS generated from metabolism are responsible for a plethora of lesions that occur on the DNA [15]. Skin cells are also exposed to UV radiation on a daily basis creating bulky adducts [11]. Evidence has shown that ATR is also vital in the absence of exogenous genotoxic stress [41]. This was initially demonstrated in cells from the blastocysts of $ATR^{-/-}$ mice where 60–65% of mitotic spreads contained fragmented chromosomes compared to the 0–2% seen in $ATR^{+/+}$ and $ATR^{+/-}$ mice [4]. Investigations have shown that ATR is associated with chromatin in normal proliferating cells to a greater extent at the S phase of the cell cycle [23] where the chromatin is most vulnerable to fragmentation. ATR has been shown to maintain fragile site stability as siRNA knockdown of ATR or expression of an ATR-KD mutant increases the average number of chromosomal breaks per cell when DNA replication is reduced using the replication inhibitor aphidicolin [42]. Furthermore, many of the proteins associated with ATR and its activation such as Rad17, TopBP1 and claspin are also associated with chromatin during unperturbed S-phase [43].

1.5 ATR and DNA Damage Repair

As well as arresting the cell cycle at the S and G2 checkpoints, ATR is pivotal to stabilising replication forks and initiating DNA damage repair (Fig. 1). This repair is primarily via HRR; however, there is considerable crosstalk between ATR and other

DNA damage repair pathways [41]. Some of this crosstalk is due to the formation of the ssDNA–dsDNA junction in repair pathway intermediates, as previously mentioned in the case of NER. ATR interacts with machinery from other pathways such as the non-homologous end joining (NHEJ) pathway, where ATR phosphorylates DNA-PK$_{CS}$ [44], and the ICL and translesion synthesis (TLS) pathways in which ATR is activated by RPA and phosphorylates FANCD2 allowing cross links to ultimately be removed [41, 45]. Removal of these cross links leaves behind a gap which must be repaired via HRR. Evidence has also shown that G2 arrest to correct the mismatch repair (MMR) substrate 6-thioguanine is ATR dependent [46], and that ATR may indirectly influence the base excision repair (BER) pathway [47].

The repair pathway in which ATR is most strongly associated is the HRR pathway which is also activated by ATM [48]. The precise role of ATR in HRR is yet to be confirmed; however, the plethora of experimental evidence surrounding ATR and other proteins implicated in HRR has demonstrated that it has a significant role in this DNA repair pathway. HRR is only active during S and G2-phases of the cell cycle as it relies on the use of the homologous sister chromatid as a template for DNA repair. It is a complex repair pathway in which the ssDNA overhang, generated at stalled replication forks or following the ATM-dependent resection of DSBs [49, 50], is rapidly coated with RPA, preventing DNA degradation and recruiting the ATRIP-ATR complex. ATM and ATR both phosphorylate BRCA1 stimulating its E3 ubiquitin ligase activity that is needed for activation of the G2 checkpoint [51]. BRCA2, which also interacts with PALB2 and BRCA1 [52] delivers RAD51 to the break and aids its displacement of RPA to form the nucleoprotein filament that can invade the complementary duplex DNA [53–55]. The sequence from this chromatid is then replicated at the point of damage, making this repair pathway error-free [18].

Phosphorylation of HRR proteins by ATR is critical to HRR proficiency. HRR assays in ATR-KD cells have shown that cells lacking ATR kinase function have reduced levels of HRR [56]. CHK1 is also vital for HRR. An HRR model system where only colonies that have undergone HRR survive was used to show that inhibition of CHK1 and ATR by UCN-01 and caffeine, respectively, reduces HRR by three- to fourfold. Inhibition of CHK1 by UCN-01 or siRNA knockdown of CHK1 also reduces RAD51 focus formation, further confirming the requirement of active CHK1 in HRR [57].

Histone H2AX is a substrate common to both ATR and ATM, and is required for DSB repair. Phosphorylation of H2AX (γ-H2AX) is a marker of DSB and occurs in an ATR-dependent manner following treatment with the replication-arresting agent hydroxyurea (HU) and also UV [58]. Investigations have indicated that H2AX facilitates HRR, as γ-H2AX foci co-localise with foci of repair proteins including RAD51, BRCA1 and 53BP1 [58, 59].

A number of other HRR-associated proteins are directly phosphorylated by ATR. These include BLM which is a RecQ helicase that is deficient in Bloom's syndrome; a condition that predisposes to cancer [60]. BLM has been shown to be directly phosphorylated by ATR, and co-localises with ATR, RAD51 and γ-H2AX in response to HU [61].

Another RecQ helicase—WRN, deficient in Werner's syndrome—also interacts with and is phosphorylated by ATR. WRN co-localises with the HRR proteins RAD51 and RAD54 in response to the DNA cross-linking agent Mitomycin C (MMC), and immunoprecipitates with ATR demonstrating a direct interaction between these proteins [62]. ATR, BLM, WRN and 53BP1 interact with one another to promote RAD51 foci formation [63, 64]. These interactions are dependent on ATR activity; siRNA knockdown of ATR prevents phosphorylation of BLM and reduces formation of 53BP1 foci in response to HU. BLM and 53BP1 no longer co-localise in cells with siRNA knockdown of CHK1 following the same treatment [63].

More recently, the role of ATR in phosphorylating proteins in the Fanconi anaemia pathway has been highlighted. These proteins are involved in removing DNA ICLs and initiating repair of the DNA via HRR [65–67]. Following treatment with MMC, cells with active ATR and ATRIP phosphorylate FANCI and FANCD2, the nuclease responsible for DNA incisions on either side of the cross link, and target it for ubiquitylation and degradation [45].

1.6 Importance of the Target in Cancer Therapy

ATR has long been thought of as a suitable target for anticancer therapy because of the variety of DNA lesions that activate it. Many of the anticancer agents that are in routine clinical use act by damaging the DNA either by causing bulky adducts that are repaired by NER, e.g. cisplatin, or by inducing DNA DSBs, e.g. IR or topoisomerase II poisons (such as doxorubicin, mitoxantrone or etoposide), or stalled replication forks. Stalled replication forks occur when nucleoside triphosphates (dNTPs) are in limited supply e.g., due to HU therapy or when unrepaired DNA lesions encounter the advancing replication fork. Such lesions may be (1) single base damage induced by DNA methylating agents such as temozolomide (TMZ) or dacarbazine (DTIC) or oxidative damage resulting from IR-induced ROS, (2) single-strand breaks induced by topoisomerase I poisons such as irinotecan or topotecan, (3) or the most difficult of all lesions to repair, ICLs, induced by cisplatin and the bifunctional alkylating agents. Since all these lesions trigger ATR to promote survival, inhibition of ATR should promote cell killing.

It is important that the enhanced cell killing is directed at tumour cells and not normal tissues otherwise there would be no patient benefit. There are two key characteristics of cancer that indicate that ATR is likely to be a highly attractive target for *selective* cancer therapy: (1) continuous proliferation and (2) dysregulation of their G1 control [68], making them reliant on their remaining S and G2 checkpoints (Fig. 2). This situation can be brought about by multiple mechanisms, for example, activation of oncogenes that drive proliferation or up-regulation of the cyclins and CDKs that promote S-phase entry and/or loss of tumour-suppressor genes such as p53 and Rb that control entry into S-phase. This means that cancerous cells are much more likely than normal cells to enter S-phase with

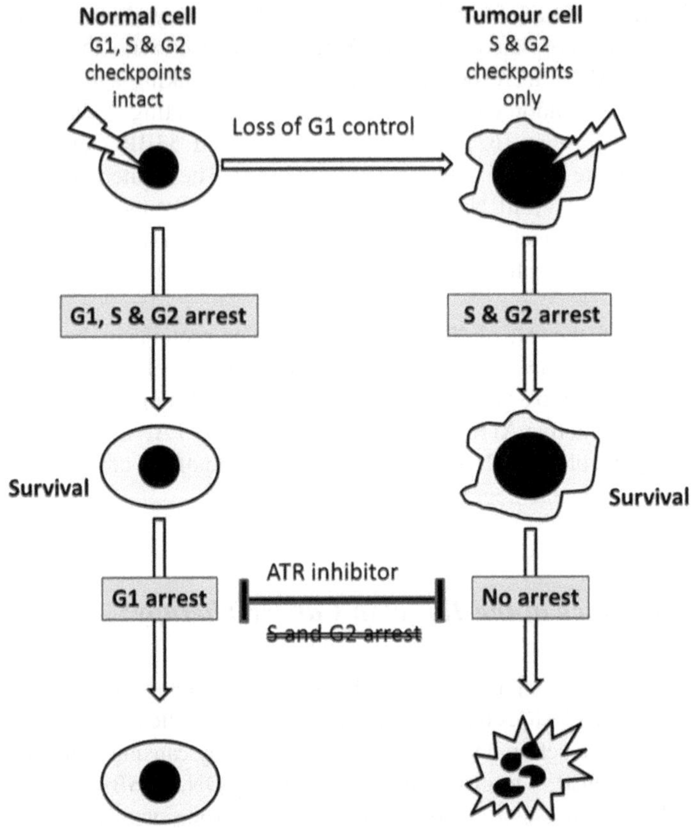

Fig. 2 Selective sensitization of cancer cells with dysfunctional G1 control by ATR inhibition. Loss of G1 control, e.g. by mutation of p53, frequently occurs during neoplastic transformation. While the normal cell (*left*) has all cell cycle checkpoints intact, the cancer cell relies on the S and G2 checkpoints it has retained. Both normal and cancer cells may be able to survive DNA damage (*lightening flash symbol*) by activation of cell cycle checkpoints. However, if ATR is inhibited the cancer cell will be unable to arrest and will die, but the normal cell may engage the G1 checkpoint and survive

damaged DNA or when appropriate biomolecules (e.g. dNTPs) are limited. Therefore, they have a high level of replicative stress, stalled replication forks and replication-associated DSBs. In addition, there are higher levels of ROS in tumours due to a number of factors, including increased metabolic activity, mitochondrial dysfunction, various oxidases [69, 70] and inflammation [71]. These give rise to approximately 100-fold higher levels of oxidative DNA lesions in tumours than in normal tissues [72]. It is apparent therefore that the DNA in a tumour cell is in a more fragile state. In response to this damage tumour cells lacking G1 control are much more likely to be dependent than normal cells on the S and G2 checkpoint function that they retain.

1.7 Validation of the Target by Genetic Inactivation

Abrogation of the S and G2 checkpoint has been the major rationale for the development of inhibitors of the ATR/CHK1 pathway. Whereas CHK1 inhibitors are in advanced clinical evaluation [73] ATR inhibitor development has lagged behind, which may reflect the difficulty of assaying an enzyme that requires a complex of co-activators and regulators [74]. Until recently, most of the studies demonstrating chemo- and radiosensitisation by ATR inactivation have used genetic manipulation. Two early independent "proof of principle" genetic studies with cells expressing ATR-KD mutants demonstrated abrogation of DNA damage-induced G2 arrest and sensitization of cells to a variety of DNA damaging chemotherapeutic agents [9, 75]. In the first study human SV40 transformed fibroblasts transfected with a doxycycline-inducible ATR-KD mutant were used. When the ATR-KD was expressed clonogenic survival following exposure to the DNA cross-linking agents cisplatin and MMC and the antimetabolite HU was profoundly reduced with a modest increase in UV sensitivity [9]. Further studies by this group also demonstrated sensitization to topoisomerase I poisons by activation of the ATR-KD in these cells [10]. In the second study, premature chromatin condensation (PCC) was used to demonstrate cell death in ATR-KD U2OS cells exposed to UV irradiation or HU [75]. These studies demonstrated that the previously observed enhancement of HU cytotoxicity by caffeine was due to ATR inhibition. Another approach used cells transfected with the Seckel mutant ATR that have very low levels of ATR activity. These cells were sixfold more sensitive to IR and the topoisomerase II poison doxorubicin, 10 to 20-fold more sensitive to the antimetabolites 5-Fluorouracil, gemcitabine, HU and methotrexate and >400-fold more sensitive to cisplatin than isogenic ATR expressing DLD1 cells [76]. ATR knockdown also caused a profound sensitization to cisplatin and gemcitabine in HeLa, HCT116 and U2OS cells [77]. ATR knockdown has also been shown to enhance the cytotoxicity of DNA methylating agents, such as TMZ, [78], and this may be dependent on a functional MMR pathway [30].

It has been postulated that targeting of the ATR/CHK1 pathway is only relevant in cells with defective G1 control through loss of the tumour suppressor gene p53. However, the selectivity of ATR inactivation may not be restricted to p53 defective cells. ATR silencing sensitised both HeLa (p53 defective) and U2OS (p53 wild-type) to topoisomerase I poisons [79]. It should be noted that U2OS cells do have an element of G1 dysfunction by virtue of p16 deletion, and following further impairment of the Rb pathway by over-expression of cyclin D, cyclin E or CDK2, there was enhanced UV-induced PCC in the ATR-KD cells, but not ATR-wt U2OS cells [75]. Inactivation of the p53 pathway in these ATR-KD U2OS cells, by MDM2 or human papilloma virus E6 expression, also increased the level of PCC threefold. This confirmed previous studies indicating that abrogation of G2 arrest and radiosensitisation by caffeine is greater in p53 null cells than p53 wt cells [80]. Sensitization was specific to replicating cells and selective to cells defective in the G1 checkpoint, although there is not a clear relationship to p53 status.

1.8 Development of Inhibitors

Caffeine (Fig. 3) was the first small molecule inhibitor of G2 arrest [81] to be identified as an inhibitor of ATR [82]; however it was weak (IC_{50} = 1.1 mM) and was a more potent inhibitor of the other PI-3K family members ATM and mTOR, and weakly active against DNA-PK. Radiosensitisation by caffeine was attributed to both ATM and ATR inhibition but UV sensitization was dependent on ATR inhibition rather than ATM inhibition [82]. Enhancement of UV-induced PCC was also specific to ATR [75]. Despite its lack of potency, its easy availability has led to several studies investigating abrogation of S and G2 checkpoints and chemo and radiosensitisation by caffeine. Caffeine also enhanced DNA damage and the cytotoxicity of TMZ in combination with IR in a glioblastoma model [83]. Schisandrin B (Fig. 3), a natural product, was identified as an inhibitor of ATR with an IC_{50} of 7.25 μM, and abrogated the UV-induced S and G2/M checkpoint and increased UV cytotoxicity in human lung cancer cells [84]. In a screen of PI-3K inhibitors PI-103 and PI-124 (Fig. 3) were identified as being more potent than previously identified ATR inhibitors with IC_{50} values of 0.9 and 2 μM, respectively [85]; however, these inhibitors have not been taken forward as ATR inhibitors due to their pan PI-3K family activity.

Recently, progress has been made on two fronts; firstly, by development of a high throughput cell-based screen to measure ATR activity and its inhibition and secondly, by the identification of novel inhibitors. The cell-based screen employed cells expressing a fusion of the ATR-activating domain of TopBP1 with a fragment of the oestrogen receptor such that ATR was activated on exposure to tamoxifen. ATR activity was then measured by phosphorylation of histone H2AX by immunofluorescence. Using this screen NVP-BEZ235 (Fig. 3), which had previously been thought to be selective for PI-3K and mTOR, was demonstrated to be a potent inhibitor of ATR (IC_{50} = 100 nM) [86]. The most potent ATR inhibitor, ETP-46464 (IC_{50} = 25 nM) (Fig. 3), identified using the screen inhibited the restart of stalled replication forks and abrogated S-phase arrest after HU exposure.

Three novel small molecules, VE-821, AZ-20 and NU6027 (Fig. 3), have recently been identified as being ATR inhibitors [34, 87–89]. All compounds inhibited CHK1 phosphorylation at Ser^{345} but there were some subtle differences in their cytotoxicity. NU6027 (K_i = 100 nM in biochemical assays and IC_{50} = 6.7 μM in cell-based assays) enhanced the sensitivity of MCF7 cells to representatives of the major classes of DNA damaging agents: IR, DNA methylating agents (TMZ), antimetabolites (HU), topoisomerase I and II poisons (camptothecin and doxorubicin), and the DNA cross-linking agent, cisplatin, but not to the antitubulin agent, paclitaxel [34]. The specificity of NU6027 for ATR was confirmed using ATR-KD cells. In line with previous suggestions that TMZ and 6-thioguanine potentiation by ATR inactivation required an intact mismatch repair system [30, 46], NU6027 potentiation of TMZ was greater in MMR-defective cells than in the parental and MMR corrected

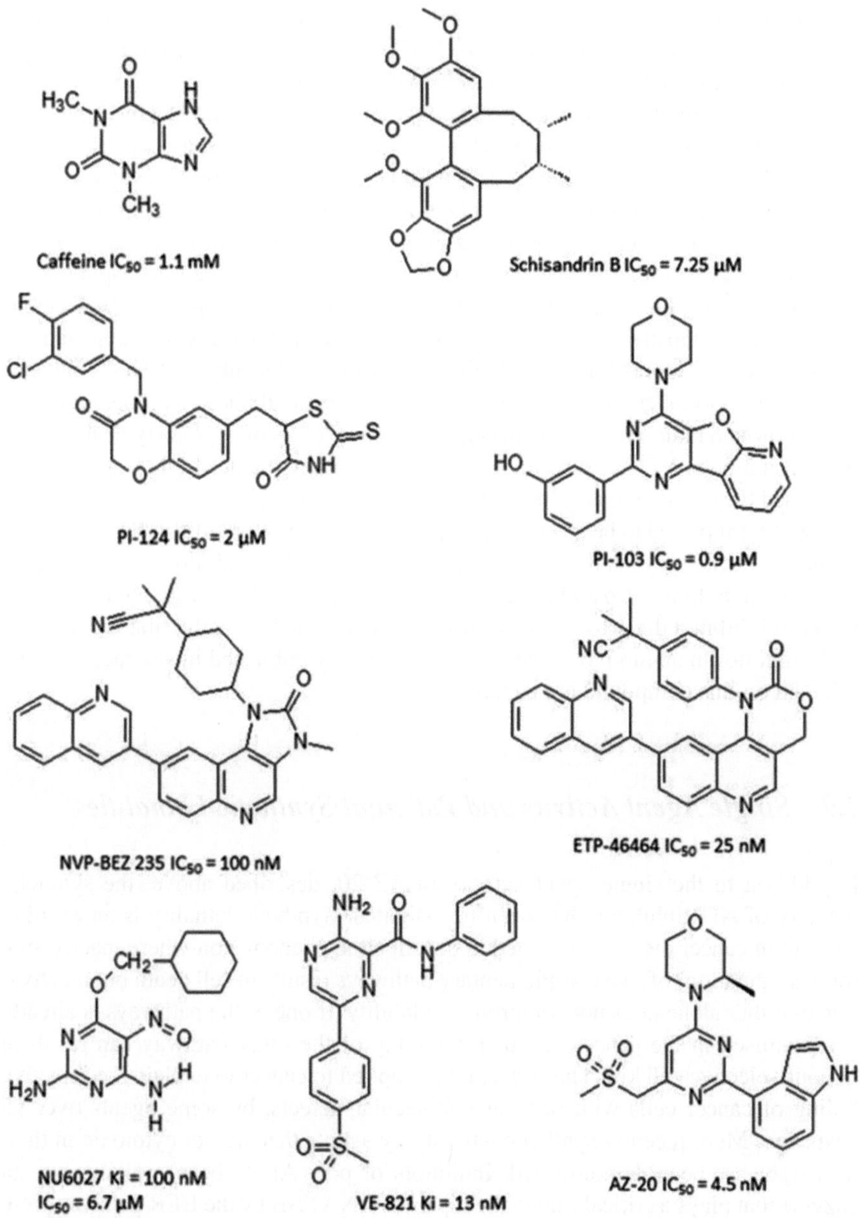

Fig. 3 Chemical structures of ATR inhibitors. Structurally diverse inhibitors of increasing potency have been identified since the prototype inhibitor, caffeine, was first described including the natural product Schisandrin B, pan PI-3K inhibitors (PI-103 and PI-124), those identified by high throughput screening (NVP-BEZ 235 and ETP-46464) and novel small molecule inhibitors (NU6027, VE-821 and AZ-20)

counterparts [34]. In the studies with NU6027 both p53 wt and mutant cell lines were potentiated with sensitization of TMZ being greatest in p53 mutant cells but cisplatin sensitization was greater in cells with functional p53.

VE-821 is a highly potent and specific ATR inhibitor (K_i = 13 nM), which enhanced the cytotoxicity of cisplatin, gemcitabine, camptothecin, etoposide and IR, but not docetaxel in HCT116 colon cancer cells. Cisplatin potentiation by VE-821 was much greater in a panel of human cancer cell lines in comparison to normal human cell lines and was more active in cells lacking p53 or ATM. In contrast to NU6027, VE-821 only caused a very modest sensitization of wild-type p53 MCF7 cells to cisplatin [88]. Recently, VE-821 has been investigated as a radiosensitiser in hypoxic cells. Regions of hypoxia develop in solid tumours because of inefficient tumour vasculature and this contributes to chemoresistance and radioresistance [90]. Not only did VE-821 enhance IR-induced cytotoxicity in a panel of 12 human cancer cell lines, but it also caused a more profound radiosensitisation in cells cultured in levels of hypoxia typically found in solid tumours. VE-821 also increased re-oxygenation-induced DNA damage and decreased the survival of cells undergoing re-oxygenation [91].

AZ-20 is reported to be an even more potent ATR inhibitor with an IC_{50} of 4.5 nM in biochemical assays and 51 nM in cellular assays. This inhibitor was active as a single agent both in vitro and in vivo, and at an oral dose of 25 mg/kg bid or 50 mg/kg qd; it inhibited the growth of LoVo xenografts [89]. This is the first report of an ATR inhibitor in an in vivo model and although only published in abstract form the full data on this compound are eagerly awaited.

1.9 Single Agent Activity and Potential Synthetic Lethalities

In addition to the single agent activity of AZ-20, described above, the synthetic lethality of ATR inhibitors has been investigated. Synthetic lethality is an exciting concept in cancer therapy; it is used to describe the phenomenon where inactivation (or dysregulation) of two complementary pathways results in cell death but inactivation of either alone does not compromise viability. If one of the pathways is already compromised in the cancer cell then targeting of the other pathway can result in tumour-selective cell kill. The concept was applied to cancer to explain the selective killing of cancer cells with particular molecular defects, by some agents over 15 years ago. More recently synthetic lethality by agents that are not cytotoxic in their own right has been demonstrated. Inhibitors of poly(ADP-ribose) polymerase, an enzyme that plays a critical role in the repair of DNA SSB by the BER pathway were profoundly cytotoxic to HRR-defective cancer cells but did not affect the viability of cells with functional HRR [92, 93]. As described above, ATR plays an important role in HRR and NU6027 inhibited RAD51 focus formation (indicative of HRR suppression), so it was a logical extension of this work to investigate ATR inhibition in cells with BER defects. NU6027 was more cytotoxic in cells lacking the BER scaffold protein, XRCC1, and also in the presence of a PARP inhibitor, suggesting the potential for synthetic lethality [34]. These findings are relevant to the molecular biology of cancer because polymorphisms in XRCC1 and other defects in DNA single-strand break repair e.g. those due to aberration in DNA polβ are also associated with cancer [94, 95] and this may be exploitable by ATR inhibition. Other recent data demonstrate that caffeine selectively radiosensitises polβ defective cells [96] implicating that ATR inhibition would have broad applicability in cancer.

It is well recognised that oncogene activation and hyperactive growth factor signalling itself causes oncogenic stress, characterised by stalled/collapsed replication forks, making such cancer cells particularly dependent on the ATR pathway for survival [97]. It was exciting to discover that inactivation of ATR or CHK1 is synthetically lethal in oncogene-activated cancer cells. Knocking down ATR to 16% of normal levels was synthetically lethal in ras-transformed cells [98]. By analogy, inhibition of both CHK1/CHK2 with AZD7762 induced cell death and significantly delayed disease progression of Myc-over-expressing lymphoma cells in vivo [99, 100]. Cyclin E, which promotes S-phase entry, is commonly over-expressed in cancer and leads to replication stress and DNA damage. The ATR inhibitor, ETP-46464 was cytotoxic to cells with induced over-expression of cyclin E [86].

1.10 Differences Between ATR Inhibitors and CHK1 Inhibitors

A number of CHK1 inhibitors are undergoing clinical evaluation and it could be argued that ATR inhibitors will have a very similar profile. However, although CHK1 is thought to be the major target of ATR, ATR inhibitors are not necessarily the same as CHK1 inhibitors and key differences have been observed. For example the inhibition of cisplatin, carboplatin and oxaliplatin by ATR siRNA seen in a variety of cell lines was not replicated when CHK1 was inhibited [77]. DLD-1 cells expressing Seckel mutant ATR were much more sensitive to cisplatin and MMC than those expressing mutant CHK1; indeed the ATR mutant cells were generally more sensitive to a range of cytotoxic drugs showing the importance of other ATR substrates [101]. The effects of the CHK1 inhibitor, PF-00477736, was compared with those of NU6027: unlike NU6027, PF-00477736 caused only modest sensitization of cisplatin and camptothecin but did enhance the cytotoxicity of paclitaxel in MCF7 cells in line with previous studies with PF-00477736 [102]. Interestingly, PF-00477736 did not inhibit the formation of RAD51 foci and, instead, increased them, suggesting that PF-00477736 stimulates HRR rather than inhibiting it [34]. Clearly therefore inhibitors of ATR will have a different spectrum of activity compared to CHK1 inhibitors.

1.11 Pharmacodynamic Biomarkers of ATR Inhibition

Since ATR inhibitors are proposed to have minimal toxicity in normal cells MTD may not be the best endpoint for clinical trials with ATR inhibitors. Pharmacodynamic biomarkers of ATR inhibition are needed to guide these trials. One possibility is

CHK1 phosphorylation, which has been used in in vitro studies. However, it may be easier to use a general marker of DNA damage signalling, such as the phosphorylation of histone H2AX (γH2AX) by ATR. Although not specific for ATR, as ATM and DNA-PK also phosphorylate H2AX in response to DNA DSB and stalled replication forks, methods already exist to determine γH2AX nuclear foci, or levels by immunofluorescence microscopy, flow cytometry or immunoblotting.

1.12 Future Prospects

The accumulating preclinical data demonstrates that the inactivation of ATR, by genetic means or small molecule inhibitors, enhances the cytotoxicity of all classes of DNA damaging anticancer agents. Moreover, these data support the premise that sensitization is greatest in cells with dysfunctional G1 control that distinguishes tumour cells from normal tissues. This evidence indicates that ATR inhibitors will have broad application as chemo- and radiosensitisers with minimal toxicities. In the last few years a number of potent and chemically diverse small molecule inhibitors of ATR have been identified and it is to be hoped that these will enter clinical evaluation in the near future.

Perhaps the most exciting aspect is the potential for synthetic lethality of ATR inhibitors in cells lacking BER or undergoing replication stress due to oncogene activation, amplification of growth factor signalling or cell cycle pathways. More work is needed to confirm these initial findings and establish biomarkers for the determinants of sensitivity to ATR inhibitors.

References

1. Durocher D, Jackson SP (2001) DNA-PK, ATM and ATR as sensors of DNA damage: variations on a theme? Curr Opin Cell Biol 13(2):225–231
2. Matsuoka S et al (2007) ATM and ATR substrate analysis reveals extensive protein networks responsive to DNA damage. Science 316(5828):1160–1166
3. Bentley NJ et al (1996) The Schizosaccharomyces pombe rad3 checkpoint gene. EMBO J 15(23):6641–6651
4. Brown EJ, Baltimore D (2000) ATR disruption leads to chromosomal fragmentation and early embryonic lethality. Genes Dev 14(4):397–402
5. Schlegel R, Pardee AB (1986) Caffeine-induced uncoupling of mitosis from the completion of DNA replication in mammalian cells. Science 232(4755):1264–1266
6. O'Driscoll M et al (2003) A splicing mutation affecting expression of ataxia-telangiectasia and Rad3-related protein (ATR) results in Seckel syndrome. Nat Genet 33(4):497–501
7. Tauchi H et al (2002) Nijmegen breakage syndrome gene, NBS1, and molecular links to factors for genome stability. Oncogene 21(58):8967–8980
8. Casper AM et al (2004) Chromosomal instability at common fragile sites in Seckel syndrome. Am J Hum Genet 75(4):654–660
9. Cliby WA et al (1998) Overexpression of a kinase-inactive ATR protein causes sensitivity to DNA-damaging agents and defects in cell cycle checkpoints. EMBO J 17(1):159–169

10. Cliby WA et al (2002) S phase and G2 arrests induced by topoisomerase I poisons are dependent on ATR kinase function. J Biol Chem 277(2):1599–1606
11. Unsal-Kacmaz K et al (2002) Preferential binding of ATR protein to UV-damaged DNA. Proc Natl Acad Sci USA 99(10):6673–6678
12. Saleh-Gohari N et al (2005) Spontaneous homologous recombination is induced by collapsed replication forks that are caused by endogenous DNA single-strand breaks. Mol Cell Biol 25(16):7158–7169
13. Burhans WC, Weinberger M (2007) DNA replication stress, genome instability and aging. Nucleic Acids Res 35(22):7545–7556
14. Paulsen RD, Cimprich KA (2007) The ATR pathway: fine-tuning the fork. DNA Repair (Amst) 6(7):953–966
15. Lindahl T (1993) Instability and decay of the primary structure of DNA. Nature 362(6422): 709–715
16. Vilenchik MM, Knudson AG (2003) Endogenous DNA double-strand breaks: production, fidelity of repair, and induction of cancer. Proc Natl Acad Sci USA 100(22):12871–12876
17. Hang B (2010) Formation and repair of tobacco carcinogen-derived bulky DNA adducts. J Nucleic Acids 2010:709521
18. Hoeijmakers JH (2001) Genome maintenance mechanisms for preventing cancer. Nature 411(6835):366–374
19. Shiotani B, Zou L (2009) Single-stranded DNA orchestrates an ATM-to-ATR switch at DNA breaks. Mol Cell 33(5):547–558
20. Symington LS, Gautier J (2011) Double-strand break end resection and repair pathway choice. Annu Rev Genet 45:247–271
21. Morishima K et al (2007) TopBP1 associates with NBS1 and is involved in homologous recombination repair. Biochem Biophys Res Commun 362(4):872–879
22. Delacroix S et al (2007) The Rad9-Hus1-Rad1 (9-1-1) clamp activates checkpoint signaling via TopBP1. Genes Dev 21(12):1472–1477
23. Dart DA et al (2004) Recruitment of the cell cycle checkpoint kinase ATR to chromatin during S-phase. J Biol Chem 279(16):16433–16440
24. Zou L, Elledge SJ (2003) Sensing DNA damage through ATRIP recognition of RPA-ssDNA complexes. Science 300(5625):1542–1548
25. Itakura E et al (2004) ATR-dependent phosphorylation of ATRIP in response to genotoxic stress. Biochem Biophys Res Commun 323(4):1197–1202
26. Zou L, Liu D, Elledge SJ (2003) Replication protein A-mediated recruitment and activation of Rad17 complexes. Proc Natl Acad Sci USA 100(24):13827–13832
27. Yoshioka K, Yoshioka Y, Hsieh P (2006) ATR kinase activation mediated by MutSalpha and MutLalpha in response to cytotoxic O6-methylguanine adducts. Mol Cell 22(4):501–510
28. Wang Y, Qin J (2003) MSH2 and ATR form a signaling module and regulate two branches of the damage response to DNA methylation. Proc Natl Acad Sci USA 100(26):15387–15392
29. Liu Y et al (2010) Interactions of human mismatch repair proteins MutSalpha and MutLalpha with proteins of the ATR-Chk1 pathway. J Biol Chem 285(8):5974–5982
30. Caporali S et al (2004) DNA damage induced by temozolomide signals to both ATM and ATR: role of the mismatch repair system. Mol Pharmacol 66(3):478–491
31. Wang X et al (2006) Rad17 phosphorylation is required for claspin recruitment and Chk1 activation in response to replication stress. Mol Cell 23(3):331–341
32. Takai H et al (2000) Aberrant cell cycle checkpoint function and early embryonic death in Chk1(–/–) mice. Genes Dev 14(12):1439–1447
33. Walker M et al (2009) Chk1 C-terminal regulatory phosphorylation mediates checkpoint activation by de-repression of Chk1 catalytic activity. Oncogene 28(24):2314–2323
34. Peasland A et al (2011) Identification and evaluation of a potent novel ATR inhibitor, NU6027, in breast and ovarian cancer cell lines. Br J Cancer 105(3):372–381
35. Liu Q et al (2000) Chk1 is an essential kinase that is regulated by Atr and required for the G(2)/M DNA damage checkpoint. Genes Dev 14(12):1448–1459

36. Parsels LA et al (2011) Assessment of chk1 phosphorylation as a pharmacodynamic biomarker of chk1 inhibition. Clin Cancer Res 17(11):3706–3715
37. Sorensen CS, Syljuasen RG (2012) Safeguarding genome integrity: the checkpoint kinases ATR, CHK1 and WEE1 restrain CDK activity during normal DNA replication. Nucleic Acids Res 40(2):477–486
38. Chen MS, Ryan CE, Piwnica-Worms H (2003) Chk1 kinase negatively regulates mitotic function of Cdc25A phosphatase through 14-3-3 binding. Mol Cell Biol 23(21):7488–7497
39. Dai Y, Grant S (2010) New insights into checkpoint kinase 1 in the DNA damage response signaling network. Clin Cancer Res 16(2):376–383
40. Lee J, Kumagai A, Dunphy WG (2001) Positive regulation of Wee1 by Chk1 and 14-3-3 proteins. Mol Biol Cell 12(3):551–563
41. Flynn RL, Zou L (2011) ATR: a master conductor of cellular responses to DNA replication stress. Trends Biochem Sci 36(3):133–140
42. Casper AM et al (2002) ATR regulates fragile site stability. Cell 111(6):779–789
43. Petermann E, Caldecott KW (2006) Evidence that the ATR/Chk1 pathway maintains normal replication fork progression during unperturbed S phase. Cell Cycle 5(19):2203–2209
44. Yajima H, Lee KJ, Chen BP (2006) ATR-dependent phosphorylation of DNA-dependent protein kinase catalytic subunit in response to UV-induced replication stress. Mol Cell Biol 26(20):7520–7528
45. Shigechi T et al (2012) ATR-ATRIP kinase complex triggers activation of the Fanconi anemia DNA repair pathway. Cancer Res 72(5):1149–1156
46. Yamane K, Taylor K, Kinsella TJ (2004) Mismatch repair-mediated G2/M arrest by 6-thioguanine involves the ATR-Chk1 pathway. Biochem Biophys Res Commun 318(1):297–302
47. Stauffer D et al (2007) p300/CREB-binding protein interacts with ATR and is required for the DNA replication checkpoint. J Biol Chem 282(13):9678–9687
48. Chen G et al (1999) Radiation-induced assembly of Rad51 and Rad52 recombination complex requires ATM and c-Abl. J Biol Chem 274(18):12748–12752
49. Di Virgilio M, Ying CY, Gautier J (2009) PIKK-dependent phosphorylation of Mre11 induces MRN complex inactivation by disassembly from chromatin. DNA Repair (Amst) 8(11): 1311–1320
50. Bolderson E et al (2010) Phosphorylation of Exo1 modulates homologous recombination repair of DNA double-strand breaks. Nucleic Acids Res 38(6):1821–1831
51. Cortez D et al (1999) Requirement of ATM-dependent phosphorylation of brca1 in the DNA damage response to double-strand breaks. Science 286(5442):1162–1166
52. Zhang F et al (2009) PALB2 links BRCA1 and BRCA2 in the DNA-damage response. Curr Biol 19(6):524–529
53. Jensen RB, Carreira A, Kowalczykowski SC (2010) Purified human BRCA2 stimulates RAD51-mediated recombination. Nature 467(7316):678–683
54. Liu J et al (2010) Human BRCA2 protein promotes RAD51 filament formation on RPA-covered single-stranded DNA. Nat Struct Mol Biol 17(10):1260–1262
55. Thorslund T et al (2010) The breast cancer tumor suppressor BRCA2 promotes the specific targeting of RAD51 to single-stranded DNA. Nat Struct Mol Biol 17(10):1263–1265
56. Wang H et al (2004) ATR affecting cell radiosensitivity is dependent on homologous recombination repair but independent of nonhomologous end joining. Cancer Res 64(19): 7139–7143
57. Sorensen CS et al (2005) The cell-cycle checkpoint kinase Chk1 is required for mammalian homologous recombination repair. Nat Cell Biol 7(2):195–201
58. Ward IM, Chen J (2001) Histone H2AX is phosphorylated in an ATR-dependent manner in response to replicational stress. J Biol Chem 276(51):47759–47762
59. Paull TT et al (2000) A critical role for histone H2AX in recruitment of repair factors to nuclear foci after DNA damage. Curr Biol 10(15):886–895
60. Chu WK et al (2010) BLM has early and late functions in homologous recombination repair in mouse embryonic stem cells. Oncogene 29(33):4705–4714

61. Davies SL et al (2004) Phosphorylation of the Bloom's syndrome helicase and its role in recovery from S-phase arrest. Mol Cell Biol 24(3):1279–1291
62. Otterlei M et al (2006) Werner syndrome protein participates in a complex with RAD51, RAD54, RAD54B and ATR in response to ICL-induced replication arrest. J Cell Sci 119(Pt 24):5137–5146
63. Tripathi V, Kaur S, Sengupta S (2008) Phosphorylation-dependent interactions of BLM and 53BP1 are required for their anti-recombinogenic roles during homologous recombination. Carcinogenesis 29(1):52–61
64. Patro BS et al (2011) WRN helicase regulates the ATR-CHK1-induced S-phase checkpoint pathway in response to topoisomerase-I-DNA covalent complexes. J Cell Sci 124(Pt 23): 3967–3979
65. Vasquez KM (2010) Targeting and processing of site-specific DNA interstrand crosslinks. Environ Mol Mutagen 51(6):527–539
66. D'Andrea AD (2010) Susceptibility pathways in Fanconi's anemia and breast cancer. N Engl J Med 362(20):1909–1919
67. Deans AJ, West SC (2011) DNA interstrand crosslink repair and cancer. Nat Rev Cancer 11(7):467–480
68. Massague J (2004) G1 cell-cycle control and cancer. Nature 432(7015):298–306
69. Storz P (2005) Reactive oxygen species in tumor progression. Front Biosci 10:1881–1896
70. Babior BM (1999) NADPH oxidase: an update. Blood 93(5):1464–1476
71. Berasain C et al (2009) Inflammation and liver cancer: new molecular links. Ann N Y Acad Sci 1155:206–221
72. Wiseman H, Halliwell B (1996) Damage to DNA by reactive oxygen and nitrogen species: role in inflammatory disease and progression to cancer. Biochem J 313(Pt 1):17–29
73. Garrett MD, Collins I (2011) Anticancer therapy with checkpoint inhibitors: what, where and when? Trends Pharmacol Sci 32(5):308–316
74. Wagner JM, Kaufmann SH (2010) Prospects for the Use of ATR Inhibitors to Treat Cancer. Pharmaceuticals 3(5):1311–1334
75. Nghiem P et al (2001) ATR inhibition selectively sensitizes G1 checkpoint-deficient cells to lethal premature chromatin condensation. Proc Natl Acad Sci USA 98(16):9092–9097
76. Wilsker D, Bunz F (2007) Loss of ataxia telangiectasia mutated- and Rad3-related function potentiates the effects of chemotherapeutic drugs on cancer cell survival. Mol Cancer Ther 6(4):1406–1413
77. Wagner JM, Karnitz LM (2009) Cisplatin-induced DNA damage activates replication checkpoint signaling components that differentially affect tumor cell survival. Mol Pharmacol 76(1):208–214
78. Caporali S et al (2008) AKT is activated in an ataxia-telangiectasia and Rad3-related-dependent manner in response to temozolomide and confers protection against drug-induced cell growth inhibition. Mol Pharmacol 74(1):173–183
79. Flatten K et al (2005) The role of checkpoint kinase 1 in sensitivity to topoisomerase I poisons. J Biol Chem 280(14):14349–14355
80. Powell SN et al (1995) Differential sensitivity of p53(−) and p53(+) cells to caffeine-induced radiosensitization and override of G2 delay. Cancer Res 55(8):1643–1648
81. Yao SL et al (1996) Selective radiosensitization of p53-deficient cells by caffeine-mediated activation of p34cdc2 kinase. Nat Med 2(10):1140–1143
82. Sarkaria JN et al (1999) Inhibition of ATM and ATR kinase activities by the radiosensitizing agent, caffeine. Cancer Res 59(17):4375–4382
83. Chalmers AJ et al (2009) Cytotoxic effects of temozolomide and radiation are additive- and schedule-dependent. Int J Radiat Oncol Biol Phys 75(5):1511–1519
84. Nishida H et al (2009) Inhibition of ATR protein kinase activity by schisandrin B in DNA damage response. Nucleic Acids Res 37(17):5678–5689
85. Knight ZA et al (2006) A pharmacological map of the PI3-K family defines a role for p110alpha in insulin signaling. Cell 125(4):733–747
86. Toledo LI et al (2011) A cell-based screen identifies ATR inhibitors with synthetic lethal properties for cancer-associated mutations. Nat Struct Mol Biol 18(6):721–727

87. Charrier JD et al (2011) Discovery of potent and selective inhibitors of ataxia telangiectasia mutated and Rad3 related (ATR) protein kinase as potential anticancer agents. J Med Chem 54(7):2320–2330
88. Reaper PM et al (2011) Selective killing of ATM- or p53-deficient cancer cells through inhibition of ATR. Nat Chem Biol 7(7):428–430
89. Jacq X et al. (2012) *AZ20,* a novel potent and selective inhibitor of ATR kinase with in vivo antitumour activity. Cancer Res. 72(8) Suppl1 Abs#1823
90. Graeber TG et al (1996) Hypoxia-mediated selection of cells with diminished apoptotic potential in solid tumours. Nature 379(6560):88–91
91. Pires IM et al (2012) Targeting radiation-resistant hypoxic tumour cells through ATR inhibition. Br J Cancer 107(2):291–299
92. Bryant HE et al (2005) Specific killing of BRCA2-deficient tumours with inhibitors of poly(ADP-ribose) polymerase. Nature 434(7035):913–917
93. Farmer H et al (2005) Targeting the DNA repair defect in BRCA mutant cells as a therapeutic strategy. Nature 434(7035):917–921
94. Starcevic D, Dalal S, Sweasy JB (2004) Is there a link between DNA polymerase beta and cancer? Cell Cycle 3(8):998–1001
95. Kiyohara C, Takayama K, Nakanishi Y (2006) Association of genetic polymorphisms in the base excision repair pathway with lung cancer risk: a meta-analysis. Lung Cancer 54(3):267–283
96. Neijenhuis S et al (2010) Targeted radiosensitization of cells expressing truncated DNA polymerase {beta}. Cancer Res 70(21):8706–8714
97. Halazonetis TD, Gorgoulis VG, Bartek J (2008) An oncogene-induced DNA damage model for cancer development. Science 319(5868):1352–1355
98. Gilad O et al (2010) Combining ATR suppression with oncogenic Ras synergistically increases genomic instability, causing synthetic lethality or tumorigenesis in a dosage-dependent manner. Cancer Res 70(23):9693–9702
99. Hoglund A et al (2011) Therapeutic implications for the induced levels of Chk1 in Myc-expressing cancer cells. Clin Cancer Res 17(22):7067–7079
100. Ferrao PT et al (2012) Efficacy of CHK inhibitors as single agents in MYC-driven lymphoma cells. Oncogene 31(13):1661–1672
101. Wilsker D et al (2012) Targeted mutations in the ATR pathway define agent-specific requirements for cancer cell growth and survival. Mol Cancer Ther 11(1):98–107
102. Zhang C et al (2009) PF-00477736 mediates checkpoint kinase 1 signaling pathway and potentiates docetaxel-induced efficacy in xenografts. Clin Cancer Res 15(14):4630–4640

Telomeres, Telomerase, and DNA Damage Response in Cancer Therapy

Raina A. Tamakawa, Helen B. Fleisig, and Judy M.Y. Wong

1 Overview

Faithful genome maintenance is essential to an organism's growth and survival. To preserve genome fidelity, the DNA Damage Response (DDR) pathway has evolved to coordinate the surveillance and repair of genomic DNA, damaged by normal metabolic or environmental insults [1]. DDR surveillance mechanisms scan for discontinuities and structural changes in the DNA double helix. Upon detection of any damage to the DNA molecule, these surveillance sensors activate signal transduction cascades to amplify the damage signal, and coordinate the arrest of proliferation for proper DNA repair [1–4]. Alternatively, apoptosis may be initiated if repair is not possible. The abrupt termini of linear eukaryotic chromosomes pose specific challenges to DDR surveillance, as these natural ends are indistinguishable from damaged double-stranded DNA. In most eukaryotic organisms with linear chromosomes, phylogenetically conserved nucleoprotein structures, known as telomeres, differentiate chromosome ends from nonspecific DNA breaks [5–7]. Telomeres mask the ends of chromosomes from DDR surveillance sensors and protect the chromosome ends from inappropriate repair by DDR mechanisms [8].

Over the past two decades, we have learned a great deal about the structure of telomeres, their homeostatic maintenance, and the cellular consequences of their dysfunction. We know that while telomeres suppress the erroneous activation of the DDR pathways by chromosome ends, the structural and functional integrity of these structures are dependent on the activities of the same DDR pathways. In this chapter, we describe the protein and nucleic acid components of telomeres, both

R.A. Tamakawa • H.B. Fleisig • J.M.Y. Wong (✉)
Faculty of Pharmaceutical Sciences, University of British Columbia,
2146 East Mall, Vancouver, BC, Canada
e-mail: judy.wong@ubc.ca

L. Panasci et al. (eds.), *Advances in DNA Repair in Cancer Therapy*,
Cancer Drug Discovery and Development 72, DOI 10.1007/978-1-4614-4741-2_11,

© Springer Science+Business Media, LLC 2013

stable and transient. We then describe the physiological mechanisms of telomere maintenance by the enzyme telomerase, its biogenesis and regulation, and how this reverse transcriptase might be utilized in anticancer chemotherapy.

2 Telomeres

2.1 Telomere Structure

At the ends of most eukaryotic chromosomes are highly conserved, tandem DNA repeats. These highly repetitive sequences are associated with their specific binding proteins, and together, these chromosome-end structures are known as telomeres. Telomeres cap chromosome ends and protect them from nonspecific nuclease digestion, as well as preventing them from being recognized as double-stranded DNA breaks. In the absence of telomeres, erroneous DNA repair can lead to chromosomal end-to-end fusions and genetic recombination [5]. The length of telomeric DNA repeats vary between species, ranging from ~300 to 600 bp in yeast [9], to ~150 kb in mice [10]. Human telomeres measure ~5–15 kb in length [11, 12]. In all vertebrate chromosomes, telomeres are made up of a G-rich hexanucleotide sequence (TTAGGG)n [13]. Telomere repeats run 5′-3′, terminating in a single-stranded 3′ overhang of the G-rich strand [14, 15]. The length of this overhang is also species-specific, measuring ~50–100 nucleotides in length in mouse and human telomeres [16].

Mammalian telomeres were previously thought to be linear. However, electron microscopy analysis of psoralene cross-linked telomeric DNA from human and mouse were visualized to end as large duplex loops [17]. At the molecular level, double-stranded telomeric DNA folds back onto itself to form a lariat structure termed the telomeric loop (Fig. 1a). This allows for the G-rich 3′ overhang to invade the duplex section of telomeric repeats, thereby forcing the formation of a single-stranded DNA displacement loop [18]. The resulting higher order chromatin structure is distinct from damaged DNA and thus serves to differentiate the normal chromosomal termini, preventing them from being recognized as double strand breaks. This differentiation mechanism is crucial in preventing the initiation of DNA damage checkpoint responses [5, 6, 16].

A six-member protein complex, termed shelterin, associates with telomeric DNA in a sequence specific manner (Fig. 1b, Table 1). This complex facilitates formation of the telomeric loop to protect chromosome ends from DNA damage surveillance mechanisms, as well as to functionally maintain telomere length. The shelterin complex is composed of six distinct proteins: telomere repeat binding factors 1 and 2 (TRF1 and TRF2), protection of telomeres 1 (POT1), TRF1-and TRF2-interacting nuclear protein 2 (TIN2), repressor/activator protein 1 (Rap1), and TPP1 (formally known as PTOP, PIP1, or TINT1) [7, 19]. TRF1 and 2 are sequence specific telomeric DNA binding proteins that recruit the other four proteins to the telomeres

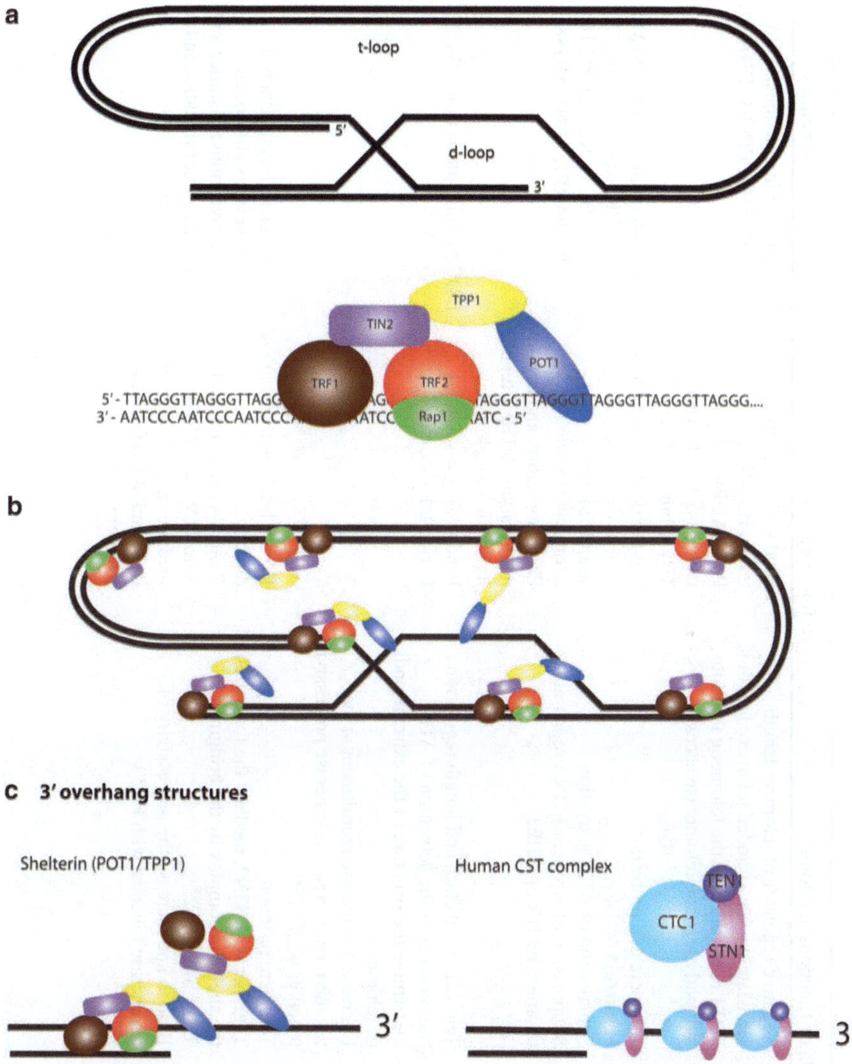

Fig. 1 Human telomeres. (**a**) Telomere repeats at chromosome ends fold back to form a lariat structure (t-loop). The 3′ telomeric DNA overhang invades the double-stranded DNA region of telomeric repeats to form a displacement-loop (d-loop). (**b**) Shelterin protein complex aids in t-loop formation and stabilization: TRF1 and TRF2 interact with double-stranded telomeric repeats, recruiting the other four shelterin proteins, POT1, TIN2, TPP1, and Rap1, to the telomere end. TIN2 links TRF1 to TRF2, contributing to the stabilization of these proteins on the telomere. POT1, which has strong binding specificity for single-stranded telomeric repeats, together with its heterodimeric partner TPP1 associates with TRF1 and TRF2 through a bridge formed by TIN2. Rap1 is recruited by TRF2, forming a TRF2-Rap complex. (**c**) The human telomere 3′ overhangs exist in two structural forms. Shelterin components POT1-TPP1 bind single-stranded telomeric DNA with high sequence specificity. Recently, the human homologs of yeast CST complex have been identified to associate with single-stranded telomeric DNA structure, with low sequence specificity and in the absence of shelterin

Table 1 Telomere Binding Proteins

	Protein	Gene name	Major functions at telomere	Telomere binding	Other biological functions
Shelterin complex	TRF 1	*TERF 1*	– Negative regulator of telomere length – Translate double-stranded telomere length to POT1 recruitment at the telomeric terminus, indirectly regulate telomerase access	– Bind to double-stranded telomeric DNA – Poly (ADP) ribosylase tankyrase 1 and 2 regulate TRF 1 binding to telomeres	DDR protein interactions: ATM, BLM, PML, PINX1, DNA-PK
	TRF2	*TERF2*	– Telomere end protection – Formation of T-loop – Suppression of ATM activation – Suppression of erroneous DNA repair at the telomere by HR or NHEJ	– Bind to double-stranded as well as double/single-stranded junction of telomeric DNA – Phosphorylation and arginine methylation of TRF2 regulate its binding to telomeres	DDR protein interactions: ATM, DNA-PK, Apollo, MRN, WRN, BLM, Ku70, ERC 1/XPF, PARP1, FEN1, MUS81
	POT1	*POT1*	– Telomere protection and length regulation – Suppression of the activation of ATR – Determine the structure of the extreme ends of telomeres – Decrease telomerase recruitment at the telomere but increase telomerase processivity together with TPP1	– Bind to single-stranded G-rich-telomeric DNA	DDR protein interactions: WRN, BLM
	TPP1	*ACD*	– Telomere proteion – Together with TIN2, mediate the formation of the shelterin complex by the bridging of TRF1 and TRF2 – Bind to and increase the processivity of telomerase together with POT1	– Do not bind telomeric DNA directly – Protein–protein interaction with TIN2 and TRF1 mediates the recruitment of itself and of POT1 to the telomere	Adrenocortical dysplasia is a spontaneous autosomal recessive mouse disorder with a splice donor mutation in the ACD gene

	TIN2	*TINF2*	– Telomere length regulation – Modulate the ribosylation of TRF1 by inhibiting TRF1 association with tankyrase – Together with TPP1, mediate the formation of the shelterin complex by the bridging of TRF1 and TRF2	– Do not bind temoleric DNA directly – Protein–protein interaction between TRF1/TRF2 to TPP1/POT1 complex – Long isoforms (TIN2L) may anchor telomers to the nuclear matrix	Genetic mutations were found in dyskeratosis congenita and Revesz syndrome patients
	RAP1	*TERF2IP*	– Telomere length regulation – Suppression of homology-directed repair at the telomere – Subtelomeric gene silencing	– Do not bind telomeric DNA directly – Protein–protein interaction with TRF2 mediates its telomere localization	Global regulation of gene expression in yeast and mouse
Human CST complex	CTC1	*C17ORF68*	– Homolog of mouse CTC1	– As a complex, these three proteins bind to telomeric G-rich single-stranded DNA in a sequence non-specific manner, analogous to the binding of ssDNA by replication protein A (RPA) – Bind to telomeric single-stranded DNA in the absence of TPP1-POT1	Also identified as subunits of DNA polymerase alpha:primase complex
	STN1	*OBFC1*	– Homolog of *S. pombe* STN1 – Bridge the association between CTC1 and TEN1		Also identified as subunits of DNA polymerase alpha:primase complex
	TEN1	*C17ORF106*	– Homolog of *S. pombe* TEN1		

[19]. Both TRF1 and TRF2 contain a C-terminal SANT/Myb-type DNA binding domain that binds to the 5′-TTAGGG-3′ sequence in duplex DNA, making the entire shelterin complex highly specific for telomeric repeats [20, 21].

TRF1 is a homodimeric protein that aids in telomeric loop formation and stabilization [22]. Its binding to arrays of telomeric repeats induces shallow bends and results in the formation of DNA loops, demonstrating the protein's architectural role on telomeres [20]. This protein has also been shown to affect telomere length. Over-expression of TRF1 results in telomere shortening while expression of a dominant negative TRF1 mutant, lacking the Myb type domain, causes telomere lengthening. This suggests a negative correlation between TRF1 function and telomere length [23, 24]. On the other hand, accumulations of TRF1 and TRF2 at telomere ends were shown to positively correlate with telomere length [23, 24]. This led to a protein counting theory of telomere length regulation, which proposed that a feedback mechanism mediated by protein interactions with TRF1 is responsible for steady-state telomere length maintenance [23].

Like TRF1, TRF2 also binds to double-stranded telomeric DNA as a homodimer [25] and plays a role in telomeric loop assembly. In contrast to TRF1, TRF2 is believed to bind near the loop-tail junction where it stabilizes the G-rich single-stranded telomeric overhang at the displacement loop by facilitating strand invasion and preventing the single-stranded sequence from being recognized as a DNA break [17, 26]. In corroboration to this model, electron microscopy of a telomere DNA track containing ~2 kb of telomeric repeats at the end of a linearized DNA plasmid and terminating in a 3′ single-stranded overhang, revealed the specific binding location of TRF2 at the telomeric loop junction [27]. Like TRF1, TRF2 also serves as a negative regulator of telomere length. TRF2 over-expression results in shortened telomeres and induces senescence in telomerase negative cells [28].

POT1 is the most highly conserved component of shelterin, and has a strong specificity for single-stranded 5′-(T)TAGGGTTAG-3′ sites [29, 30]. Following DNA replication, single-stranded telomeric overhangs initially associate with replication protein A (RPA). Heterogeneous nuclear ribonucleoprotein A1 (hnRNPA1) binding displaces RPA binding, while the increase in TERRA expression levels (see below) following S phase removes hnRNPA1 from telomeric DNA, allowing for sequence-specific binding by POT1 [31]. POT1 accumulation at chromosome ends is believed to regulate telomerase activity by relaying telomere length information from the double-stranded region of the telomeric loop to the single-stranded region through its interaction with TRF1 [32]. Studies have also demonstrated that POT1 plays a positive role in telomere length maintenance, as ectopic expression of POT1 results in an increase in telomeric DNA [33, 34].

TPP1, the heterodimeric partner of POT1 [35, 36], enhances POT1 affinity for single-stranded telomeric DNA [36]. Most of the POT1-TPP1 complexes are associated with TRF1 and TRF2 through a bridge formed by TIN2, which functions to stabilize the interactions between these proteins [37]. In addition to the protection of telomere ends, the TPP1-POT1 complex also serves as a regulator of telomere length maintenance. Through its oligonucleotide- and oligosaccharide-binding fold, TPP1 has been suggested to regulate telomerase activity and the enzyme's access to single-stranded telomeric DNA, both negatively and positively in a context dependent manner [36].

TIN2 co-localizes with TRF1 on metaphase chromosomes [38]. TIN2 forms bridges that join POT1 to TRF1 and TRF2 and also TRF1 to TRF2, contributing to the stabilization of these proteins at telomeres [39, 40]. The binding of TRF1 to TIN2 leads to the compaction of telomeric DNA and telomeric loop stabilization. Both events limit the accessibility of telomerase to telomere ends and thereby functions as a negative regulator of telomere length [38]. Genetic lesions of TIN2 underlie a subpopulation of autosomal dominant form of dyskeratosis congenita, representing the only shelterin protein associated with this genetic disease of telomere dysfunction [41].

Rap1 is recruited to the telomere by protein interactions with TRF2, forming a TRF2-Rap1 complex [42]. Rap1 affects telomere length homeostasis through its interactions with telomere length regulator proteins Rif1 and Rif2. Like the other shelterin proteins, Rap1 is a negative regulator of telomere length. Over-expression of Rap1 leads to telomere shortening, while expression of dominant negative mutants results in the gain of telomere length [43]. In addition, RAP1 binds to non-telomeric sequences and is implicated in diverse cellular activities including gene silencing and the transcriptional regulation of gene targets involved in adhesion, metabolism, and cancer [44].

In *Saccharomyces cerevisiae*, the cdc13-stn1-ten1 (CST) complex binds single-stranded telomeric DNA in place of POT1-TPP1 [45, 46]. Recently, the human version of this protein complex has been identified. The human ctc1-stn1-ten1 (CST) complex contains two human homologs of the ScCST complex (Stn1 and Ten1), and a third component, the conserved telomere maintenance component 1 (ctc1) [47]. Similar to the ScCST complex, human CST binds single-stranded G-rich telomeric DNA, in the absence of POT1-TPP1. Unlike ScCST, human CST does not exhibit sequence specificity for telomeric repeats, and likely associates with other single-stranded DNA in a manner analogous to the binding of single-stranded DNA by replication protein A (Fig. 1c, Table 1). A significant increase in the G-strand overhang was observed in Stn1 depleted human cells, indicating a role of the CST complex in single-strand telomeric DNA regulation [47]. Whether this newly identified CST complex functionally interacts with the shelterin complex is currently under investigation.

In addition to the binding of telomere-specific shelterin and the single-stranded DNA-specific CST complexes, heterochromatin formation via the epigenetic regulation of telomeric chromatin is also observed. DNA methylation of subtelomeric regions [48], together with histone methylation of the telomeric chromatin [49], are postulated to negatively regulate gene transcription, suppress homologous recombination and prevent telomerase access for telomere elongation [50]. Telomere-repeat containing RNAs (TERRA) are long UUAGGG-repeat containing noncoding RNA transcripts that have been recently identified [51, 52]. TERRAs are transcribed starting from the subtelomeric region, using the C-rich strand of the telomere as a template. TERRAs are found to be stably associated with telomeric chromatin and cellular machinery responsible for the nonsense mediated decay of dysfunctional RNA transcripts. Current models of these noncoding RNA functions prescribe a role in the induction of heterochromatin formation, serving as a negative regulator of telomerase access to telomeres [53]. Cell-cycle phase-specific changes in TERRA expression levels are also postulated to mediate the switch from RPA to POT1 binding at single-stranded telomeric DNA termini [31].

2.2 Telomere Function: End Replication Problem and the Hayflick Limit

Besides structurally protecting the ends of chromosomes, telomeres also serve as a solution to the end-replication problem. Because DNA polymerases fail to completely copy chromosomes to the very end, the placement of telomeres at the extreme ends of chromosomes allows them to buffer gene coding sequence from being eroded [11, 54]. Instead, telomeric DNA is lost after every round of DNA replication. Telomeric DNA loss is cumulative and with continual proliferation; telomeres will eventually reach a critical short length. At this point, genome surveillance mechanisms will trigger replicative senescence, an irreversible cellular growth arrest state where cells can no longer divide into daughter progeny, but remain metabolically active (Fig. 2) [5, 55–57]. This short telomere checkpoint serves as a "mitotic clock" which counts down the number of cell divisions in each cell lineage. Leonard Hayflick first described this relationship by observing the replicative potential of human primary fibroblasts in culture in 1965 [58]. Termed the Hayflick Limit, the number of times a cell lineage could divide before short telomere-induced proliferative arrest was determined by the structural integrity of telomeres and the activities of biological pathways responsible for maintaining the length of these specialized DNA tracts [59, 60]. Incidentally, this process can be viewed as a tumor suppressive mechanism: by limiting the number of cell divisions that can occur in a particular cell lineage, one can reduce the accumulation of deleterious mutations that precede cellular transformation [61].

In rare cases, some somatic cells are able to bypass this short telomere checkpoint by inactivating the genome surveillance mechanisms mediated by the tumor suppressor genes p53 and retinoblastoma protein (Rb) (Fig. 2). Further cell divisions in p53/Rb-inactivated cells continue to deplete telomeric DNA, leading to the disruption of the telomere structure [62]. Uncapped telomeres are recognized by cellular repair mechanisms as damaged DNA, resulting in cells attempting to repair these damages. Erroneous repair leads to chromosome end fusions and rampant genomic instability. When this happens, a second checkpoint termed "crisis" [57] is activated and cells are triggered to undergo apoptosis. Under this extreme selective pressure, most cells will die. In extremely rare cases (~1 in 10 million cells), genomic instability can lead to the reactivation of a specialized cellular reverse transcriptase, termed telomerase, which is capable of adding telomeric repeats to chromosome ends [63]. Telomerase expression

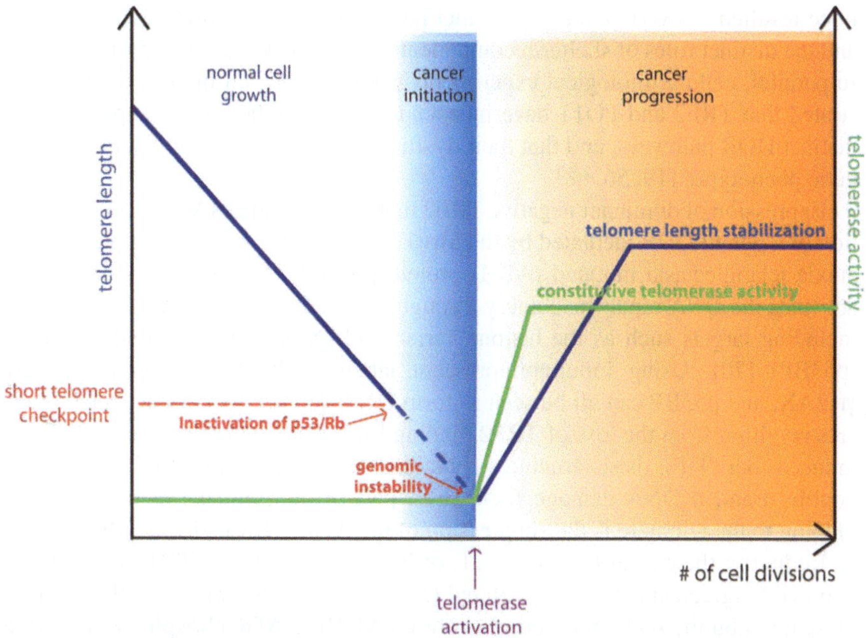

Fig. 2 Telomere dynamics and cancer development. With each cell division, approximately 50–100 bp of telomeric DNA is lost from chromosome ends. With continual proliferation, telomeres will eventually reach a critical short length and are triggering replicative senescence. Inactivation of genome surveillance mechanisms mediated by the tumor suppressor genes p53 and Rb allow continual cell divisions, further depleting telomeric DNA leading to rampant genomic instability and the induction of apoptosis. A rare cell (~1 in 10 million) can be forced to reactivate telomerase, allowing the cell to replace lost telomeric repeats, prevent further genomic instability and confer the unlimited proliferative capacity required for the formation of a malignant tumor cell

allows cells to replace lost telomeric repeats and prevent further chromosome instability. In these cases of forced reactivation of telomerase enzyme expression, constitutive telomerase activity confers the unlimited proliferative capacity required for the formation of a malignant tumor cell (Fig. 2).

2.3 *Dysfunctional Telomere Capping Is Recognized as DNA Damage*

Dysfunctional telomeres are created when telomeric sequences are shortened beyond a critical length allowing for the formation of a higher order chromatin structure [5], or by the genetic deletion or protein dysfunction of key shelterin components [28, 56]. Uncapped telomeres expose the ends of chromosomes, thereby inducing the DDR and resulting in the cascade of genomic instability through erroneous chromosome end repair. Genetic deletion of different shelterin components in mice resulted in overlapping yet distinct phenotypes, underscoring the complexity and the distinct roles of shelterin components in chromosome end capping [64–67]. In parallel, cellular biological experiments using human cell models have demonstrated that TRF2 and POT1 have independent roles in the normal suppression of distinct DDR pathways, and that their dysfunctions cause severe molecular cytogenetic phenotypes [19, 56, 68].

Expression of dominant negative TRF2 that cannot bind DNA leads to the induction of a potent DDR mediated by the Mre11-Rad50-NBS (MRN) complex and the ataxia telangiectasia mutated (ATM) protein [26, 69]. As a protein kinase, ATM activates the DNA repair machinery through a cascade of phosphorylation activity, including targets such as the histone variant H2AX and the p53 binding protein (p53BP) [70]. Using immunofluorescent labeling, the MRN complex, ATM, γH2AX, and p53BP can all be seen to form DNA damage foci at uncapped telomeres resulting from the loss of TRF2 binding [56]. Known as telomere dysfunction induced foci (TIF), these structures contain DDR factors similar to those found in double-stranded DNA damage foci. Thus, part of the normal function of TRF2 binding to the telomere is the suppression of the ATM DDR pathway [19, 69].

Reducing the expression of POT1, or its binding partner TPP1, activates the ataxia telangiectasia and Rad3-related (ATR) kinases [69, 71]. Together with its obligate subunit, ATR interacting protein (ATRIP), ATR phosphorylates DDR effectors responsible for diverse DNA damages, such as those induced by UV exposure, exposure to nucleophilic crosslinking agents or resulting from collapsed replication forks [72]. *In vitro* biochemical experiments have shown that ATR is activated by replication protein A (RPA)-coated single-stranded DNA. This is reminiscent of the single-stranded telomeric G-rich overhang left vacant by the removal of POT1 binding. Thus, part of the normal function of POT1 binding to the telomeric terminai is the suppression of the ATR DDR pathway [19, 64, 69].

2.4 Key DDR Players Are Required for Normal Telomere Maintenance

The relationship between telomeric chromatin and the DDR machinery extends beyond simple antagonism. DDR components are known to play positive roles in the normal homeostatic maintenance of telomeres, in the absence or presence of telomerase activity. Evidence that key DDR components have important roles in normal telomere homeostasis comes from studies of inherited human diseases and from animal models of these diseases. Genetic disorders such as Ataxia telangiectasia, Nijmegen break syndrome, Bloom syndrome and Werner syndrome all exhibit molecular phenotypes of accelerated telomere shortening [70, 73]. Initially believed to be the function of an increased rate of telomere attrition due to the higher cellular turnover, animal models and cellular biology studies later revealed the normal functions of these proteins in telomere homeostasis.

DDR mediators are activated each time telomeric DNA undergoes replication [78, 79]. Normal replication through telomere ends requires the resolution of the telomere loop, followed by leading and lagging strand synthesis through the ends. This requires the action of RecQ helicases, such as Bloom and Werner, the elective de novo synthesis of G-rich telomeric DNA when telomerase is active, nuclease trimming by Apollo, XPF or the MRN complex to create the correct telomeric DNA terminus and overhangs, and the reformation of the telomeric loop structures through the actions of Rad51D, RPA and other homologous recombination pathway effectors. In addition, the steps necessary to "open up" a telomere for DNA replication machinery access predicts that the transient recognition of these "open" telomere structures by DDR sensors. Indeed, both ATM-MRN and ATR-ATRIP complexes are found at functional telomeres during the DNA replication phase of the cell cycle [78, 79]. Despite the data supporting these models, detailed mechanisms of how telomeric binding proteins coordinate with transient DDR signals to direct telomere formation, instead of promoting erroneous DNA repair at these sites, still need to be elucidated.

3 Telomerase

3.1 Telomerase Structure and Biogenesis

The human telomere terminal transferase enzyme, more commonly referred to as telomerase, is a ribonucleoprotein (RNP) responsible for the de novo synthesis of telomere repeats. This unique reverse transcriptase extends chromosome ends by utilizing an integral RNA subunit as a template to synthesize the TTAGGG telomeric DNA repeats. The core components of this enzyme complex consist of the telomerase reverse transcriptase catalytic subunit (TERT) and the telomerase RNA (TER), which contains the template sequence for telomere synthesis. In the human enzyme, RNA-binding proteins such as the H/ACA proteins dyskerin, Nop10, and Nhp2 are also found to associate with the core enzyme complex (Fig. 3, Table 2). Other proteins transiently associate with the core enzyme complex and play important roles in the regulation of the catalytic activity, enzyme stability, cellular localization and intracellular trafficking of the enzyme (Table 3) [80–82].

TER and TERT were identified as the catalytic core of this complex by virtue of their ability to form a complex and elongate telomeres *in vitro*, in the absence of other protein factors [83]. However, *in vivo*, telomerase employs an intricate biogenesis pathway involving specific factors for enzyme assembly, trafficking and the subcellular localization of the holoenzyme complex. TER transcription is ubiquitous in all human cells. The stability of TER is dependent on biogenesis protein factors Shq1 and NAF-1 mediated complex formation with the H/ACA proteins (dyskerin, Nhp2, and Nop10). TER association with the H/ACA complex results in the formation of a stable but inactive telomerase RNP. Assembly of this

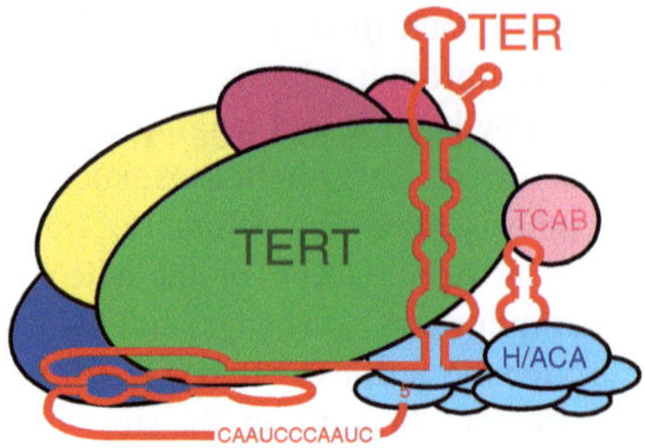

Fig. 3 Human Telomerase. Schematic depiction of the human telomerase enzyme. Telomerase is a specialized reverse transcriptase carrying its own RNA template (TER). Telomerase RNA serves multiple functions. The template domain allows sequence specific alignment of the linear chromosome ends into the catalytic site and provides the 6nt template sequence for RT. Other domains of the RNA serve structural and catalytic functions, RT activity is provided by the protein subunit, telomerase reverse transcriptase (TERT). Together, TER and TERT comprise the minimal functional unit that can be reconstituted *in vitro* for telomerase activity. The *in vivo* accumulation and stability of TER requires the association of RNA with two sets of H/ACA proteins. Other protein factors involved in the regulation of enzyme functions through cellular localization (TCAB1), assembly (Pontin and Reptin) and other mechanisms associate with the holoenzyme complex in a transient manner

inactive telomerase RNP with TERT is required for catalytic activity [84]. Telomerase enzyme assembly is cell cycle and subcellular localization dependent [85]. Numerous biogenesis factors, including staufen, L22, SmB/SmD3, PinX1, 14-3-3, nucleolin, pontin, reptin, Hsp90, p23, and telomerase Cajal body protein 1 (TCAB1) have all been demonstrated to play important roles in TER/TERT subcellular localization and enzyme assembly. Finally, following the formation of a functional telomerase enzyme, additional trafficking factors, such as TCAB, hEST1A, and hnRNPs, are required for proper transport of the active enzyme to chromosome ends.

3.2 Telomerase RNA

TER is a noncoding RNA that serves as a template for TERT-dependent addition of telomeric repeats. Ubiquitously expressed, human TER is synthesized by RNA polymerase II (pol II) and processed into a mature 451-nucleotide (nt) product with a 5′ trimethyl guanosine cap that lacks a polyadenosine tail at its 3′ end [86, 87]. TER contains a 341-nt pol II-type promoter region upstream of the transcription start site [88]. Nuclear factor-Y (NF-Y), Sp1 and Sp3 are essential regulators of TER promoter function.

Primary and secondary structural elements of TER contain many motifs that are essential for telomerase activity as well as cellular accumulation of mature TER. The 11-nt telomeric repeat template sequence is contained within the 5′ portion of TER in the pseudoknot domain (nt 1–209). The H/ACA motif (nt 275–451)

Table 2 Core Telomerase Holoenzyme Subunits

Holoenzyme component	Role in telomerase	Other suggested biological functions	Effects of functional/genetic changes to telomere maintenance
TERT	– Reverse transcriptase – Expression levels is transcriptionally regulated – Expression levels control temporal pattern of telomerase activity in human cells	– Promotes double-stranded DNA damage repair – Suggested to act as a transcription factor in mouse, in the absence of TER	– Knock down results in the lost of telomerase activity – Haploinsufficient in humans – Genetic mutations are associated with autosorral dominant dyskeratosis congenita
TER	– RNA template – Ubiquitously expressed	– Inhibit ATR activation, in the absence of TERT	– Knockdown results in the lost of telomerase activity – Haploinsufficient in humans – Genetic mutations are associated with autosomal dominant dyskeratosis congenita
DKC1	– Member of H/ACA protein complex – Binds to TER and direct correct processing, localization and promotes in vivo stability of TER	– Pseudourdine synthase – Use H/ACA RNA as a guide to modify rRNA and snoRNA	– Knockdown results in decreased TER accumulation – Genetic mutations in DKC1 are associated with X-linked dyskeratosis congenita
Nop10	– H/ACA protein – Necessary for the correct processing, localization and in vivo stability of TER	– Together with DKC1 and Nhp2, forms the RNA bining surface for association with H/ACA RNAs	– Genetic mutations in NOP10 are associated with autosomal recessive dyskeratosis congenita
Nhp2	– H/ACA protein – Necessary for correct processing, localization and in vivo stability of TER	– Together with DKC1 and NOP10, forms the RNA bining surface for association with H/ACA RNAs	– Genetic mutations in Nhp2 are associated with autosomal recessive dyskeratosis congenita

Table 3 Telomerase Biogenesis and Regulation Factors

TER biogenesis and regulation	Protein components	Roles in telomerase pathways	Other biological functions	Effects of changes in expression levels in telomere maintenance
	NAF1	Early assembly factors for H/ACA RNA (including TER)	H/ACA RNPs biogenesis	ND
	Shq1			Knockdown of SHQ1 causes a decrease in TER accumulation
	hSaufen	TER associating complex	Proper RNA localization	ND
	L22		Component of 60S ribosomal unit	
	SmB and SmD3	TER associating comples, binding to the CAB box of the CR7 region	SMN complex	ND
TERT biogenesis and regulation	Hsp90 and p23	TERT chaperones	Chaperone functions in nuclear and ligand activated receptors	Inhibition of Hsp90 or p23 block the assembly of telomerase
	14-3-3	Nuclear localization of TERT by masking of the TERT nuclear export signal	Controlling intracellular signaling, molecular chaperone functions and intracellular localization	Mutant 14-3-3 leads to the decrease of TERT nuclear accumulation but does not reduce telomerase activity *in vitro*
	Nucleolin	TERT nucleolar localization	Synthesis and maturation of ribosome RNAs	Nucleolin mutants prevent TERT nucleolar localizarion but have no effects on telomerase activity
	PinX1	TERT nucleolar localization	rRNA and snoRNA maturation	Over-expression of PinX1 inhibits telomerase activity and leads to telomere shortening
	c-Abl	c-Abl phosphorylates TERT and inhibits its activity	Cell differentiation, cell division, cell adhesion and stress response	c-Abl$^{-/-}$ causes increase in telomere length

	Akt	Akt phosphorylates TERT and inhibits its activity	Angiogenesis, cell survival and cell division	Over-expression of Akt decreases telomere length in HTC cells
	MKRN	Regulates ubiquitination of TERT	Lysine ubiquitination, targeting proteins for proteasome degradation	Over-expression of MKRN1 decreases telomerase activity and telomere length
	SMURF	SMURF2 activated in presence of shortened telomeres	Lysine ubiquitination, targeting proteins for proteasome degradation	SMURF2 is increased by telomere shortening
Telomerase assembly and localization factors	T-CAB	TER associating protein, binding to the CAB box of the CR7 region	Small Cajal body RNAs biogenesis and localization	Mutant TCAB is associated with AR Dyskeratosis congenita TCAB knockdown leads to telomere shortening
	pontin and reptin	Assembly of TER, dyskerin and TERT	AAA + ATPase chromatin remodeling, transcription co-regulation, DNA damage responses	Pontin knockdown decreases dyskerin steady state levels and telomerase activity by 10–20%
	NAT10	Telomerase assembly, disassembly and localization	Nucleolar assembly, cytokinesis and microtubule stabilization	NAT10 over-expression causes a decrease in telomere length but does not effect telomerase activity
	GNL3L	Telomerase assembly, disassembly and localization	Nucleolar GTPase	GNL3L over-expression causes a decrease in telomere length but does not affect telomerase activity
Telomere tethering/ binding	EST1A	Recruitment of TERT to telomeric terminus. Bind TERT independent of TER.	Involved in nonsense mediated mRNA decay. Implicated in TERRA-mediated telomere chromatin formation.	Recombinant expression of EST1A in the presence of TERT increases telomere length when compared to cells receiving TERT alone. EST1A expression alone causes a loss of telomere repeats.

(continued)

Table 3 (continued)

	EST1B	Associate with telomerase activity. Bind TERT independent of TER.	Involved in nonsense mediated mRNA decay. Implicated in TERRA-mediated telomere chromatin formation.	Recombinant expression of EST1B did not significantly change telomere length
	hnRNPs	Telomere telomerase interaction	Regulation of alternative splicing and mRNA transport and packaging	Over-expression of hnRNP C1 and U leads to a decrease in telomere length
	Pif1	DNA helicase negatively regulate telomerse by unwinding DNA/RNA hybrid	In yeast, the mitochondrial form of PIF1 is involved in mitochondrial DNA repair	Expression of wildtype PIF1 in human fibrosarcoma cells resulted in telomere shortening

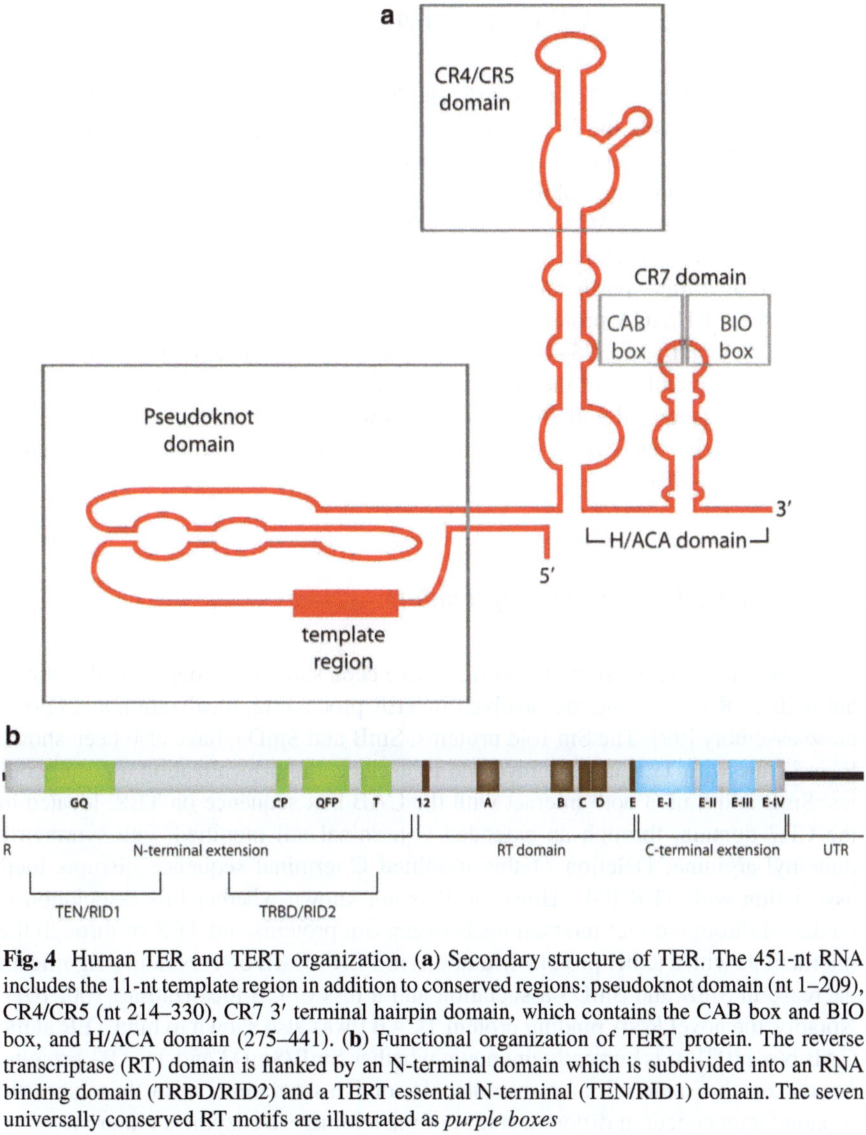

Fig. 4 Human TER and TERT organization. (**a**) Secondary structure of TER. The 451-nt RNA includes the 11-nt template region in addition to conserved regions: pseudoknot domain (nt 1–209), CR4/CR5 (nt 214–330), CR7 3′ terminal hairpin domain, which contains the CAB box and BIO box, and H/ACA domain (275–441). (**b**) Functional organization of TERT protein. The reverse transcriptase (RT) domain is flanked by an N-terminal domain which is subdivided into an RNA binding domain (TRBD/RID2) and a TERT essential N-terminal (TEN/RID1) domain. The seven universally conserved RT motifs are illustrated as *purple boxes*

is essential for TER association with the chaperone H/ACA protein complex. Association with the H/ACA proteins is crucial for cellular accumulation and 3′ end processing of TER. The 3′ terminal hairpin domain (CR7; nt 408–422) contains a Cajal body specific localization signal (CAB box) necessary for the accumulation of TER to the Cajal bodies (CBs), as well as a biogenesis box (BIO box), which is necessary for *in vivo* accumulation of TER (Fig. 4a) [89, 90].

3.3 *The Hinge/ACA Proteins (H/ACA)*

The H/ACA proteins dyskerin, Nop10, and Nhp2 form the core trimer that acts as a chaperone to promote the *in vivo* accumulation of TER. The binding of these proteins with TER immediately following transcription is essential for its cellular accumulation, processing and stability [93]. In contrast to other protein factors described in the later sections, H/ACA proteins associate with TER throughout the enzyme's life span and are considered stable components of the telomerase holoenzyme, as illustrated by affinity purification experiments [92, 93].

Two sets of H/ACA proteins bind to the distal and proximal stem loops of the TER H/ACA motif (nt 275–441) [80]. Mutations in the H/ACA motif in TER, as well as in the members of the H/ACA core trimer complex (dyskerin, Nhp2, and Nop10), are associated with genetic diseases with the common etiology of telomerase deficiencies and overlapping clinical presentations of premature tissue aging phenotypes [94–98].

3.4 *Other TER-Associating Factors*

RNA binding proteins, staufen and L22, have been shown to independently associate with TER *in vivo* and are involved in TER processing, localization and telomerase assembly [99]. The Sm-fold proteins, SmB and SmD3, have also been shown to associate with TER and are involved in its subcellular localization to Cajal bodies. SmB and SmD3 both interact with the CAB box sequence on TER, located in the CR7 domain, through an extended C-terminal tail modified with symmetric dimethyl-arginine. Deletion of this modified C-terminal sequence disrupts their association with TER [94]. However, it is not known whether this association is mediated through direct interactions between Sm proteins and TER or through the interactions with a tether protein. Mutations in TER's CAB box result in a significant decrease in SmB and SmD3 association and a loss of CB localization [100, 101]. Notably, the novel RNA binding protein TCAB1 was also shown to bind TER at the CAB box [103]. It is currently unknown whether SmB/SmD3 and TCAB1 proteins coexist on the same telomerase molecule, or if associations with these specific protein factors occur at different stages of the telomerase enzyme's maturity.

3.5 *Telomerase Reverse Transcriptase*

Catalytic activation of the telomerase complex requires the transcriptional activation of TERT. The TERT gene, located on chromosome 5p15.33, is composed of 16 exons and encompasses more than 37 kb [103, 104]. The GC-rich promoter region is located 1,100 bp upstream from the ATG start codon [104, 105]. This region lacks both TATA and CAAT boxes [103] and was found to be hypermethylated in somatic cells which correlates with its transcriptional inactive state. The TERT promoter contains numerous c-myc, as well as other oncogenic transcription factors, such as c-Jun and c-fos binding sites, which have been demonstrated to mediate TERT transcriptional activation in transformed cells [105]. Transcription activation of the TERT locus produces a full length TERT-mRNA, as well as a variety of alternative spliced forms. TERT alternative splicing is believed to regulate the levels of functional telomerase in a development stage specific manner [107]. Following protein translation of the full length 125 kDa polypeptide [108, 109], TERT associates with chaperones Hsp90 and p23, and is transported to the nucleus via its nuclear localization signal where it is assembled with the TER-H/ACA complex to form the fully functional telomerase enzyme [110].

TERT contains a central reverse transcriptase (RT) domain that is flanked by a N-terminal region and a C-terminal domain. The TERT N-terminal region is further subdivided into two domains: an RNA binding domain (TRBD) and a TERT essential N-terminal (TEN) domain. A large non-conserved linker region separates the two N-terminal domains (Fig. 4b) [112].

The RT domain contains the seven universally conserved RT motifs (1, 2, A, B′, C, D, and E) [113]. An invariant trio of aspartic acids (found in motifs A and C) is directly involved in catalysis, as mutations of these residues results in abolished catalytic activity *in vitro* and *in vivo* [84, 114–117]. Mutations of other amino acid residues in any of the conserved RT motifs were also found to reduce or eliminate telomerase reverse transcriptase activity (Fig. 4b) [84, 115, 117].

The high affinity RNA binding domain (TRBD), also known as the RNA interacting domain 2 (RID 2), contains telomerase specific motifs CP, QFP, and T, also referred to as domains II, III, and IV, respectively [118–120]. These motifs mediate TER recognition and have a relatively high binding affinity to structured RNA stem loops, interacting with the CR4/CR5 domain of TER [121]. This domain plays a role in promoting stable enzyme assembly, as mutations in these motifs result in severe defects in TER–TERT association (Fig. 4b) [122].

The TERT essential N-terminal (TEN) domain or RNA interacting domain 1 (RID 1), contains the non-conserved extreme N-terminus motif [123] and moderately conserved GQ motif (also referred to as domain I) [112, 122]. The GQ motif is further divided into domains IA and IB, separated by a DAT (dissociates activities of telomerase) domain [124]. The TEN domain interacts with the TER pseudoknot-template domain [121], but is not considered a major TER binding surface as mutations in this region only result in modest reductions of TER–TERT association [122]. This region also displays high single-stranded telomeric DNA binding affinity, suggesting an important role in substrate recognition and primer binding (Fig. 4b) [121, 124–126].

The smaller, less-conserved C-terminal domain (TEC or CDAT) plays several roles in telomerase function: it contributes to telomerase catalytic activity [121, 127], regulates the cellular localization of the enzyme, and plays a role in polymerase processivity [128, 129]. However, this domain is not essential for RNA binding, as mutations in this region were not found to impair TER–TERT association (Fig. 4b) [129].

3.6 TERT Chaperones and Localization Factors

Molecular chaperone proteins p23 and Hsp90 were identified as key factors in the assembly and functionality of the telomerase holoenzyme. Both were found to associate with TERT and aid in its nuclear import and localization. They were also demonstrated to be required for the assembly of active telomerase enzyme both *in vitro* and *in vivo*, as inhibition of either chaperone protein disrupts telomerase assembly leading to a reduction in enzyme activity ([110]; see geldanamycin mechanism below).

The nuclear retention of TERT is dependent on its association with the 14-3-3 proteins, a protein family involved in intracellular trafficking/targeting, cell cycle regulation, cytoskeleton structure, and transcription [129]. TERT and 14-3-3 interact via their respective C-termini. This interaction is required for the nuclear accumulation of TERT, as 14-3-3 proteins promote the nuclear retention of TERT by masking the nuclear export signal (NES)-like motif in the C-terminal region of TERT. Binding of 14-3-3 inhibits the binding of CRM1/exportin 1 to TERT NES, resulting in the nuclear accumulation of the reverse transcriptase.

Nucleolin is a phosphorprotein that binds to TERT through its RNA binding domain 4 and the carboxyl terminal RGG domain. RNA binding domain 1 may also be involved in the nucleolar localization of telomerase holoenzyme through its interactions with TER. Biochemical experiments had shown that the binding of TERT with the nucleolin-4R fragment, which lacks a nucleolar localization signal, resulted in the mislocalization of TERT in the cytoplasm, thereby implicating this protein in the subnuclear localization of TERT [130].

PINX1, a PIN2/TRF1 interacting protein, is involved in TERT nucleolar localization and has also been characterized as an inhibitor of telomerase activity and a negative regulator of telomere length. Inhibition of endogenous PINX1 resulted in an increase in telomerase activity, whereas over-expression of PINX1 decreases telomerase activity and shortens telomeres [131]. PINX1 was found to bind directly with TERT at its RNA binding domain and indirectly associate with TER through TERT [132].

3.7 TERT Post-translational Modifications

Telomerase activity is regulated via post-translational modifications of TERT. Several studies have demonstrated that the phosphorylation of TERT is required for the catalytic activity of the enzyme [133–136]. Protein kinase B (Akt) and protein kinase Cα have both been shown to interact with and phosphorylate TERT *in vitro* and *in vivo* [133–135], resulting in the increase in telomerase activity. Conversely, protein phosphatase 2A inhibits telomerase activity via the dephosphorylation of TERT directly [133, 137] or indirectly, through the dephosphorylation and inhibition of Akt. c-Abl protein tyrosine kinase associates with TERT and mediates TERT phosphorylation *in vitro* and *in vivo*. In contrast to the activation models above, c-Abl phosphorylation of TERT resulted in the inhibition of telomerase activity, making this kinase a negative regulator of TERT [138].

The E3 ubiquitin ligase MKRN1 was shown to have a negative role on telomere length homeostasis. MKRN1 is responsible for the ubiquitination of TERT, targeting TERT for protease degradation. Over-expression of MKRN1 results in the decrease of telomerase activity and subsequently in the shortening of telomere length [139].

3.8 TER–TERT Biogenesis/Assembly Factors

Pontin and reptin, members of the AAA+ family of DNA helicases [140], play pivotal roles in telomerase assembly. These helicases are found to bind to dyskerin and play a role in the formation of the TER-dyskerin complex. Subsequently, these helicases bind to endogenous TERT and mediate the assembly with TER-dyskerin complex to form the catalytically active telomerase enzyme [141]. The formation of the TERT-pontin-reptin complex is regulated by cell cycle stages, with the highest level of complex formation occurring during S-phase, providing evidence for another level of cell cycle dependent regulation of TERT.

Nucleoplasmic Cajal bodies (CBs) have been suggested as one of the sites for telomerase assembly. The novel RNA binding protein TCAB1 was shown to be required for telomerase localization to these sites. Knockdown of TCAB using retroviral shRNA and RNA interference resulted in a significant reduction in the percentage of cells with TER staining in CBs by microscopic analysis [102], indicating its role in CBs localization of telomerase. TCAB1 was found to associate with TER by binding specifically to the CAB-box sequence (CR7 motif). Inhibition of TCAB1 by shRNA also reduced the amount of TER at telomeres during S phase of the cell cycle, resulting in telomere shortening. This data suggested that TCAB1 plays a role in controlling the access of telomerase complex to telomeres, representing an additional level of enzyme activity regulation [102] and see below).

3.9 Targeting Telomerase Holoenzyme to Telomere

Newly assembled, catalytically active telomerase enzyme must travel to and associate with the limited number of telomere ends for its proper function. As illustrated with the earlier discussion on TCAB, the Cajal Bodies were suggested as sites where the delivery of the active enzyme to the telomeres occurs [142–144]. TER is found localized at CBs in cancer cells throughout the cell cycle [101, 142]. Mutations in the CAB box motif decrease the accumulation of TER in CBs as well as the frequency of TER association with telomeres, resulting in shorter telomere length [101, 144]. Recent analysis of genetic lesions responsible for the rare autosomal recessive isoforms of dyskeratosis congenita (AR-DC) identified TCAB1 compound heterozygous mutations in a small subpopulation of AR-DC. While TER accumulations were within the normal range, telomerase RNA was found to accumulate at nucleoli instead of Cajal bodies. Mislocalization of the telomerase holoenzyme prevented telomere access leading to a loss of telomere length maintenance. This data identified TCAB1 as a critical telomerase regulation factor, which recruits the holoenzyme complex to Cajal bodies for proper telomere access and synthesis [145].

The presence of TERT was also found to be necessary for the localization and accumulation of TER in CBs as well as trafficking of telomerase to telomeres during S phase of the cell cycle [142, 146, 147]. However, outside of S phase, TERT resides in subnuclear foci, termed TERT foci [143], indicating that these two components are not transported to CBs as an assembled complex. Inhibition of TERT resulted in a decrease of TER colocalized with CBs and telomeres without affecting the levels of TER in cells. Additionally, expression of TERT in telomerase negative cells resulted in the accumulation of TER at both sites [146]. These observations again suggest that CB localization of telomerase is connected to enzyme biogenesis and catalytic activity in transformed cells.

hEst1A has also been suggested to play an important role in telomere maintenance in a manner similar to its yeast homologue Est1p. Yeast Est1p interacts with TLC (yeast TER) and the yeast telomere binding protein Cdc13, thereby recruiting telomerase to the proximity of the telomeres [148, 149]. Using in silico methods, three human homologs of yeast telomerase telomere-recruitment factor Est1p were identified. Of these three proteins, Est1A shows the highest sequence homology with ScEst1p [150]. Over-expression of Est1A reduced the steady state telomere length, but co-expression of TERT and Est1A increases telomere length substantially, suggesting that Est1A's role in telomere length regulation is completely telomerase dependent. The human Est1p homologs have recently been implicated in TERRA and telomeric chromatin regulation [52, 151].

hnRNPs are also implicated in the localization of telomerase to telomeric ends for the de novo synthesis of telomeric repeats. *In vitro* studies demonstrated that hnRNPs A1/UP1, A2, A3, C1/C2, and D bind to TER and single-stranded telomeric DNA [153–156], suggesting possible roles in the bridging and recruitment of telomerase holoenzyme to the telomeres. In parallel, hnRNP A1/UP1 was found at telomere ends *in vivo* and was suggested to stimulate telomerase activity through the disruption of G-quadruplex structures formed during telomere synthesis by telomerase [156].

The human homolog of yeast DNA helicase Pif1 negatively influences the regulation of telomere length, by modulating telomerase activity [157]. hPif1 reduces telomerase processivity at the telomere by binding to and unwinding the DNA substrate and RNA template hybrid, resulting in the removal of telomerase from chromosome ends. hPif1 expression is regulated by cell cycle progression, peaking at late S/G2 [158]. Over-expression of hPIF1 induces telomere shortening in human HT1080 cells through telomerase activity modulation [157].

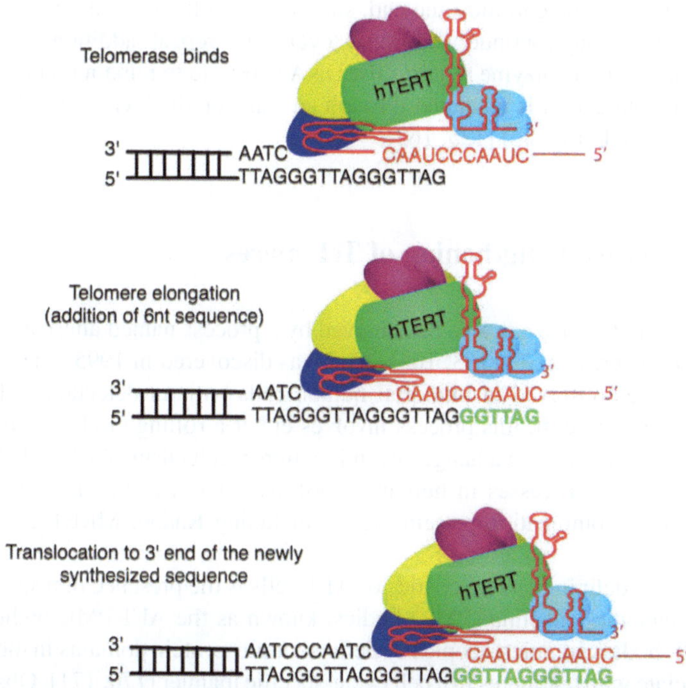

Fig. 5 Telomere repeat synthesis. Due to its short RNA template sequence, telomerase relies on two movement behaviors to add multiple 6-nuleotide (nt) telomeric sequences to chromosome ends. Addition of each 6-nt repeat to the 3′ end of the template is followed by telomerase translocation. This mediates realignment of the chromosome end from the 5′ end to the 3′ end of the template to enable subsequent rounds of repeat addition. Telomerase's ability to carry out these two movement behaviors is termed nucleotide addition processivity and repeat addition processivity, respectively

4 Telomerase Catalytic Cycle

TERT directs the addition of deoxynucleotide triphosphates (dNTPs) to the ends of the G-rich strand of the chromosome by copying the last six nucleotide of the 11-nt telomere repeat template sequence of TER [159, 160]. This activity results in the *de novo* synthesis of a single, 6 nt repeat. Because the TER RNA template region is quite short, to generate multiple repeats within a single catalytic event, telomerase holoenzyme undergoes multiple rounds of transient dissociation from the DNA substrate, to reposition the enzyme-substrate complex. Telomerase relies on its unique ability to transiently move away from the active site after the addition of a single 6-nt repeat, translocate towards the 3′ end of the newly synthesized chromosome and mediate the realignment of the new chromosome end with the TER RNA template, in order to continue subsequent rounds of multiple telomeric repeat addition (Fig. 5). Following the addition of each telomeric repeat, the enzyme may either disassociate from the chromosome end, stay bound without continuing elongation, or translocate and continue additional cycles of repeat addition [124, 161]. Translocation of the enzyme requires the DNA substrate to remain bound to telomerase. This interaction is mediated through an "anchor site" within the N-terminal domain of TERT (Fig. 4b) [125, 162].

5 Alternative Lengthening of Telomeres

Telomere maintenance can also be achieved by a process named alternative lengthening of telomeres (ALT) [163, 164]. ALT was discovered in 1995 when telomere elongation was observed in immortal human cells without detectable telomerase activity [165]. In yeast, this process involves either a rolling circle recombination mechanism or a strand exchange recombination mechanism. ALT is believed to occur by similar processes in humans [166], as it requires the activity of many homologous recombination protein factors including Rad50, MRE11, and NBS1 [167, 168].

One of the defining characteristics of ALT cells is the presence of a special class of promyelocytic leukemia (PML) bodies, known as the ALT-PML bodies [169]. ALT-PML bodies are microscopically defined, multi-protein domains in the nucleus that associate with telomeres in a cell cycle-specific manner [170, 171]. Observation of the ALT cell line U2OS revealed that TRF1 and FANCD2, a member of the Fanconi anemia protein family, colocalized with ALT-PML bodies at the same stages of the cell cycle. Monoubiquitination of FANCD2 is essential for this association, as depletion of FANCA, a member of the ubiquitination complex, or FANCL, the E3 ubiquitin-protein ligase, resulted in the loss of FANCD2 signals in ALT-PML bodies [172]. Depletion of FANCA or FANCD2 also resulted in an increase in telomere-signal-free chromosome ends in ALT cells. Due to the heterogeneity of telomere length in ALT cells, there were no significant changes in the average telomere length corresponding to these events. However, examination of newly synthesized telomere ends revealed that in the absence of FANCA or FANCD2 there is a significant decrease in sister chromatid exchange, supporting a role for monoubiquitination of FANCD2 in the ALT-mechanism of telomere maintenance through homologous recombination. Mutations in the chromatin remodeling proteins, the alpha thalassemia/mental retardation syndrome X-linked protein (ATRX) and death-associated protein 6 (DAXX) were found to associate with the ALT phenotype in a panel of pancreatic neuroendocrine tumors. Loss of ATRX-DAXX function is postulated to compromise heterochromatin states at telomeres, leading to the development of ALT by providing a permissive environment for nonreciprocal homologous recombination [173].

Although detected in human cells, ALT is not considered to be the normal physiological process for the maintenance of telomeres in humans. It has only been observed in a small number of human tumors (carcinoma and osteocarcinoma) and some transformed cell types in culture (mainly fibroblasts) [165, 174]. Long-term telomerase inhibition could potentially select tumor cells for ALT activation, as recently described in a mouse model of inducible TERT expression [175]. The under-representation of ALT-positive tumors was puzzling, until a 2002 study proved that the ALT mechanism cannot fully substitute telomerase in tumorigenesis: expression of the oncogenic H-Ras allele in the immortal human fibroblast ALT cell line GM00847 did not result in malignant transformation when injected into nude mice. In contrast, the co-expression of TERT in these cells imparted a tumorigenic phenotype [176]. This tumorigenic phenotype was again observed with the introduction of a mutant TERT, TERT-HA, which retains its enzymatic activity *in vitro* but is incapable of maintaining telomere length *in vivo*. Additionally, recombinant telomerase expression in ALT models accelerates cell growth and promotes anchorage-independent growth. Telomerase-positive ALT cells pass through cell-division phases of the cell cycle more quickly, implying that the observed cell-growth advantage is cumulative over cycles of proliferation [177]. The ALT recombination mechanism was not able to completely replace telomerase in the process of cellular transformation, implicating an additional, tumor growth-promoting role of TERT, independent of it role in telomere length maintenance.

6 Telomerase and Cancer

Most normal human somatic cells do not express detectable levels of telomerase activity as TERT expression is rapidly down-regulated following embryonic development [178]. In some human cell types, such as germline cells and stem cells, where there is a high demand for proliferation, TERT transcription is periodically activated to allow for transient expression of the enzyme. In contrast, more than 85% of human tumors surveyed harbor robust telomerase activity [179]. In almost all cases, the transcriptional up-regulation of TERT is responsible for the increase in ectopic telomerase activity in tumor cells [180]. Proof-of-concept experiments showed that the inhibition of telomerase in human cancer cells resulted in telomere-induced crisis and apoptosis in cell culture models [181, 182].

Telomerase expression is not considered to be oncogenic, as it alone does not lead to the development of cancer [183]. Additionally, it has been shown that TERT expression alone is not sufficient for the immortalization of human mammary epithelial cells, keratinocytes [184], prostate epithelial cells [185], or airway epithelial cells [186]. Cooperation between TERT and other oncogenic factors are essential for the transformed phenotype [187].

Paradoxically, early neoplastic lesions typically have undetectable or low telomerase activity, when compared to advanced malignant lesions that over-express the enzyme [188, 189]. This suggests that initiation of tumor development may require the absence of telomerase activity. Indeed, data from tumor cytogenetic studies have demonstrated that telomere length from precancerous lesions are much shorter than in normal tissues [12, 190, 191]. Several studies have reported critically short telomeres as a common early feature of many human cancers, such as colon [189], lung [192], breast [193], pancreatic [194], and prostate [195]. The telomere dysfunction model of carcinogenesis suggested that rampant chromosome instability following the uncapping of dysfunctional, short telomeres contributes to the eventual development of aneuploidy, a genetic signature of cellular transformation and carcinogenesis [57, 196]. Telomere dysfunction is thus recognized as a late event in the process of cancer initiation. After which, telomerase activity has to be induced to prevent further chromosome instability that hinder cancer growth, and provide a mechanism for the indefinite proliferation and immortality phenotype in malignant tumors [197] (Fig. 2).

7 NON-telomere Maintenance Roles of Telomerase

Besides its role in telomere maintenance, there is growing evidence pointing to telomerase's additional role in the cancer biology. Higher mRNA levels of several DNA repair and chromatin modifying genes, as well as better double-stranded break repair kinetics, were observed in human foreskin fibroblast cells expressing TERT as compared to cells lacking ectopic expression of the enzyme [198]. Importantly, these effects occurred rapidly before any significant telomere lengthening was observed. A transcriptome study done by Smith and colleagues [199] demonstrated that the ectopic expression of telomerase in human mammary epithelial cells reduced the need for exogenous mitogens for cellular proliferation, correlating with the telomerase dependent induction of gene expression that promotes cell growth and survival. This latter study provided evidence for a role of telomerase in cellular proliferation by affecting the expression profiles of growth and survival-related genes. In corroboration of this model, TERT is shown to act as a transcriptional co-activator of the beta-catenin transcriptional complex in mice [200], a function that is independent of its reverse transcriptase activity and its association with the telomerase RNA [200, 201]. These data have been recently corroborated by Masutomi and Hahn's model implicating human TERT in the promotion of TWIST expression and the resultant epithelial–mesenchymal transition. In conformity with the mouse models, TERT was found to complex with the BRG1, a SWI/SNF-related chromatin-remodeling factor, in transformed human cells.. Distinct from mouse models, human BRG1-TERT complexes with additional nucleolar proteins nucleostemin (NS) and/or GNL3L [202].

Aside from transcriptional regulation, TERT activity is also implicated in optimal mitochondrial function, independently of TER [203, 204]. Although direct molecular proof of TERT's functionality in the mitochondrion has not yet been established, TERT has intriguingly been shown to exhibit a RNA-dependent RNA polymerase activity when partnered with a mitochondrial RNA, RMRP [203]. DePinho's group showed that a switch to the ALT mechanism of telomere maintenance in mouse T-cell lymphoma, through the inhibition of TERT expression, was accompanied by a specific induction of mitochondrial enzymes that reduce oxidative damage. This supports the hypothesis that TERT harbors mitochondrial functions, independent of TER [175].

In addition to transcription co-activator functions, constitutive TERT expression is also involved in enhanced DNA repair. Normal, diploid human fibroblasts over-expressing TERT were found to be more resistant to apoptosis and necrosis induced by DNA damages, but equally susceptible to the cytotoxic effects of oxidative agents as normal fibroblasts without TERT expression [205]. This suggested that telomerase is involved in enhancing cellular survival following genotoxic stress. Direct evidence implicating telomerase's role as a regulator of the DNA damage response pathway was provided by a cell biology study [206]. By suppressing endogenous TERT expression in diploid human fibroblasts using either an TERT-coding sequence specific shRNA or an TERT 3′ untranslated region-specific shRNA (TERT 3′ UTR shRNA), it was shown that TERT participates in DNA damage responses and chromatin maintenance in a manner that is separate from its role in telomere length maintenance. Following ionizing radiation (IR), irinotecan, or etoposide treatment, phosphorylation of H2AX and the ataxia telangiectasia mutated (ATM) protein was greatly impaired in telomerase knock-down cells as compared to control cells expressing normal levels of TERT. As a direct consequence, the phosphorylation of BRCA1 tumor suppressor proteins was not observed and protein levels of p53 were not up-regulated. These results indicate impaired DNA damage responses in cells lacking TERT. Telomerase knock-down cells also exhibited increased sensitivity to IR as shown by the decreased relative survival in clonogenic growth assays. When wildtype recombinant TERT was introduced into cells expressing the TERT 3′ UTR shRNA, which does not target these recombinant copies of TERT, the cells ability to respond to DNA damage was restored. The molecular mechanism of how TERT perform this role to promote DNA damage survival remains unclear, but is suggested to be associated with TERT's chromatin remodeling activities. In agreement with these data, our laboratory showed that transient telomerase inhibition synergistically increased the cytotoxicity of double-stranded DNA-damaging agents, in a cell-cycle phase-specific manner. This short-term telomerase inhibition was not predicted to significantly reduce telomere length, and the synergistic cellular toxicity may be ascribed to the inhibition of a non-telomere-related telomerase function in tumor cell growth [207].

8 Targeting Telomeres and Telomerase in Anticancer Chemotherapy

Uncapped telomeres induce a dramatic DDR response culminating in cell cycle arrest and programmed cell death [19, 56, 69]. While targeted disruptions of the telomere structure could have been a viable strategy in anticancer therapy, the therapeutic index would be extremely low, considering that the same DDR activation will be induced in cancer and normal cells alike. Conversely, the apparent lack of TERT expression in normal somatic cells, and the growing evidence for TERT's additional roles in cancer biology, makes telomerase an ideal target for anticancer therapies. Telomerase is constitutively over-expressed in over 85% of all human cancers [179]. Early proof-of-principle experiments demonstrated that the expression of a dominant negative form of TERT completely inhibited telomerase activity and substantially reduced telomere length in several cancer models [180]. The resulting telomere dysfunctions led to the formation of dicentric chromosomes and other types of chromosome fusions, resulting in the loss of cellular viability and apoptosis. This inhibition of TERT was demonstrated to limit tumorigenicity of mouse xenograft models of cancer [181].

Following these proof-of-principle experiments, numerous strategies targeting the telomerase holoenzyme components are described. In the following sections, we discuss some of the more notable strategies of telomerase inhibition in targeted therapy against cancers.

9 Telomerase Catalytic Activity Inhibitors

9.1 BIBR1532

BIBR1532 is a small molecule, non-nucleoside inhibitor that interferes with telomeric DNA repeat addition by telomerase through the targeting of the catalytic component TERT [208]. Treatment of cancer cells with low doses BIBR1532 reduces their growth capacity and sensitizes them to other chemotherapeutic drugs, in a telomere length-dependent manner [209]. At high doses of BIBR1532, cells exhibited off target cytotoxic effects independent of telomerase's catalytic function [210, 211]. Leukemia cells, but not normal hematopoietic stem cells, treated with 30–80 μM BIBR1532 displayed an immediate reduction in proliferative capacity. In particular, telomere dysfunctions are manifested as increases in telomere signal free ends, formation of chromosome end-to-end fusions, and an increase in phosphorylation of p53 and a loss of TRF2 signals at the telomere. However, BIBR1532 induced cytotoxic effects may not be confined to the formation of dysfunctional telomeres and this off-target effect hampers its further development as an anticancer therapeutic agent.

9.2 3′-Azido-2′, 3′-Dideoxythymidine (AZT)

AZT is a reverse transcriptase inhibitor used in the highly active antiretroviral therapy (HAART) against HIV infection and in the treatment of virus-associated cancers. As a thymidine analog, AZT has been shown to inhibit telomerase *in vitro* and *in vivo*. Upon its activation through phosphorylation by thymidine kinase, this nucleoside analog is incorporated into telomeric DNA as a chain terminator, blocking further reverse transcription and telomere elongation [212, 213]. Prolonged treatment of adult T-cell leukemic cells with AZT results in telomere attrition, accompanied by increased expression of $p14^{ARF}$ and activation of the p53-dependent apoptotic pathway [214]. This leads to an increase in the p53 target $p21^{WAF}$ expression and the accumulation of $p27^{KIP}$, to induce cell cycle arrest or program cell death of the tumor cells. In combination with chemotherapy agents such as 5-fluorouracil, AZT has been shown to increase treatment toxicity in colorectal cancer cell model, most likely in a synergistic manner.

9.3 Oligonucleotide-Based Specific Inhibitors of Telomerase

Oligonucleotide-based inhibitors of telomerase designed to target the TER template may provide a highly specific, telomerase-based antitumor therapy [86, 215, 216]. GRN163L is a 13-base, lipid modified N3′-P5′ thiophosphoramidate oligomer, complementary to the template region of TER. GRN163L binds with high affinity to telomerase [217, 218] and has been demonstrated to effectively inhibit the enzyme, resulting in telomere length shortening and subsequent growth arrest. The 5′-lipid palmitoyl domain facilitates cellular and tissue penetration, as well as makes this agent more acid resistant than other anti-telomerase oligonucleotides, thereby increasing the cellular uptake and bioavailability of the drug [220]. GRN163L shows antitumor effects in several cancers, including breast, liver, lung, and multiple myeloma, both *in vitro* and *in vivo* [219–223]. This drug is currently undergoing clinical trials in patients with chronic lymphocytic leukemia, multiple myeloma, solid tumor malignancies, locally recurrent or metastatic breast cancer and advanced or metastatic non-small cell lung cancer [224].

10 Telomerase Expression, Biogenesis and Assembly Inhibitors

10.1 Costunolide

Costunolide is a sesquiterpene lactone isolated from *Magnolia sieboldii*. Reported to harbor anti-inflammatory, antifungal, and antiviral properties [225–228], it was also shown to suppress cell proliferation and induce apoptosis in several tumor cell lines, including breast cancer and leukemia cells [229, 230]. Costunolide exerts its anticancer properties through transcription regulation of TERT. A decrease in c-Myc or Sp1 binding to their cognate DNA binding sites on the TERT promoter was observed after costunolide treatment, in a dose-dependent manner [229]. Corresponding to the decrease in TERT mRNA levels, there is a reduction in telomerase activity resulting in an inhibition of cell growth and an increase in apoptosis.

10.2 Geldanamycin

Geldanamycin is a benzoquinone ansamycin antibiotic and inhibits the binding of cofactor ATP and partner p23 to the molecular chaperone Hsp90 [231, 232]. The Hsp90-p23 complex is a molecular chaperone that binds to and stabilizes cytoplasmic TERT at intermediate stages for folding, assembly and movement across nuclear membranes. Geldanamycin blocks the assembly of active telomerase both *in vitro* and *in vivo* [110] by disrupting the Hsp90-p23-telomerase interaction. Geldanamycin actions result in the ubiquitination and proteosome degradation of TERT and the reduction of telomerase activity [139]. However, since Hsp90 and p23 form chaperone complexes that have integral roles in numerous biological processes, geldanamycin mediated inhibition of Hsp90 function lacks specificity for the telomerase pathway. Given that many of the Hsp90-p23 binding partners are key players in cancer progression, such as v-Src, Bcr-Abl, Raf-1, and ErbB2 [233–236], geldanamycin promiscuous activities might be beneficial in anticancer chemotherapy. The utility of geldanamycin disruption of Hsp90-p23 formation should be revisited in specific cancer types, based on the molecular etiology of the disease.

11 Telomerase Immunotherapy

Telomerase is tested as a novel target for cancer immunotherapy. In telomerase-positive cancers, TERT peptides are presented as epitopes on the tumor cell surface by the major histocompatibility complex (MHC) class I pathway. TERT antigen presentation was demonstrated to produce cytotoxic T lymphocyte responses [237–239]. Two first-generation vaccines have been developed: GRNVAC1 and GV1001. Telomerase cancer vaccine, GRNVAC1, uses an *ex vivo* process where mature dendritic cells are isolated from the patient's blood and transfected with TERT mRNA. These cells are then returned to the body where they stimulate the production of $CD4^+$ and $CD8^+$ T-cells specific for TERT [240]. GV1001 is a peptide vaccine derived from the active functional domain of telomerase. GV1001 binds multiple human leukocyte antigen (HLA) class II molecules and harbors putative HLA class I epitopes, and also illicit $CD4^+$ and $CD8^+$ T-cell responses specific for TERT [240]. Both vaccines were test successful in phase I/II clinical trials for efficacy in producing telomerase specific $CD4^+$ and $CD8^+$ T-lymphocytes [240, 241]. GV1001 is currently in two phase III clinical trials for the treatments of pancreatic cancer while GRNVAC1 is being investigated in a phase II clinical trial in patients with acute myeloid leukemia [242].

12 Telomerase-Telomere Recruitment Inhibitors

12.1 Tankyrase1 Inhibitors

Poly(ADP-ribose) polymerase (PARPs) is a large family of enzymes that use NAD+ as a substrate to generate ADP-ribose polymers onto glutamic acid residues on protein acceptors [243–245]. Tankyrase 1 and 2 are PARP family members specifically known for their telomeric poly (ADP-ribosyl) polymerase activities. Tankyrase 1 and 2 ribosylates TRF1, preventing TRF1 from binding to telomeric DNA, and leading to TRF1's proteolytic degradation [246]. Over-expression of tankyrase 1 reduces TRF1 binding to the telomere, enables telomerase access at the telomere ends and the corresponding telomere elongation. Conversely, inhibition of tankyrase 1 induces telomere shortening and cell death through a telomere length independent mechanism: in the absence of tankyrase 1, cells undergoing mitosis are unable to resolve sister telomeres cohesion and were arrested at the mitotic phase [248, 249].

Small molecule inhibitors of tankyrase 1's PARP activities have been shown to complement telomerase inhibition to enhance the rate of telomere attrition [249]. However, given PARPs are known to mediate the ribosylation of multiple protein acceptors, the likelihood of off-target effects by these small molecules PARP inhibitors is high.

12.2 G-Quadruplex Stabilizers

G-quadruplexes are stable 4-stranded DNA structures made up of G-rich sequences where the guanine residues form square arrangements. The 3′ telomeric DNA overhang is guanine rich and can form these higher order molecular structures, in addition to the normal telomeric DNA structures. Small molecule, non-nucleoside compounds such as telomestatin, BRACO-19, TMPyP4, and carbcynanine dyes, are predicted to bind within the grooves [250] or intercalate [251] G-quadruplex DNA, to stabilize these structures. Compounds that intercalate into the DNA to stabilize the G-quadruplex tend to have large, flat aromatic surfaces and are cationically charged to allow for π-stacking interactions. Examples of such molecules are porphyrins and cisplatin [252]. These older platinum containing complexes are shown to potently inhibit telomerase, leading to telomere shortening, arrested cell growth and subsequent cell death. Newer platinum (II) containing structures are also reported to inhibit telomerase *in vitro*, with distinct covalent linkage that could lock the G-quadruplex structure irreversibly [253].

Several different G-quadruplex inhibitors have been shown to disrupt the binding of telomere-associated proteins, inhibit telomerase activity and induce apoptosis *in vitro* [254–258]. However, as G-quadruplex binding agents, these compounds are predicted to bind elsewhere in the genome and disrupt their local structure, leading to altered functions. For example, telomestatin can bind to non-telomeric G-rich DNA found in the promoter region of the c-myc oncogene [259, 260]. Expression of myc is reduced by telomestatin binding, which stabilizing the G-quartet structure in its promoter and prevent transcription factor access. In addition to these off-target effects, another major problem with G-quadruplex stabilizer is their inability to penetrate the cell membrane. Optimal delivery protocol for these types of drugs has yet to be developed.

13 Genetic Therapy Against Telomerase

13.1 TER with Mutant Template

The expression of mutant-template human telomerase RNA (MT-TER), in telomerase positive cells, has been tested as an anticancer gene therapy. MT-TER assemble with endogenous TERT and the recombinant enzyme then erroneously adds DNA repeats with mutant sequence to chromosome ends. A few copies of mutant DNA repeats are enough to disrupt the binding of telomeric proteins. The resulting compromised telomere structure leads to a loss in cellular viability by inducing apoptosis [260–262]. Even though mutant TER is dominant over endogenously expressed wild-type TER, it can only be expressed at low levels, thereby limiting its cytotoxic efficiency in cancer cells. To overcome this deficiency, co-expression of siRNA against endogenous TER, as well as lentiviral expression of mutant TER, has proven to increase the therapeutic efficacy of MT-TER [262].

13.2 TERT-Promoter Driven Suicidal Gene Therapy

Based on the selective activation of the TERT promoter in cancer cells, several groups reported the use of recombinant DNA vectors, with TERT promoter driving the expression of cytotoxic transgenes, including the herpes simplex virus thymidine kinase, Bcl2-associating X protein, caspase 8 and bacterial nitro-reductase, delivering suicidal enzymatic activities in a cancer cell specific manner [263–271]. While these proof-of-principle experiments provided the framework for a cancer specific targeting strategy, more work is still needed for the development, delivery, and clinical validity of these cancer gene therapies.

13.3 Hammerhead Ribozyme

Hammerhead ribozymes targeting either the RNA component or reverse transcriptase component of telomerase are shown to be effective strategies in several cancer models. Colon and gastric carcinoma cells treated with retrovirus delivered ribozyme targeted TERT displayed a significant decrease in telomerase activity and rapid induction of apoptosis [272]. In endometrial and hepatocellular carcinoma cells, ribozyme targeting of TER resulted in a dose dependent decrease in telomerase activity [273, 274]. Up to 90% inhibition of telomerase activity could be achieved at relatively low concentrations of the ribozyme. As with other genetic means of telomerase activity inhibition, the current lack of efficient delivery protocols hamper their use in clinical settings.

13.4 Zinc Finger Proteins

Zinc Finger Proteins are synthetic peptides designed to target specific chromosomal loci and alter their functionality or sequence identity. Transcription activation of TERT in tumor cells relies on the activation at multiple transcription factor binding sites on TERT's promoter, including that for SP1, c-MYC, ER, E2F-1, WT-1 and MZF-2 [105, 275, 276]. Conceivably, ZFP designed to target these chromosomal loci will interfere with TERT transcription activities. Recently, a ZFP that recognizes a 12 bp sequence within the core TERT promoter fused to a KRAB repressor domain has been described [277]. *In vitro* expression of this ZFP resulted in >80% reduction of TERT expression. Cancer cell lines engineered to express this ZFP are shown to have significantly lower endogenous TERT mRNA levels, a decrease in telomerase activity and inhibition of cell proliferation within 8–12 days. Longer-term repression of endogenous TERT transcription in human cancer cell lines expressing this ZFP in a stable fashion mirrored these results and displayed shortened telomeres. Despite the positive laboratory data, several issues such as ZFP's treatment efficacy, target efficiency and specificity, as well as the availability of appropriate delivery protocols will need to be addressed before the adoption of these novel therapeutic options into clinical applications.

14 Therapeutic Considerations

14.1 Combination Chemotherapies

Despite the demonstrations of several successful strategies targeting telomeres and telomerase in cancer cells, their usefulness in the clinics has been marred by several deficiencies. The timeline of inducing cytotoxicity by telomerase inhibition relies completely on the kinetics of telomere shortening to a critically short length. As telomere length decreases at a rate of 50–100 bp per cell division, this process can be quite long, and tumor specific. This time lag can range from weeks to months of continual telomerase inhibition therapy. However, prolonged inhibition of the telomerase enzyme could affect normal human cells that are also dependent on transient telomerase activity for their functionality [61]. In these cases, telomere erosion in off-target cells from telomerase inhibition therapy could precipitate adverse treatment effects in these normal cell types. Premature telomere shortening translate to the accelerated rate of tissue aging. If these cells were allowed to divide beyond the short telomere check point, due to the inactivation of tumor suppressive mechanism, new rounds of chromosome instability cycles could trigger the development of secondary tumors. This paradox, in addition to the lack of proper delivery methods for genetic-based inhibition of TERT function, argues that telomerase inhibition on its own is not efficacious as an anticancer therapy.

On the other hand, telomerase inhibition has been demonstrated to increase the sensitivity to chemotherapeutic agents by overwhelming the DNA repair mechanism, with the creation of unprotected chromosome ends. For example, telomere dysfunction in late generation $TERC^{-/-}$ mice, lacking the mouse telomerase RNA gene, resulted in decreased cellular survival after exposure to IR [278]. At the cellular level, the rate of apoptosis in gastrointestinal crypt cells and primary thymocytes was higher in telomerase deficient mice as compared to control. These $TER^{-/-}$ cells also displayed delayed DNA break repair kinetics, as well as persistent chromosomal breaks, complex chromosomal aberrations and massive fragmentation.

Reduction of telomerase activity also resulted in increased cell sensitivity to topoisomerase inhibitors. The MCF-7 breast cancer cell line and HBL-100 immortal breast cell line expressing an anti-TERT ribozyme, which cleaves human telomerase mRNA, resulted in inhibition of telomerase activity, decreased telomere length and induced apoptosis. Additionally, an increased sensitivity to the topoisomerase II inhibitor doxorubicin was also observed in these cell lines. In parallel, when exogenous TERT was introduced into telomerase-negative human fibroblasts, there was a decrease in the sensitivity of these cell lines to doxorubicin, as well as two other topoisomerase inhibitors: mitoxantrone and etoposide [279].

Telomerase inhibition via the ectopic expression of dominant negative-TERT (DN-TERT) in human cancer cells resulted in telomere shortening, growth arrest and apoptosis [181, 182]. Expression of recombinant DN-TERT in BCR-ABL positive leukemia cells completely inhibited endogenous telomerase activity and resulted in an increase in apoptosis following treatment with the tyrosine kinase inhibitor imatinib [280].

Telomerase inhibition was also demonstrated to increase telomerase positive pharynx Fadu tumor cell's sensitivity to paclitaxel [281]. Telomerase inhibition was achieved using either antisense TER, which blocks the template for telomere synthesis, or 3′-azido-3′deoxythymidine (AZT), a nucleoside analog reverse transcriptase inhibitor. The combination of AZT and paclitaxel resulted in decreased tumor size, increased apoptosis, and prolonged survival in FaDu xenograft tumor mice models. This effect was not observed in telomerase negative human osteocarcinoma Saos-2 cells, indicating that the increase in sensitivity to paclitaxel was due to telomerase inhibition [282].

Knockdown of telomerase activity in human cells can also be achieved via retroviral transfer of siRNA targeting TERT. These telomerase knockdown cells displayed increased sensitivity to IR and chemotherapeutic agents etoposide, bleomycin, and doxorubicin [283]. In addition, the combination therapy using the TERT siRNA increased the apoptotic effect of cisplatin, a platinum-based chemotherapeutic agent, on the hepatocellular cell line SMMC7721 in vitro and also greatly reduced SMMC7721 and HepG2 tumor growth in the mouse xenograft model as compared to cisplatin monotherapy [284].

In 2005, Ward and Autexier reported the effects of telomerase inhibition on drug resistant leukemia and breast cancer cells by the non-nucleosidic small molecule inhibitor BIBR1532, a proprietary formulation from Boehringer Ingelheim [285]. This drug impairs telomere elongation by affecting telomerase translocation or promoting the disassociation of the enzyme from the telomere end [208]. They observed an increase in chemotherapy sensitivity when drug resistant leukemia and breast cancer cells were concurrently treated with BIBR1532. Continuous BIBR1532 treatment was found to decrease the proliferative capacity of these cells. As the number of population doublings with BIBR1532 increased these cells are progressively sensitized cells to the chemotherapeutic agents. This observation suggested that the effects of BIBR1532 treatment were telomere length dependent [211].

Combination chemotherapy studies demonstrated synergistic effects of GRN163L in combination with ionizing radiation [286]. Enhanced radiation sensitivity by GRN163L application was observed following long-term (42 days) drug treatment, with no significant differences in short-term (2 and 9 days) and intermediate inhibition (20 days) [286]. Accordingly, this synergistic effect was attributed to the generation of critically short telomeres following long-term telomerase inhibition. With breast cancer models, previous studies have also demonstrated that GRN163L in combination with the microtubule stabilizing agent paclitaxel [287], and tratsuzumab, a monoclonal antibody against the HER-2 receptor [288], has synergistic treatment effects, in a telomere-length dependent manner.

Combination studies have provided genetic and biological evidence linking telomere dysfunction and increased sensitivity to chemotherapeutic agents, making telomerase inhibition an effective therapeutic option for many different types of cancers. Many of these studies concluded that the observed increase in sensitivity of cancer cells to cytotoxic agents was telomere length dependent [209, 222, 286]. However, telomere shortening caused by the continuous inhibition of telomerase may affect normal human cell types that also require telomerase activity for growth and proliferation [61]. Recent data from our laboratory demonstrated that a transient inhibition of telomerase activity, at the time of the induction of DNA damage, also elicit a synergistic cytotoxicity response in breast and colon cancer cells. This potentiation of cytotoxicity is dependent on the timing and mode of action of the genotoxic agents, as only S/G2 specific DNA damage inducers are observed with increased cytotoxicity in combination with telomerase inhibition [207]. Even though the exact mechanism by which telomerase inhibition increases cellular toxicity in this manner, independent of telomere length, is not known, our work also demonstrated that inhibiting the ATM kinase, in conjunction with telomerase inhibition, synergistically increases the cytotoxicity of these S/G2 specific double-stranded DNA-damaging agents, suggesting a role for telomerase in DNA repair [207].

14.2 Transient Telomerase Activation as a Genome Maintenance Mechanism

Higher telomere attrition rates are often seen with chronic inflammation [289]. Accelerated telomere shortening in these conditions is associated with increased cellular turnover, leading to premature loss of tissue renewal capacity, and an increased risk of genomic instability. Proof-of-principle experiments showed that telomerase activation extends the replication lifespan of tissues with high turnover [290, 291]. Spurred by these early reports, telomerase activation strategies using small-molecule transcription activators are being heavily pursued. One of these agents, TAT2 (cycloastragenol) is extracted from the root of *Astragalus membranaceus*, a flowering plant used extensively in traditional Chinese herbal medicine [292]. TAT2 has been shown to activate telomerase in cell culture models, through the induction of TERT transcription. Short-term TAT2 treatment (12–18 days) improves the proliferative capacity of CD8+ T-lymphocytes from HIV-infected individuals by moderately increasing telomerase activity. By delaying the onset of immunosenescence of these T-cell models, TAT2 treatment increases the cytokine/chemokine production and antiviral (HIV) activity of T-cells in vitro. This effect is blocked by the addition of a specific telomerase inhibitor, GRN163L, confirming that the improvement of immune function by TAT2 is mediated by telomerase activation [291].

There is a strong negative correlation between mean telomere length and chronological age in humans [293, 294]. Conceivably, strategies to stimulate telomerase-dependent telomere maintenance in later life not only will contribute to boost tissue renewal capacity but will also help preserve the stability and integrity of the genome. This protection against genomic alterations which are frequently associated with cancers could be invaluable to older individuals, as age is one of the biggest risk factors for cancer [197]. Several studies have been initiated to test this. In one such study, TA-65, a compound related to TAT2, also isolated from *Astragalus membranaceus*, is being given as one of the active components of a dietary supplement. Other active ingredients include standard vitamins and trace minerals. An interim (1-year) report of this study revealed moderate improvement in participants' immune system profiles with a continuous regimen of TA-65 in low doses [295]. The long-term utility of these strategies in health promotion including an improved tissue renewal capacity and cancer prevention, as well as the off-target/untoward effects of such therapies require further investigation.

14.3 Concluding Remarks

The integrity of telomere function has a paramount role in promoting chromosome stability. Loss of telomere function is implicated in the replicative aging of human tissues, and also has a major effect on cellular transformation related to carcinogenesis [1, 197]. The relationship between telomere structural maintenance and DDR pathways is an illustration of functional adaptation. Telomeres exist to protect the ends of chromosomes from being recognized by DDR sensors and undergoing erroneous repair by DDR mechanisms [7, 19, 69]. Yet, normal homeostatic maintenance of this nucleoprotein structure relies on many of the same DDR factors that need to be kept in check [78, 79]. How do these simple 6nt DNA repeats and their protein-binding partners accomplish these conflicting tasks? What are the regulatory/signal transduction events that allow the same DDR machinery to adapt its functions for the specific requirement of maintaining telomeres?

Interindividual variations in telomere maintenance capacity are an understudied area of telomere biology [289]. TERT A1062T non-synonymous single nucleotide polymorphism (SNP) has recently been discovered to associate, with a high prevalence, to acute myeloid leukemia (AML). Patients diagnosed with AML had three times higher prevalence of the 1062T-TERT isoforms compared to controls. 1062T-TERT was found to exhibit decreased telomerase activity compared to wild type (1062A) [296]. Conceivably, a decreased telomere maintenance capacity could accelerate the rate of telomeric DNA loss leading to the premature exhaustion of replicative cell pools and the precipitation of genomic instability. With increased availability of data from large-scale epidemiology studies, it is expected that genetic variations in telomerase and other telomere pathway components could also be named as risk factors in other types of malignancies, as well as idiosyncratic and orphaned tissue failure syndromes [289].

To maintain an immortal phenotype, cancer cells need to replenish lost telomere repeats. Accordingly, high telomerase activity is observed in more than 85% of all human cancers [179]. Even though telomerase inhibition alone has limited clinical efficacy as an anticancer treatment, chemotherapy regimens targeting telomerase, when combined with other cytotoxic stress, are reported to be effective against multiple types of malignancies. Several Phase I and II clinical studies with non-small cell lung cancer and breast cancer patients are currently underway [297, 298]. The results of these trials may provide new clinical anticancer strategies.

Acknowledgment We thank Dragony Fu, Suzanne Lee, and Naresh Thumati for reading this chapter and for their helpful comments. Research in Judy Wong's laboratory is supported by the Canadian Institutes of Health Research, the Canadian Cancer Society, and the Leukemia and Lymphoma Society of Canada. JMYW is supported by the Canada Research Chair and Michael Smith Foundation of Health Research career development programs.

References

1. Jackson SP, Bartek J (2009) The DNA-damage response in human biology and disease. Nature 461:1071–1078
2. You Z, Bailis JM (2010) DNA damage and decisions: CtIP coordinates DNA repair and cell cycle checkpoints. Trends Cell Biol 20:402–409
3. Morrison C, Sonoda E, Takao N, Shinohara A, Yamamoto K, Takeda S (2000) The controlling role of ATM in homologous recombinational repair of DNA damage. EMBO J 19:463–471
4. Paull TT, Rogakou EP, Yamazaki V, Kirchgessner CU, Gellert M, Bonner WM (2000) A critical role for histone H2AX in recruitment of repair factors to nuclear foci after DNA damage. Curr Biol 10:886–895
5. de Lange T (2002) Protection of mammalian telomeres. Oncogene 21:532–540
6. de Lange T (2004) T-loops and the origin of telomeres. Nat Rev Mol Cell Biol 5:323–329
7. de Lange T (2005) Shelterin: the protein complex that shapes and safeguards human telomeres. Genes Dev 19:2100–2110

8. Blackburn EH (2001) Switching and signaling at the telomere. Cell 106:661–673
9. Chan CS, Tye BK (1983) Organization of DNA sequences and replication origins at yeast telomeres. Cell 33:563–573
10. Kipling D, Cooke HJ (1990) Hypervariable ultra-long telomeres in mice. Nature 347:400–402
11. Harley CB, Futcher AB, Greider CW (1990) Telomeres shorten during ageing of human fibroblasts. Nature 345:458–460
12. de Lange T, Shiue L, Myers RM, Cox DR, Naylor SL, Killery AM et al (1990) Structure and variability of human chromosome ends. Mol Cell Biol 10:518–527
13. Moyzis RK, Buckingham JM, Cram LS, Dani M, Deaven LL, Jones MD et al (1988) A highly conserved repetitive DNA sequence, (TTAGGG)n, present at the telomeres of human chromosomes. Proc Natl Acad Sci USA 85:6622–6626
14. Makarov VL, Hirose Y, Langmore JP (1997) Long G tails at both ends of human chromosomes suggest a C strand degradation mechanism for telomere shortening. Cell 88:657–666
15. McElligott R, Wellinger RJ (1997) The terminal DNA structure of mammalian chromosomes. EMBO J 16:3705–3714
16. Greider CW (1999) Telomeres do D-loop-T-loop. Cell 97:419–422
17. Griffith JD, Comeau L, Rosenfield S, Stansel RM, Bianchi A, Moss H et al (1999) Mammalian telomeres end in a large duplex loop. Cell 97:503–514
18. Klobutcher LA, Swanton MT, Donini P, Prescott DM (1981) All gene-sized DNA molecules in four species of hypotrichs have the same terminal sequence and an unusual 3′ terminus. Proc Natl Acad Sci USA 78:3015–3019
19. Palm W, de Lange T (2008) How shelterin protects mammalian telomeres. Annu Rev Genet 42:301–334
20. Bianchi A, Stansel RM, Fairall L, Griffith JD, Rhodes D, de Lange T (1999) TRF1 binds a bipartite telomeric site with extreme spatial flexibility. EMBO J 18:5735–5744
21. Court R, Chapman L, Fairall L, Rhodes D (2005) How the human telomeric proteins TRF1 and TRF2 recognize telomeric DNA: a view from high-resolution crystal structures. EMBO Rep 6:39–45
22. Shay JW (1999) At the end of the millennium, a view of the end. Nat Genet 23:382–383
23. van Steensel B, de Lange T (1997) Control of telomere length by the human telomeric protein TRF1. Nature 385:740–743
24. Smogorzewska A, van Steensel B, Bianchi A, Oelmann S, Schaefer MR, Schnapp G et al (2000) Control of human telomere length by TRF1 and TRF2. Mol Cell Biol 20:1659–1668
25. Broccoli D, Smogorzewska A, Chong L, de Lange T (1997) Human telomeres contain two distinct Myb-related proteins, TRF1 and TRF2. Nat Genet 17:231–235
26. Karlseder J, Broccoli D, Dai Y, Hardy S, de Lange T (1999) p53- and ATM- dependent apoptosis induced by telomeres lacking TRF2. Science 283:1321–1325
27. Stansel RM, de Lange T, Griffith JD (2001) T-loop assembly in vitro involves binding of TRF2 near the 3′ telomeric overhang. EMBO J 20:5532–5540
28. Karlseder J, Smogorzewska A, de Lange T (2002) Senescence induced by altered telomere state, not telomere loss. Science 295:2446–2449
29. Lei M, Podell ER, Cech TR (2004) Structure of human POT1 bound to telomeric single-stranded DNA provides a model for chromosome end-protection. Nat Struct Mol Biol 11:1223–1229
30. Loayza D, Parsons H, Donigian J, Hoke K, de Lange T (2004) DNA binding features of human POT1: a nonamer 5′-TAGGGTTAG-3′ minimal binding site, sequence specificity, and internal binding to multimeric sites. J Biol Chem 279:13241–13248
31. Flynn RL, Centore RC, O'Sullivan RJ, Rai R, Tse A, Songyang Z et al (2011) TERRA and hnRNPA1 orchestrate an RPA-to-POT1 switch on telomeric single-stranded DNA. Nature 471:532–536
32. Loayza D, De Lange T (2003) POT1 as a terminal transducer of TRF1 telomere length control. Nature 423:1013–1018
33. Colgin LM, Baran K, Baumann P, Cech TR, Reddel RR (2003) Human POT1 facilitates telomere elongation by telomerase. Curr Biol 13:942–946

34. Armbruster BN, Linardic CM, Veldman T, Bansal NP, Downie DL, Counter CM (2004) Rescue of an hTERT mutant defective in telomere elongation by fusion with hPot1. Mol Cell Biol 24:3552–3561
35. Wang F, Podell ER, Zaug AJ, Yang Y, Baciu P, Cech TR et al (2007) The POT1-TPP1 telomere complex is a telomerase processivity factor. Nature 445:506–510
36. Xin H, Liu D, Wan M, Safari A, Kim H, Sun W et al (2007) TPP1 is a homologue of ciliate TEBP-beta and interacts with POT1 to recruit telomerase. Nature 445:559–562
37. O'Connor MS, Safari A, Xin H, Liu D, Songyang Z (2006) A critical role for TPP1 and TIN2 interaction in high-order telomeric complex assembly. Proc Natl Acad Sci USA 103: 11874–11879
38. Kim SH, Kaminker P, Campisi J (1999) TIN2, a new regulator of telomere length in human cells. Nat Genet 23:405–412
39. Liu D, O'Connor MS, Qin J, Songyang Z (2004) Telosome, a mammalian telomere-associated complex formed by multiple telomeric proteins. J Biol Chem 279:51338–51342
40. Ye JZ, de Lange T (2004) TIN2 is a tankyrase 1 PARP modulator in the TRF1 telomere length control complex. Nat Genet 36:618–623
41. Savage SA, Giri N, Baerlocher GM, Orr N, Lansdorp PM, Alter BP (2008) TINF2, a component of the shelterin telomere protection complex, is mutated in dyskeratosis congenita. Am J Hum Genet 82:501–509
42. Li B, Oestreich S, de Lange T (2000) Identification of human Rap1: implications for telomere evolution. Cell 101:471–483
43. Li B, de Lange T (2003) Rap1 affects the length and heterogeneity of human telomeres. Mol Biol Cell 14:5060–5068
44. Martinez P, Blasco MA (2011) Telomeric and extra-telomeric roles for telomerase and the telomere-binding proteins. Nat Rev Cancer 11:161–176
45. Surovtseva YV, Churikov D, Boltz KA, Song X, Lamb JC, Warrington R et al (2009) Conserved telomere maintenance component 1 interacts with STN1 and maintains chromosome ends in higher eukaryotes. Mol Cell 36:207–218
46. Wellinger RJ, The CST (2009) complex and telomere maintenance: the exception becomes the rule. Mol Cell 36:168–169
47. Miyake Y, Nakamura M, Nabetani A, Shimamura S, Tamura M, Yonehara S et al (2009) RPA-like mammalian Ctc1-Stn1-Ten1 complex binds to single-stranded DNA and protects telomeres independently of the Pot1 pathway. Mol Cell 36:193–206
48. Gonzalo S, Jaco I, Fraga MF, Chen T, Li E, Esteller M et al (2006) DNA methyltransferases control telomere length and telomere recombination in mammalian cells. Nat Cell Biol 8:416–424
49. Garcia-Cao M, O'Sullivan R, Peters AH, Jenuwein T, Blasco MA (2004) Epigenetic regulation of telomere length in mammalian cells by the Suv39h1 and Suv39h2 histone methyltransferases. Nat Genet 36:94–99
50. Blasco MA (2007) The epigenetic regulation of mammalian telomeres. Nat Rev Genet 8: 299–309
51. Redon S, Reichenbach P, Lingner J (2010) The non-coding RNA TERRA is a natural ligand and direct inhibitor of human telomerase. Nucleic Acids Res 38:5797–5806
52. Luke B, Lingner J (2009) TERRA: telomeric repeat-containing RNA. EMBO J 28:2503–2510
53. Feuerhahn S, Iglesias N, Panza A, Porro A, Lingner J (2010) TERRA biogenesis, turnover and implications for function. FEBS Lett 584:3812–3818
54. Olovnikov AM (1973) A theory of marginotomy. The incomplete copying of template margin in enzymic synthesis of polynucleotides and biological significance of the phenomenon. J Theor Biol 41:181–190
55. Smogorzewska A, de Lange T (2002) Different telomere damage signaling pathways in human and mouse cells. EMBO J 21:4338–4348
56. Takai H, Smogorzewska A, de Lange T (2003) DNA damage foci at dysfunctional telomeres. Curr Biol 13:1549–1556

57. Wright WE, Shay JW (1992) The two-stage mechanism controlling cellular senescence and immortalization. Exp Gerontol 27:383–389
58. Hayflick L (1965) The limited in vitro lifetime of human diploid strains. Exptl Cell Res 37:614–636
59. Effros RB, Pawelec G (1997) Replicative senescence of T cells: does the Hayflick limit lead to immune exhaustion? Immunol Today 18:450–454
60. Shay JW, Wright WE (2000) Hayflick, his limit, and cellular ageing. Nat Rev Mol Cell Biol 1:72–76
61. Fleisig HB, Wong JM (2007) Telomerase as a clinical target: current strategies and potential applications. Exp Gerontol 42:102–112
62. Rajaraman S, Choi J, Cheung P, Beaudry V, Moore H, Artandi SE (2007) Telomere uncapping in progenitor cells with critical telomere shortening is coupled to S-phase progression in vivo. Proc Natl Acad Sci USA 104:17747–17752
63. Wright WE, Shay JW (1992) Telomere positional effects and the regulation of cellular senescence. Trends Genet 8:193–197
64. Chan SS, Chang S (2010) Defending the end zone: studying the players involved in protecting chromosome ends. FEBS Lett 584:3773–3778
65. Tejera AM, Alcontres M, Thanasoula M, Marion RM, Martinez P, Liao C et al (2010) TPP1 is required for TERT recruitment, telomere elongation during nuclear reprogramming, and normal skin development in mice. Dev Cell 18:775–789
66. Sfeir A, Kabir S, van Overbeek M, Celli GB, de Lange T (2010) Loss of Rap1 induces telomere recombination in the absence of NHEJ or a DNA damage signal. Science 327:1657–1661
67. Hockemeyer D, Daniels JP, Takai H, de Lange T (2006) Recent expansion of the telomeric complex in rodents: two distinct POT1 proteins protect mouse telomeres. Cell 126:63–77
68. Hockemeyer D, Sfeir AJ, Shay JW, Wright WE, de Lange T (2005) POT1 protects telomeres from a transient DNA damage response and determines how human chromosomes end. EMBO J 24:2667–2678
69. Denchi EL, de Lange T (2007) Protection of telomeres through independent control of ATM and ATR by TRF2 and POT1. Nature 448:1068–1071
70. Lavin MF (2008) Ataxia-telangiectasia: from a rare disorder to a paradigm for cell signalling and cancer. Nat Rev Mol Cell Biol 9:759–769
71. He H, Wang Y, Guo X, Ramchandani S, Ma J, Shen MF et al (2009) Pot1b deletion and telomerase haploinsufficiency in mice initiate an ATR-dependent DNA damage response and elicit phenotypes resembling dyskeratosis congenita. Mol Cell Biol 29:229–240
72. Cimprich KA, Cortez D (2008) ATR: an essential regulator of genome integrity. Nat Rev Mol Cell Biol 9:616–627
73. Pandita TK (2001) The role of ATM in telomere structure and function. Radiat Res 156: 642–647
74. Pandita TK (2002) ATM function and telomere stability. Oncogene 21:611–618
75. Hickson I, Zhao Y, Richardson CJ, Green SJ, Martin NM, Orr AI et al (2004) Identification and characterization of a novel and specific inhibitor of the ataxia-telangiectasia mutated kinase ATM. Cancer Res 64:9152–9159
76. Chang S, Multani AS, Cabrera NG, Naylor ML, Laud P, Lombard D et al (2004) Essential role of limiting telomeres in the pathogenesis of Werner syndrome. Nat Genet 36:877–882
77. Crabbe L, Jauch A, Naeger CM, Holtgreve-Grez H, Karlseder J (2007) Telomere dysfunction as a cause of genomic instability in Werner syndrome. Proc Natl Acad Sci USA 104:2205–2210
78. Verdun RE, Crabbe L, Haggblom C, Karlseder J (2005) Functional human telomeres are recognized as DNA damage in G2 of the cell cycle. Mol Cell 20:551–561
79. Verdun RE, Karlseder J (2006) The DNA damage machinery and homologous recombination pathway act consecutively to protect human telomeres. Cell 127:709–720
80. Mitchell JR, Cheng J, Collins K (1999) A box H/ACA small nucleolar RNA-like domain at the human telomerase RNA 3′ end. Mol Cell Biol 19:567–576

81. Dragon F, Pogacic V, Filipowicz W (2000) In vitro assembly of human H/ACA small nucleolar RNPs reveals unique features of U17 and telomerase RNAs. Mol Cell Biol 20:3037–3048
82. Wang C, Meier UT (2004) Architecture and assembly of mammalian H/ACA small nucleolar and telomerase ribonucleoproteins. EMBO J 23:1857–1867
83. Weinrich SL, Pruzan R, Ma L, Ouellette M, Tesmer VM, Holt SE et al (1997) Reconstitution of human telomerase with the template RNA component hTR and the catalytic protein subunit hTRT. Nat Genet 17:498–502
84. Harrington L (2003) Biochemical aspects of telomerase function. Cancer Lett 194:139–154
85. Wong JM, Kusdra L, Collins K (2002) Subnuclear shuttling of human telomerase induced by transformation and DNA damage. Nat Cell Biol 4:731–736
86. Feng J, Funk WD, Wang SS, Weinrich SL, Avilion AA, Chiu CP et al (1995) The RNA component of human telomerase. Science 269:1236–1241
87. Zaug AJ, Linger J, Cech TR (1996) Method for determining RNA 3′ ends and application to human telomerase RNA. Nucleic Acids Res 24:532–533
88. Zhao JQ, Hoare SF, McFarlane R, Muir S, Parkinson EK, Black DM et al (1998) Cloning and characterization of human and mouse telomerase RNA gene promoter sequences. Oncogene 16:1345–1350
89. Fu D, Collins K (2003) Distinct biogenesis pathways for human telomerase RNA and H/ACA small nucleolar RNAs. Mol Cell 11:1361–1372
90. Chen JL, Blasco MA, Greider CW (2000) Secondary structure of vertebrate telomerase RNA. Cell 100:503–514
91. Dez C, Henras A, Faucon B, Lafontaine DLJ, Caizergues-Ferrer M, Henry Y (2001) Stable expression in yeast of the mature form of human telomerase RNA depends of its association with the box H/ACA small nucleolar RNP proteins Cbf5p, Nhp2p and Nop10p. Nucleic Acids Res 29:598–603
92. Cohen SB, Graham ME, Lovrecz GO, Bache N, Robinson PJ, Reddel RR (2007) Protein composition of catalytically active human telomerase from immortal cells. Science 315: 1850–1853
93. Fu D, Collins K (2007) Purification of human telomerase complexes identifies factors involved in telomerase biogenesis and telomere length regulation. Mol Cell 28:773–785
94. Collins K, Mitchell JR (2002) Telomerase in the human organism. Oncogene 21:564–579
95. Heiss NS, Knight SW, Vulliamy TJ, Klauck SM, Wiemann S, Mason PJ et al (1998) X-linked dyskeratosis congenita is caused by mutations in a highly conserved gene with putative nucleolar functions. Nat Genet 19:32–38
96. Mitchell JR, Wood E, Collins K (1999) A telomerase component is defective in the human disease dyskeratosis congenita. Nature 402:551–555
97. Vulliamy T, Marrone A, Goldman F, Dearlove A, Bessler M, Mason PJ et al (2001) The RNA component of telomerase is mutated in autosomal dominant dyskeratosis congenita. Nature 413:432–435
98. Wong JM, Collins K (2003) Telomere maintenance and disease. Lancet 362:983–988
99. Le S, Sternglanz R, Greider CW (2000) Identification of two RNA-binding proteins associated with human telomerase RNA. Mol Biol Cell 11:999–1010
100. Fu D, Collins K (2006) Human telomerase and Cajal body ribonucleoproteins share a unique specificity of Sm protein association. Genes Dev 20:531–536
101. Jady BE, Bertrand E, Kiss T (2004) Human telomerase RNA and box H/ACA scaRNAs share a common Cajal body-specific localization signal. J Cell Biol 164:647–652
102. Venteicher AS, Abreu EB, Meng Z, McCann KE, Terns RM, Veenstra TD et al (2009) A human telomerase holoenzyme protein required for Cajal body localization and telomere synthesis. Science 323:644–648
103. Cong YS, Wen J, Bacchetti S (1999) The human telomerase catalytic subunit hTERT: organization of the gene and characterization of the promoter. Hum Mol Genet 8:137–142
104. Wick M, Zubov D, Hagen G (1999) Genomic organization and promoter characterization of the gene encoding the human telomerase reverse transcriptase (hTERT). Gene 232:97–106

105. Horikawa I, Cable PL, Afshari C, Barrett JC (1999) Cloning and characterization of the promoter region of human telomerase reverse transcriptase gene. Cancer Res 59:826–830
106. Wu KJ, Grandori C, Amacker M, Simon-Vermot N, Polack A, Lingner J et al (1999) Direct activation of TERT transcription by c-MYC. Nat Genet 21:220–224
107. Yi X, White DM, Aisner DL, Baur JA, Wright WE, Shay JW (2000) An alternate splicing variant of the human telomerase catalytic subunit inhibits telomerase activity. Neoplasia 2:433–440
108. Meyerson M, Counter CM, Eaton EN, Ellisen LW, Steiner P, Caddle SD et al (1997) hEST2, the putative human telomerase catalytic subunit gene, is up-regulated in tumor cells and during immortalization. Cell 90:785–795
109. Yang J, Chang E, Cherry AM, Bangs CD, Oei Y, Bodnar A et al (1999) Human endothelial cell life extension by telomerase expression. J Biol Chem 274:26141–26148
110. Holt SE, Aisner DL, Baur J, Tesmer VM, Dy M, Ouellette M et al (1999) Functional requirement of p23 and Hsp90 in telomerase complexes. Genes Dev 13:817–826
111. Kelleher C, Teixeira MT, Forstemann K, Lingner J (2002) Telomerase: biochemical considerations for enzyme and substrate. Trends Biochem Sci 27:572–579
112. Xiong Y, Eickbush TH (1990) Origin and evolution of retroelements based upon their reverse transcriptase sequences. EMBO J 9:3353–3362
113. Harrington L, Zhou W, McPhail T, Oulton R, Yeung DS, Mar V et al (1997) Human telomerase contains evolutionarily conserved catalytic and structural subunits. Genes Dev 11:3109–3115
114. Lingner J, Hughes TR, Shevchenko A, Mann M, Lundblad V, Cech TR (1997) Reverse transcriptase motifs in the catalytic subunit of telomerase. Science 276:561–567
115. Beattie TL, Zhou W, Robinson MO, Harrington L (1998) Reconstitution of human telomerase activity in vitro. Curr Biol 8:177–180
116. Nakayama J, Tahara H, Tahara E, Saito M, Ito K, Nakamura H et al (1998) Telomerase activation by hTRT in human normal fibroblasts and hepatocellular carcinomas. Nat Genet 18:65–68
117. Xia J, Peng Y, Mian IS, Lue NF (2000) Identification of functionally important domains in the N-terminal region of telomerase reverse transcriptase. Mol Cell Biol 20:5196–5207
118. Bryan TM, Goodrich KJ, Cech TR (2000) A mutant of *Tetrahymena* telomerase reverse transcriptase with increased processivity. J Biol Chem 275:24199–24207
119. Bosoy D, Peng Y, Mian IS, Lue NF (2003) Conserved N-terminal motifs of telomerase reverse transcriptase required for ribonucleoprotein assembly in vivo. J Biol Chem 278: 3882–3890
120. Lai CK, Mitchell JR, Collins K (2001) RNA binding domain of telomerase reverse transcriptase. Mol Cell Biol 21:990–1000
121. Moriarty TJ, Huard S, Dupuis S, Autexier C (2002) Functional multimerization of human telomerase requires an RNA interaction domain in the N terminus of the catalytic subunit. Mol Cell Biol 22:1253–1265
122. Lue NF (2004) Adding to the ends: what makes telomerase processive and how important is it? Bioessays 26:955–962
123. Armbruster BN, Banik SS, Guo C, Smith AC, Counter CM (2001) N-terminal domains of the human telomerase catalytic subunit required for enzyme activity in vivo. Mol Cell Biol 21: 7775–7786
124. Lee SR, Wong JM, Collins K (2003) Human telomerase reverse transcriptase motifs required for elongation of a telomeric substrate. J Biol Chem 278:52531–52536
125. Moriarty TJ, Ward RJ, Taboski MA, Autexier C (2005) An anchor site-type defect in human telomerase that disrupts telomere length maintenance and cellular immortalization. Mol Biol Cell 16:3152–3161
126. Bachand F, Autexier C (2001) Functional regions of human telomerase reverse transcriptase and human telomerase RNA required for telomerase activity and RNA-protein interactions. Mol Cell Biol 21:1888–1897
127. Hossain S, Singh S, Lue NF (2002) Functional analysis of the C-terminal extension of telomerase reverse transcriptase. A putative "thumb" domain. J Biol Chem 277:36174–36180
128. Huard S, Moriarty TJ, Autexier C (2003) The C terminus of the human telomerase reverse transcriptase is a determinant of enzyme processivity. Nucleic Acids Res 31:4059–4070

129. Aitken A (2006) 14-3-3 proteins: a historic overview. Semin Cancer Biol 16:162–172
130. Khurts S, Masutomi K, Delgermaa L, Arai K, Oishi N, Mizuno H et al (2004) Nucleolin interacts with telomerase. J Biol Chem 279:51508–51515
131. Zhou XZ, Lu KP (2001) The Pin2/TRF1-interacting protein PinX1 is a potent telomerase inhibitor. Cell 107:347–359
132. Banik SS, Counter CM (2004) Characterization of interactions between PinX1 and human telomerase subunits hTERT and hTR. J Biol Chem 279:51745–51748
133. Li H, Zhao L, Yang Z, Funder JW, Liu JP (1998) Telomerase is controlled by protein kinase Calpha in human breast cancer cells. J Biol Chem 273:33436–33442
134. Kang SS, Kwon T, Kwon DY, Do SI (1999) Akt protein kinase enhances human telomerase activity through phosphorylation of telomerase reverse transcriptase subunit. J Biol Chem 274:13085–13090
135. Breitschopf K, Zeiher AM, Dimmeler S (2001) Pro-atherogenic factors induce telomerase inactivation in endothelial cells through an Akt-dependent mechanism. FEBS Lett 493:21–25
136. Haendeler J, Hoffmann J, Rahman S, Zeiher AM, Dimmeler S (2003) Regulation of telomerase activity and anti-apoptotic function by protein-protein interaction and phosphorylation. FEBS Lett 536:180–186
137. Li H, Zhao LL, Funder JW, Liu JP (1997) Protein phosphatase 2A inhibits nuclear telomerase activity in human breast cancer cells. J Biol Chem 272:16729–16732
138. Kharbanda S, Kumar V, Dhar S, Pandey P, Chen C, Majumder P et al (2000) Regulation of the hTERT telomerase catalytic subunit by the c-Abl tyrosine kinase. Curr Biol 10: 568–575
139. Kim JH, Park SM, Kang MR, Oh SY, Lee TH, Muller MT et al (2005) Ubiquitin ligase MKRN1 modulates telomere length homeostasis through a proteolysis of hTERT. Genes Dev 19:776–781
140. Huber O, Menard L, Haurie V, Nicou A, Taras D, Rosenbaum J (2008) Pontin and reptin, two related ATPases with multiple roles in cancer. Cancer Res 68:6873–6876
141. Venteicher AS, Meng Z, Mason PJ, Veenstra TD, Artandi SE (2008) Identification of ATPases pontin and reptin as telomerase components essential for holoenzyme assembly. Cell 132: 945–957
142. Zhu Y, Tomlinson RL, Lukowiak AA, Terns RM, Terns MP (2004) Telomerase RNA accumulates in Cajal bodies in human cancer cells. Mol Biol Cell 15:81–90
143. Tomlinson RL, Ziegler TD, Supakorndej T, Terns RM, Terns MP (2006) Cell cycle-regulated trafficking of human telomerase to telomeres. Mol Biol Cell 17:955–965
144. Cristofari G, Adolf E, Reichenbach P, Sikora K, Terns RM, Terns MP et al (2007) Human telomerase RNA accumulation in Cajal bodies facilitates telomerase recruitment to telomeres and telomere elongation. Mol Cell 27:882–889
145. Zhong F, Savage SA, Shkreli M, Giri N, Jessop L, Myers T et al (2011) Disruption of telomerase trafficking by TCAB1 mutation causes dyskeratosis congenita. Genes Dev 25:11–16
146. Tomlinson RL, Abreu EB, Ziegler T, Ly H, Counter CM, Terns RM et al (2008) Telomerase reverse transcriptase is required for the localization of telomerase RNA to cajal bodies and telomeres in human cancer cells. Mol Biol Cell 19:3793–3800
147. Jady BE, Richard P, Bertrand E, Kiss T (2006) Cell cycle-dependent recruitment of telomerase RNA and Cajal bodies to human telomeres. Mol Biol Cell 17:944–954
148. Pennock E, Buckley K, Lundblad V (2001) Cdc13 delivers separate complexes to the telomere for end protection and replication. Cell 104:387–396
149. Evans SK, Lundblad V (2002) The Est1 subunit of Saccharomyces cerevisiae telomerase makes multiple contributions to telomere length maintenance. Genetics 162:1101–1115
150. Snow BE, Erdmann N, Cruickshank J, Goldman H, Gill RM, Robinson MO et al (2003) Functional conservation of the telomerase protein Est1p in humans. Curr Biol 13:698–704
151. Azzalin CM, Reichenbach P, Khoriauli L, Giulotto E, Lingner J (2007) Telomeric repeat containing RNA and RNA surveillance factors at mammalian chromosome ends. Science 318:798–801

152. Ding J, Hayashi MK, Zhang Y, Manche L, Krainer AR, Xu R (1999) Crystal structure of the two-RRM domain of hnRNP A1 (UP1) complexed with single-stranded telomeric DNA. Genes Dev 13:1102–1115
153. Eversole A, Maizels N (2000) In vitro properties of the conserved mammalian protein hnRNP D suggest a role in telomere maintenance. Mol Cell Biol 20:5425–5432
154. Ford LP, Suh JM, Wright WE, Shay JW (2000) Heterogeneous nuclear ribonucleoproteins C1 and C2 associate with the RNA component of human telomerase. Mol Cell Biol 20:9084–9091
155. Fiset S, Chabot B (2001) hnRNP A1 may interact simultaneously with telomeric DNA and the human telomerase RNA in vitro. Nucleic Acids Res 29:2268–2275
156. Dallaire F, Dupuis S, Fiset S, Chabot B (2000) Heterogeneous nuclear ribonucleoprotein A1 and UP1 protect mammalian telomeric repeats and modulate telomere replication in vitro. J Biol Chem 275:14509–14516
157. Zhang DH, Zhou B, Huang Y, Xu LX, Zhou JQ (2006) The human Pif1 helicase, a potential Escherichia coli RecD homologue, inhibits telomerase activity. Nucleic Acids Res 34: 1393–1404
158. Mateyak MK, Zakian VA (2006) Human PIF helicase is cell cycle regulated and associates with telomerase. Cell Cycle 5:2796–2804
159. Shippen-Lentz D, Blackburn EH (1990) Functional evidence for an RNA template in telomerase. Science 247:546–552
160. Blackburn EH (1992) Telomerases. Annu Rev Biochem 61:113–129
161. Fulton TB, Blackburn EH (1998) Identification of Kluyveromyces lactis telomerase: discontinuous synthesis along the 30-nucleotide-long templating domain. Mol Cell Biol 18:4961–4970
162. Wyatt HD, Lobb DA, Beattie TL (2007) Characterization of physical and functional anchor site interactions in human telomerase. Mol Cell Biol 27:3226–3240
163. Reddel RR, Bryan TM, Colgin LM, Perrem KT, Yeager TR (2001) Alternative lengthening of telomeres in human cells. Radiat Res 155:194–200
164. Henson JD, Neumann AA, Yeager TR, Reddel RR (2002) Alternative lengthening of telomeres in mammalian cells. Oncogene 21:598–610
165. Bryan TM, Englezou A, Gupta J, Bacchetti S, Reddel RR (1995) Telomere elongation in immortal human cells without detectable telomerase activity. EMBO J 14:4240–4248
166. Cerone MA, Londono-Vallejo JA, Bacchetti S (2001) Telomere maintenance by telomerase and by recombination can coexist in human cells. Hum Mol Genet 10:1945–1952
167. Bechter OE, Zou Y, Shay JW, Wright WE (2003) Homologous recombination in human telomerase-positive and ALT cells occurs with the same frequency. EMBO Rep 4:1138–1143
168. Jiang WQ, Zhong ZH, Henson JD, Neumann AA, Chang AC, Reddel RR (2005) Suppression of alternative lengthening of telomeres by Sp100-mediated sequestration of the MRE11/ RAD50/NBS1 complex. Mol Cell Biol 25:2708–2721
169. Yeager TR, Neumann AA, Englezou A, Huschtscha LI, Noble JR, Reddel RR (1999) Telomerase-negative immortalized human cells contain a novel type of promyelocytic leukemia (PML) body. Cancer Res 59:4175–4179
170. Grobelny JV, Godwin AK, Broccoli D (2000) ALT-associated PML bodies are present in viable cells and are enriched in cells in the G(2)/M phase of the cell cycle. J Cell Sci 113(Pt 24):4577–4585
171. Wu G, Lee WH, Chen PL (2000) NBS1 and TRF1 colocalize at promyelocytic leukemia bodies during late S/G2 phases in immortalized telomerase-negative cells. Implication of NBS1 in alternative lengthening of telomeres. J Biol Chem 275:30618–30622
172. Fan Q, Zhang F, Barrett B, Ren K, Andreassen PR (2009) A role for monoubiquitinated FANCD2 at telomeres in ALT cells. Nucleic Acids Res 37:1740–1754
173. Heaphy CM, de Wilde RF, Jiao Y, Klein AP, Edil BH, Shi C et al (2011) Altered telomeres in tumors with ATRX and DAXX mutations. Science 333:425
174. Heaphy CM, Subhawong AP, Hong SM, Goggins MG, Montgomery EA, Gabrielson E et al (2011) Prevalence of the alternative lengthening of telomeres telomere maintenance mechanism in human cancer subtypes. Am J Pathol 179:1608–1615

175. Hu J, Hwang SS, Liesa M, Gan B, Sahin E, Jaskelioff M et al (2012) Antitelomerase therapy provokes ALT and mitochondrial adaptive mechanisms in cancer. Cell 148:651–663
176. Stewart SA, Hahn WC, O'Connor BF, Banner EN, Lundberg AS, Modha P et al (2002) Telomerase contributes to tumorigenesis by a telomere length-independent mechanism. Proc Natl Acad Sci USA 99:12606–12611
177. Fleisig HB, Wong JM (2012) Telomerase promotes efficient cell cycle kinetics and confers growth advantage to telomerase-negative transformed human cells. Oncogene 31: 954–965
178. Wright WE, Piatyszek MA, Rainey WE, Byrd W, Shay JW (1996) Telomerase activity in human germline and embryonic tissues and cells. Dev Genet 18:173–179
179. Shay JW, Bacchetti S (1997) A survey of telomerase activity in human cancer. Eur J Cancer 33:787–791
180. Kim NW, Piatyszek MA, Prowse KR, Harley CB, West MD, Ho PL et al (1994) Specific association of human telomerase activity with immortal cells and cancer. Science 266: 2011–2015
181. Hahn WC, Stewart SA, Brooks MW, York SG, Eaton E, Kurachi A et al (1999) Inhibition of telomerase limits the growth of human cancer cells. Nat Med 5:1164–1170
182. Zhang X, Mar V, Zhou W, Harrington L, Robinson MO (1999) Telomere shortening and apoptosis in telomerase-inhibited human tumor cells. Genes Dev 13:2388–2399
183. Harley CB (2002) Telomerase is not an oncogene. Oncogene 21:494–502
184. Kiyono T, Foster SA, Koop JI, McDougall JK, Galloway DA, Klingelhutz AJ (1998) Both Rb/p16^{INK4a} inactivation and telomerase activity are required to immortalize human epithelial cells. Nature 396:84–88
185. Berger R, Febbo PG, Majumder PK, Zhao JJ, Mukherjee S, Signoretti S et al (2004) Androgen-induced differentiation and tumorigenicity of human prostate epithelial cells. Cancer Res 64:8867–8875
186. Lundberg AS, Randell SH, Stewart SA, Elenbaas B, Hartwell KA, Brooks MW et al (2002) Immortalization and transformation of primary human airway epithelial cells by gene transfer. Oncogene 21:4577–4586
187. Hahn WC, Counter CM, Lundberg AS, Beijersbergen RL, Brooks MW, Weinberg RA (1999) Creation of human tumour cells with defined genetic elements. Nature 400:464–468
188. Chadeneau C, Hay K, Hirte HW, Gallinger S, Bacchetti S (1995) Telomerase activity associated with acquisition of malignancy in human colorectal cancer. Cancer Res 55: 2533–2536
189. Engelhardt M, Drullinsky P, Guillem J, Moore MA (1997) Telomerase and telomere length in the development and progression of premalignant lesions to colorectal cancer. Clin Cancer Res 3:1931–1941
190. Miura N, Horikawa I, Nishimoto A, Ohmura H, Ito H, Hirohashi S et al (1997) Progressive telomere shortening and telomerase reactivation during hepatocellular carcinogenesis. Cancer Genet Cytogenet 93:56–62
191. Rudolph KL, Millard M, Bosenberg MW, DePinho RA (2001) Telomere dysfunction and evolution of intestinal carcinoma in mice and humans. Nat Genet 28:155–159
192. Lantuejoul S, Soria JC, Morat L, Lorimier P, Moro-Sibilot D, Sabatier L et al (2005) Telomere shortening and telomerase reverse transcriptase expression in preinvasive bronchial lesions. Clin Cancer Res 11:2074–2082
193. Meeker AK, Hicks JL, Gabrielson E, Strauss WM, De Marzo AM, Argani P (2004) Telomere shortening occurs in subsets of normal breast epithelium as well as in situ and invasive carcinoma. Am J Pathol 164:925–935
194. van Heek NT, Meeker AK, Kern SE, Yeo CJ, Lillemoe KD, Cameron JL et al (2002) Telomere shortening is nearly universal in pancreatic intraepithelial neoplasia. Am J Pathol 161: 1541–1547
195. Meeker AK, Hicks JL, Platz EA, March GE, Bennett CJ, Delannoy MJ et al (2002) Telomere shortening is an early somatic DNA alteration in human prostate tumorigenesis. Cancer Res 62:6405–6409

196. Gisselsson D, Jonson T, Petersen A, Strombeck B, Dal Cin P, Hoglund M et al (2001) Telomere dysfunction triggers extensive DNA fragmentation and evolution of complex chromosome abnormalities in human malignant tumors. Proc Natl Acad Sci USA 98:12683–12688
197. DePinho RA (2000) The age of cancer. Nature 408:248–254
198. Sharma GG, Gupta A, Wang H, Scherthan H, Dhar S, Gandhi V et al (2003) hTERT associates with human telomeres and enhances genomic stability and DNA repair. Oncogene 22:131–146
199. Smith LL, Coller HA, Roberts JM (2003) Telomerase modulates expression of growth-controlling genes and enhances cell proliferation. Nat Cell Biol 5:474–479
200. Park JI, Venteicher AS, Hong JY, Choi J, Jun S, Shkreli M et al (2009) Telomerase modulates Wnt signalling by association with target gene chromatin. Nature 460:66–72
201. Choi J, Southworth LK, Sarin KY, Venteicher AS, Ma W, Chang W et al (2008) TERT promotes epithelial proliferation through transcriptional control of a Myc- and Wnt-related developmental program. PLoS Genet 4:e10
202. Okamoto N, Yasukawa M, Nguyen C, Kasim V, Maida Y, Possemato R et al (2011) Maintenance of tumor initiating cells of defined genetic composition by nucleostemin. Proc Natl Acad Sci USA 108:20388–20393
203. Maida Y, Yasukawa M, Furuuchi M, Lassmann T, Possemato R, Okamoto N et al (2009) An RNA-dependent RNA polymerase formed by TERT and the RMRP RNA. Nature 461:230–235
204. Rosenbluh J, Nijhawan D, Chen Z, Wong KK, Masutomi K, Hahn WC (2011) RMRP is a non-coding RNA essential for early murine development. PLoS One 6:e26270
205. Gorbunova V, Seluanov A, Pereira-Smith OM (2002) Expression of human telomerase (hTERT) does not prevent stress-induced senescence in normal human fibroblasts but protects the cells from stress-induced apoptosis and necrosis. J Biol Chem 277:38540–38549
206. Masutomi K, Possemato R, Wong JM, Currier JL, Tothova Z, Manola JB et al (2005) The telomerase reverse transcriptase regulates chromatin state and DNA damage responses. Proc Natl Acad Sci USA 102:8222–8227
207. Tamakawa RA, Fleisig HB, Wong JM (2010) Telomerase inhibition potentiates the effects of genotoxic agents in breast and colorectal cancer cells in a cell cycle-specific manner. Cancer Res 70:8684–8694
208. Pascolo E, Wenz C, Lingner J, Hauel N, Priepke H, Kauffmann I et al (2002) Mechanism of human telomerase inhibition by BIBR1532, a synthetic, non-nucleosidic drug candidate. J Biol Chem 277:15566–15572
209. Ward RJ, Autexier C (2005) Pharmacological telomerase inhibition can sensitize drug-resistant and drug-sensitive cells to chemotherapeutic treatment. Mol Pharmacol 68:779–786
210. El Daly H, Martens UM (2007) Telomerase inhibition and telomere targeting in hematopoietic cancer cell lines with small non-nucleosidic synthetic compounds (BIBR1532). Methods Mol Biol 405:47–60
211. El-Daly H, Kull M, Zimmermann S, Pantic M, Waller CF, Martens UM (2005) Selective cytotoxicity and telomere damage in leukemia cells using the telomerase inhibitor BIBR1532. Blood 105:1742–1749
212. Furman PA, Fyfe JA, St Clair MH, Weinhold K, Rideout JL, Freeman GA et al (1986) Phosphorylation of 3′-azido-3′-deoxythymidine and selective interaction of the 5′-triphosphate with human immunodeficiency virus reverse transcriptase. Proc Natl Acad Sci USA 83:8333–8337
213. Mitsuya H, Weinhold KJ, Furman PA, St Clair MH, Lehrman SN, Gallo RC et al (1985) 3′-Azido-3′-deoxythymidine (BW A509U): an antiviral agent that inhibits the infectivity and cytopathic effect of human T-lymphotropic virus type III/lymphadenopathy-associated virus in vitro. Proc Natl Acad Sci USA 82:7096–7100
214. Datta A, Bellon M, Sinha-Datta U, Bazarbachi A, Lepelletier Y, Canioni D et al (2006) Persistent inhibition of telomerase reprograms adult T-cell leukemia to p53-dependent senescence. Blood 108:1021–1029
215. Geary RS, Yu RZ, Levin AA (2001) Pharmacokinetics of phosphorothioate antisense oligodeoxynucleotides. Curr Opin Investig Drugs 2:562–573

216. Folini M, Brambilla C, Villa R, Gandellini P, Vignati S, Paduano F et al (2005) Antisense oligonucleotide-mediated inhibition of hTERT, but not hTERC, induces rapid cell growth decline and apoptosis in the absence of telomere shortening in human prostate cancer cells. Eur J Cancer 41:624–634
217. Akiyama M, Hideshima T, Shammas MA, Hayashi T, Hamasaki M, Tai YT et al (2003) Effects of oligonucleotide N3′->P5′ thio-phosphoramidate (GRN163) targeting telomerase RNA in human multiple myeloma cells. Cancer Res 63:6187–6194
218. Asai A, Oshima Y, Yamamoto Y, Uochi TA, Kusaka H, Akinaga S et al (2003) A novel telomerase template antagonist (GRN163) as a potential anticancer agent. Cancer Res 63:3931–3939
219. Herbert BS, Gellert GC, Hochreiter A, Pongracz K, Wright WE, Zielinska D et al (2005) Lipid modification of GRN163, an N3′->P5′ thio-phosphoramidate oligonucleotide, enhances the potency of telomerase inhibition. Oncogene 24:5262–5268
220. Dikmen ZG, Gellert GC, Jackson S, Gryaznov S, Tressler R, Dogan P et al (2005) In vivo inhibition of lung cancer by GRN163L: a novel human telomerase inhibitor. Cancer Res 65:7866–7873
221. Gellert GC, Dikmen ZG, Wright WE, Gryaznov S, Shay JW (2006) Effects of a novel telomerase inhibitor, GRN163L, in human breast cancer. Breast Cancer Res Treat 96:73–81
222. Hochreiter AE, Xiao H, Goldblatt EM, Gryaznov SM, Miller KD, Badve S et al (2006) Telomerase template antagonist GRN163L disrupts telomere maintenance, tumor growth, and metastasis of breast cancer. Clin Cancer Res 12:3184–3192
223. Djojosubroto MW, Chin AC, Go N, Schaetzlein S, Manns MP, Gryaznov S et al (2005) Telomerase antagonists GRN163 and GRN163L inhibit tumor growth and increase chemosensitivity of human hepatoma. Hepatology 42:1127–1136
224. Brower V (2010) Telomerase-based therapies emerging slowly. J Natl Cancer Inst 102:520–521
225. Chen HC, Chou CK, Lee SD, Wang JC, Yeh SF (1995) Active compounds from Saussurea lappa Clarks that suppress hepatitis B virus surface antigen gene expression in human hepatoma cells. Antiviral Res 27:99–109
226. Kang JS, Yoon YD, Lee KH, Park SK, Kim HM (2004) Costunolide inhibits interleukin-1beta expression by down-regulation of AP-1 and MAPK activity in LPS-stimulated RAW 264.7 cells. Biochem Biophys Res Commun 313:171–177
227. Park HJ, Jung WT, Basnet P, Kadota S, Namba T (1996) Syringin 4-O-beta-glucoside, a new phenylpropanoid glycoside, and costunolide, a nitric oxide synthase inhibitor, from the stem bark of Magnolia sieboldii. J Nat Prod 59:1128–1130
228. Wedge DE, Galindo JC, Macias FA (2000) Fungicidal activity of natural and synthetic sesquiterpene lactone analogs. Phytochemistry 53:747–757
229. Choi SH, Im E, Kang HK, Lee JH, Kwak HS, Bae YT et al (2005) Inhibitory effects of costunolide on the telomerase activity in human breast carcinoma cells. Cancer Lett 227:153–162
230. Kanno S, Kitajima Y, Kakuta M, Osanai Y, Kurauchi K, Ujibe M et al (2008) Costunolide-induced apoptosis is caused by receptor-mediated pathway and inhibition of telomerase activity in NALM-6 cells. Biol Pharm Bull 31:1024–1028
231. Grenert JP, Sullivan WP, Fadden P, Haystead TA, Clark J, Mimnaugh E et al (1997) The amino-terminal domain of heat shock protein 90 (hsp90) that binds geldanamycin is an ATP/ADP switch domain that regulates hsp90 conformation. J Biol Chem 272:23843–23850
232. Prodromou C, Roe SM, O'Brien R, Ladbury JE, Piper PW, Pearl LH (1997) Identification and structural characterization of the ATP/ADP-binding site in the Hsp90 molecular chaperone. Cell 90:65–75
233. Gorre ME, Ellwood-Yen K, Chiosis G, Rosen N, Sawyers CL (2002) BCR-ABL point mutants isolated from patients with imatinib mesylate-resistant chronic myeloid leukemia remain sensitive to inhibitors of the BCR-ABL chaperone heat shock protein 90. Blood 100:3041–3044
234. Hartson SD, Matts RL (1994) Association of Hsp90 with cellular Src-family kinases in a cell-free system correlates with altered kinase structure and function. Biochemistry 33:8912–8920
235. Solit DB, Basso AD, Olshen AB, Scher HI, Rosen N (2003) Inhibition of heat shock protein 90 function down-regulates Akt kinase and sensitizes tumors to Taxol. Cancer Res 63: 2139–2144

236. Stancato LF, Chow YH, Hutchison KA, Perdew GH, Jove R, Pratt WB (1993) Raf exists in a native heterocomplex with hsp90 and p50 that can be reconstituted in a cell-free system. J Biol Chem 268:21711–21716
237. Vonderheide RH, Hahn WC, Schultze JL, Nadler LM (1999) The telomerase catalytic subunit is a widely expressed tumor-associated antigen recognized by cytotoxic T lymphocytes. Immunity 10:673–679
238. Minev B, Hipp J, Firat H, Schmidt JD, Langlade-Demoyen P, Zanetti M (2000) Cytotoxic T cell immunity against telomerase reverse transcriptase in humans. Proc Natl Acad Sci USA 97:4796–4801
239. Nair SK, Heiser A, Boczkowski D, Majumdar A, Naoe M, Lebkowski JS et al (2000) Induction of cytotoxic T cell responses and tumor immunity against unrelated tumors using telomerase reverse transcriptase RNA transfected dendritic cells. Nat Med 6:1011–1017
240. Su Z, Dannull J, Yang BK, Dahm P, Coleman D, Yancey D et al (2005) Telomerase mRNA-transfected dendritic cells stimulate antigen-specific CD8+ and CD4+ T cell responses in patients with metastatic prostate cancer. J Immunol 174:3798–3807
241. Bernhardt SL, Gjertsen MK, Trachsel S, Moller M, Eriksen JA, Meo M et al (2006) Telomerase peptide vaccination of patients with non-resectable pancreatic cancer: a dose escalating phase I/II study. Br J Cancer 95:1474–1482
242. Shay JW, Keith WN (2008) Targeting telomerase for cancer therapeutics. Br J Cancer 98:677–683
243. Cook BD, Dynek JN, Chang W, Shostak G, Smith S (2002) Role for the related poly(ADP-Ribose) polymerases tankyrase 1 and 2 at human telomeres. Mol Cell Biol 22:332–342
244. Smith S, de Lange T (2000) Tankyrase promotes telomere elongation in human cells. Curr Biol 10:1299–1302
245. Smith S, Giriat I, Schmitt A, de Lange T (1998) Tankyrase, a poly(ADP-ribose) polymerase at human telomeres. Science 282:1484–1487
246. Chang W, Dynek JN, Smith S (2003) TRF1 is degraded by ubiquitin-mediated proteolysis after release from telomeres. Genes Dev 17:1328–1333
247. Dynek JN, Smith S (2004) Resolution of sister telomere association is required for progression through mitosis. Science 304:97–100
248. Seimiya H, Muramatsu Y, Ohishi T, Tsuruo T (2005) Tankyrase 1 as a target for telomere-directed molecular cancer therapeutics. Cancer Cell 7:25–37
249. Mergny JL, Duval-Valentin G, Nguyen CH, Perrouault L, Faucon B, Rougee M et al (1992) Triple helix-specific ligands. Science 256:1681–1684
250. Escude C, Nguyen CH, Kukreti S, Janin Y, Sun JS, Bisagni E et al (1998) Rational design of a triple helix-specific intercalating ligand. Proc Natl Acad Sci USA 95:3591–3596
251. Shi DF, Wheelhouse RT, Sun D, Hurley LH (2001) Quadruplex-interactive agents as telomerase inhibitors: synthesis of porphyrins and structure-activity relationship for the inhibition of telomerase. J Med Chem 44:4509–4523
252. Bertrand H, Bombard S, Monchaud D, Teulade-Fichou MP (2008) New platinum(II) complexes targeting the loops of the human telomeric G-quadruplex. Nucleic Acids Symp Ser (Oxf) 52:163–164
253. Sun D, Thompson B, Cathers BE, Salazar M, Kerwin SM, Trent JO et al (1997) Inhibition of human telomerase by a G-quadruplex-interactive compound. J Med Chem 40:2113–2116
254. Perry PJ, Gowan SM, Reszka AP, Polucci P, Jenkins TC, Kelland LR et al (1998) 1,4- and 2,6-disubstituted amidoanthracene-9,10-dione derivatives as inhibitors of human telomerase. J Med Chem 41:3253–3260
255. Perry PJ, Gowan SM, Read MA, Kelland LR, Neidle S (1999) Design, synthesis and evaluation of human telomerase inhibitors based upon a tetracyclic structural motif. Anticancer Drug Des 14:373–382
256. Perry PJ, Read MA, Davies RT, Gowan SM, Reszka AP, Wood AA et al (1999) 2,7-Disubstituted amidofluorenone derivatives as inhibitors of human telomerase. J Med Chem 42:2679–2684
257. Gowan SM, Heald R, Stevens MF, Kelland LR (2001) Potent inhibition of telomerase by small-molecule pentacyclic acridines capable of interacting with G-quadruplexes. Mol Pharmacol 60:981–988

258. Simonsson T, Pecinka P, Kubista M (1998) DNA tetraplex formation in the control region of c-myc. Nucleic Acids Res 26:1167–1172
259. Siddiqui-Jain A, Grand CL, Bearss DJ, Hurley LH (2002) Direct evidence for a G-quadruplex in a promoter region and its targeting with a small molecule to repress c-MYC transcription. Proc Natl Acad Sci USA 99:11593–11598
260. Guiducci C, Cerone MA, Bacchetti S (2001) Expression of mutant telomerase in immortal telomerase-negative human cells results in cell cycle deregulation, nuclear and chromosomal abnormalities and rapid loss of viability. Oncogene 20:714–725
261. Kim MM, Rivera MA, Botchkina IL, Shalaby R, Thor AD, Blackburn EH (2001) A low threshold level of expression of mutant-template telomerase RNA inhibits human tumor cell proliferation. Proc Natl Acad Sci USA 98:7982–7987
262. Li S, Rosenberg JE, Donjacour AA, Botchkina IL, Hom YK, Cunha GR et al (2004) Rapid inhibition of cancer cell growth induced by lentiviral delivery and expression of mutant-template telomerase RNA and anti-telomerase short-interfering RNA. Cancer Res 64:4833–4840
263. Abdul-Ghani R, Ohana P, Matouk I, Ayesh S, Ayesh B, Laster M et al (2000) Use of transcriptional regulatory sequences of telomerase (hTER and hTERT) for selective killing of cancer cells. Mol Ther 2:539–544
264. Gu J, Kagawa S, Takakura M, Kyo S, Inoue M, Roth JA et al (2000) Tumor-specific transgene expression from the human telomerase reverse transcriptase promoter enables targeting of the therapeutic effects of the Bax gene to cancers. Cancer Res 60:5359–5364
265. Koga S, Hirohata S, Kondo Y, Komata T, Takakura M, Inoue M et al (2000) A novel telomerase-specific gene therapy: gene transfer of caspase-8 utilizing the human telomerase catalytic subunit gene promoter. Hum Gene Ther 11:1397–1406
266. Boyd M, Mairs RJ, Mairs SC, Wilson L, Livingstone A, Cunningham SH et al (2001) Expression in UVW glioma cells of the noradrenaline transporter gene, driven by the telomerase RNA promoter, induces active uptake of [131I]MIBG and clonogenic cell kill. Oncogene 20:7804–7808
267. Komata T, Kondo Y, Kanzawa T, Hirohata S, Koga S, Sumiyoshi H et al (2001) Treatment of malignant glioma cells with the transfer of constitutively active caspase-6 using the human telomerase catalytic subunit (human telomerase reverse transcriptase) gene promoter. Cancer Res 61:5796–5802
268. Koga S, Hirohata S, Kondo Y, Komata T, Takakura M, Inoue M et al (2001) FADD gene therapy using the human telomerase catalytic subunit (hTERT) gene promoter to restrict induction of apoptosis to tumors in vitro and in vivo. Anticancer Res 21:1937–1943
269. Koga S, Kondo Y, Komata T, Kondo S (2001) Treatment of bladder cancer cells in vitro and in vivo with 2-5A antisense telomerase RNA. Gene Ther 8:654–658
270. Majumdar AS, Hughes DE, Lichtsteiner SP, Wang Z, Lebkowski JS, Vasserot AP (2001) The telomerase reverse transcriptase promoter drives efficacious tumor suicide gene therapy while preventing hepatotoxicity encountered with constitutive promoters. Gene Ther 8:568–578
271. Plumb JA, Bilsland A, Kakani R, Zhao J, Glasspool RM, Knox RJ et al (2001) Telomerase-specific suicide gene therapy vectors expressing bacterial nitroreductase sensitize human cancer cells to the pro-drug CB1954. Oncogene 20:7797–7803
272. Hao ZM, Luo JY, Cheng J, Li L, He D, Wang QY et al (2005) Intensive inhibition of hTERT expression by a ribozyme induces rapid apoptosis of cancer cells through a telomere length-independent pathway. Cancer Biol Ther 4:1098–1103
273. Kanazawa Y, Ohkawa K, Ueda K, Mita E, Takehara T, Sasaki Y et al (1996) Hammerhead ribozyme-mediated inhibition of telomerase activity in extracts of human hepatocellular carcinoma cells. Biochem Biophys Res Commun 225:570–576
274. Yokoyama Y, Takahashi Y, Shinohara A, Lian Z, Wan X, Niwa K et al (1998) Attenuation of telomerase activity by a hammerhead ribozyme targeting the template region of telomerase RNA in endometrial carcinoma cells. Cancer Res 58:5406–5410
275. Lin SY, Elledge SJ (2003) Multiple tumor suppressor pathways negatively regulate telomerase. Cell 113:881–889
276. Takakura M, Kyo S, Kanaya T, Hirano H, Takeda J, Yutsudo M et al (1999) Cloning of human telomerase catalytic subunit (hTERT) gene promoter and identification of proximal

core promoter sequences essential for transcriptional activation in immortalized and cancer cells. Cancer Res 59:551–557
277. Sohn JH, Yeh BI, Choi JW, Yoon J, Namkung J, Park KK et al (2010) Repression of human telomerase reverse transcriptase using artificial zinc finger transcription factors. Mol Cancer Res 8:246–253
278. Wong KK, Chang S, Weiler SR, Ganesan S, Chaudhuri J, Zhu C et al (2000) Telomere dysfunction impairs DNA repair and enhances sensitivity to ionizing radiation. Nat Genet 26:85–88
279. Ludwig A, Saretzki G, Holm PS, Tiemann F, Lorenz M, Emrich T et al (2001) Ribozyme cleavage of telomerase mRNA sensitizes breast epithelial cells to inhibitors of topoisomerase. Cancer Res 61:3053–3061
280. Tauchi T, Nakajima A, Sashida G, Shimamoto T, Ohyashiki JH, Abe K et al (2002) Inhibition of human telomerase enhances the effect of the tyrosine kinase inhibitor, imatinib, in BCR-ABL-positive leukemia cells. Clin Cancer Res 8:3341–3347
281. Multani AS, Li C, Ozen M, Imam AS, Wallace S, Pathak S (1999) Cell-killing by paclitaxel in a metastatic murine melanoma cell line is mediated by extensive telomere erosion with no decrease in telomerase activity. Oncol Rep 6:39–44
282. Mo Y, Gan Y, Song S, Johnston J, Xiao X, Wientjes MG et al (2003) Simultaneous targeting of telomeres and telomerase as a cancer therapeutic approach. Cancer Res 63:579–585
283. Nakamura M, Masutomi K, Kyo S, Hashimoto M, Maida Y, Kanaya T et al (2005) Efficient inhibition of human telomerase reverse transcriptase expression by RNA interference sensitizes cancer cells to ionizing radiation and chemotherapy. Hum Gene Ther 16:859–868
284. Guo X, Wang W, Zhou F, Lu Z, Fang R, Jia F et al (2008) siRNA-mediated inhibition of hTERT enhances chemosensitivity of hepatocellular carcinoma. Cancer Biol Ther 7:1555–1560
285. Damm K, Hemmann U, Garin-Chesa P, Hauel N, Kauffmann I, Priepke H et al (2001) A highly selective telomerase inhibitor limiting human cancer cell proliferation. EMBO J 20:6958–6968
286. Gomez-Millan J, Goldblatt EM, Gryaznov SM, Mendonca MS, Herbert BS (2007) Specific telomere dysfunction induced by GRN163L increases radiation sensitivity in breast cancer cells. Int J Radiat Oncol Biol Phys 67:897–905
287. Goldblatt EM, Gentry ER, Fox MJ, Gryaznov SM, Shen C, Herbert BS (2009) The telomerase template antagonist GRN163L alters MDA-MB-231 breast cancer cell morphology, inhibits growth, and augments the effects of paclitaxel. Mol Cancer Ther 8:2027–2035
288. Goldblatt EM, Erickson PA, Gentry ER, Gryaznov SM, Herbert BS (2009) Lipid-conjugated telomerase template antagonists sensitize resistant HER2-positive breast cancer cells to trastuzumab. Breast Cancer Res Treat 118:21–32
289. Trudeau M, Wong J (2010) Genetic variations in telomere maintenance, with implications on tissue renewal capacity and chronic disease pathologies. Curr Pharmacogenomics Person Med 8:7–24
290. Bodnar AG, Ouellette M, Frolkis M, Holt SE, Chiu CP, Morin GB et al (1998) Extension of life-span by introduction of telomerase into normal human cells. Science 279:349–352
291. Fauce SR, Jamieson BD, Chin AC, Mitsuyasu RT, Parish ST, Ng HL et al (2008) Telomerase-based pharmacologic enhancement of antiviral function of human CD8+ T lymphocytes. J Immunol 181:7400–7406
292. Zhu J, Lee S, Ho MK, Hu Y, Pang H, Ip FC et al (2010) In vitro intestinal absorption and first-pass intestinal and hepatic metabolism of cycloastragenol, a potent small molecule telomerase activator. Drug Metab Pharmacokinet 25:477–486
293. Benetos A, Okuda K, Lajemi M, Kimura M, Thomas F, Skurnick J et al (2001) Telomere length as an indicator of biological aging: the gender effect and relation with pulse pressure and pulse wave velocity. Hypertension 37:381–385
294. Cawthon RM, Smith KR, O'Brien E, Sivatchenko A, Kerber RA (2003) Association between telomere length in blood and mortality in people aged 60 years or older. Lancet 361:393–395
295. Harley CB, Liu W, Blasco M, Vera E, Andrews WH, Briggs LA et al (2011) A natural product telomerase activator as part of a health maintenance program. Rejuvenation Res 14:45–56

296. Calado RT, Regal JA, Hills M, Yewdell WT, Dalmazzo LF, Zago MA et al (2009) Constitutional hypomorphic telomerase mutations in patients with acute myeloid leukemia. Proc Natl Acad Sci USA 106:1187–1192
297. Gryaznov SM (2010) Oligonucleotide n3′–>p5′ phosphoramidates and thio-phoshoramidates as potential therapeutic agents. Chem Biodivers 7:477–493
298. Harley CB (2008) Telomerase and cancer therapeutics. Nat Rev Cancer 8:167–179

RAD51 Is a Key Protein of DNA Repair and Homologous Recombination in Humans

Alexander V. Mazin and Olga M. Mazina

1 Introduction

Ionizing radiation and various chemotherapeutic agents kill cancer cells by inducing DNA double-strand breaks (DSBs) or interstrand DNA cross links. Cells however can resist the killing effect by repairing these lesions using the homologous recombination (HR) pathway [1–3]. HR achieves high fidelity of repairing DNA breaks through the unique mechanism that employs homologous DNA as a template [4]. The initial step of HR involves exonucleolytic processing of the DNA ends into a resected DNA duplex with protruding 3′-ssDNA tails (Fig. 1) [5]. Then, RAD51 protein loads onto the ssDNA to form a contiguous helical nucleoprotein filament that promotes a search for the homologous dsDNA [6, 7]. Once the homologous sequence is found, RAD51 promotes the exchange of DNA strands that resulted in formation of joint molecules [8, 9]. Joint molecules provide both a template and a primer for the DNA synthesis that is required for retrieving the information lost at the site of the break and for the consequent restoration of a contiguous DNA structure.

It is thought that the joint molecules continue down one of two pathways [4, 10, 11]. They can proceed through the Synthesis Dependent Strand Annealing (SDSA) mechanism in which the joint molecules dissociate by branch migration, leading to rejoining of the broken chromosome (Fig. 1a). Alternatively, joint molecules can proceed through the Double-Strand Break Repair (DSBR) mechanism, which includes the formation of stable double Holliday junctions that are later resolved by structure-specific endonucleases (Fig. 1b). Whereas, the SDSA pathway occurs in both mitotically and meiotically dividing cells and produces non-crossovers, the DSBR takes place primarily in meiosis and produces crossovers that have a crucial role in the proper segregation of chromosomes [12].

A.V. Mazin, Ph.D. (✉) • O.M. Mazina
Department of Biochemistry and Molecular Biology, Drexel University College of Medicine, 245 N 15th Street, MS 497, NCB, Room 10103, Philadelphia, PA 19102-1192, USA
e-mail: amazin@drexelmed.edu

L. Panasci et al. (eds.), *Advances in DNA Repair in Cancer Therapy*, Cancer Drug Discovery and Development 72, DOI 10.1007/978-1-4614-4741-2_12,
© Springer Science+Business Media, LLC 2013

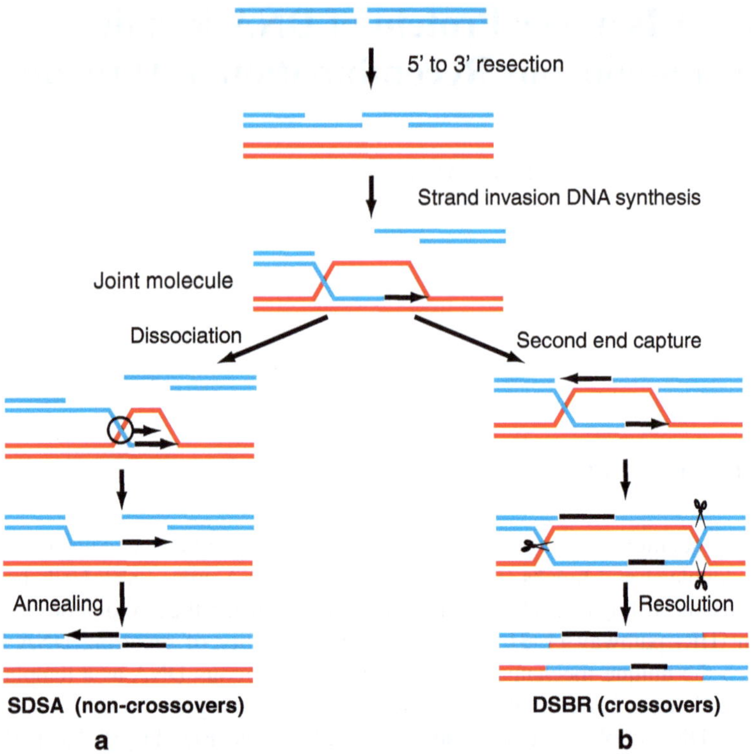

Fig. 1 Pathways of HR. Repair of a DSB proceeds either through the *S*ynthesis *D*ependent *S*trand *A*nnealing (SDSA) mechanism (**a**), which results in non-crossovers and is more common in mitosis than in meiosis; or through the *D*ouble-*S*tranded *B*reak *R*epair (DSBR) mechanism (**b**), which ultimately results in crossovers and occurs mainly in meiosis. Red and blue represent DNA copies of two different homologous chromosomes

2 The Homologous Recombination Machinery

In eukaryotes, the genes that constitute the core of the HR pathway were discovered by genetic screens of *Saccharomyces cerevisiae* for mutants conferring strong ionizing radiation sensitivity and moderate ultraviolet (UV) light sensitivity [13–16]. The core of HR, known as the Rad52 group of genes, includes *RAD50*, *RAD51*, *RAD52*, *RAD54*, *RDH54*/*TID1*, *RAD55*, *RAD57*, *RAD59*, *MRE11*, and *XRS2* [17]. The *rad51* is one of the three most IR-sensitive single mutants in *S. cerevisiae*; two other mutants are *rad52* and *rad54* [18]. Rad51 is highly evolutionarily conserved protein; it shares homology with a key bacterial HR protein known as RecA [19–21]; human RAD51 shows 68% amino acid identity with the *S. cerevisiae* homologue [22]. Rad51/RecA protein is ubiquitous in all kingdoms of life: Eukaryota, Archaea, and Bacteria. Most other members of the Rad52 group including Rad50, Rad52, Rad54, and Mre11are also conserved in eukaryotes, but have no obvious bacterial homologues. Rad55 and Rad57 are Rad51 paralogs that share homology and distant evolutionarily origin with Rad51 and with each other [22, 23]. In mammals, the core of HR includes five RAD51 paralogs (XRCC2, XRCC3, Rad51B, Rad51C, and Rad51D) that share approximately 20–30% amino acid identity with RAD51 and each other [22, 24, 25]; one RAD54 paralog (RAD54B) [26, 27]; BRCA1 and BRCA2, breast cancer susceptibility proteins 1 and 2; NBS1, Nijmegen breakage syndrome protein that forms a heterotrimeric complex (MRN) with MRE11 and RAD50 [28, 29]. In addition to the core proteins, a large number of other proteins participate in specific HR events, particularly the proteins of the Fanconi anemia (FA) group [30] and the RecQ family [31].

3 The Biological Function of Rad51

In *Escherichia coli, S. cerevisiae, and S. pombe recA* and rad51 mutants (rhp51, in *S. pombe*) are viable, albeit show reduced viability, but extremely sensitive to DNA-damaging agents such as ionizing radiation (IR) and methyl methanesulfonate (MMS) [3, 32]. In contrast, in vertebrates Rad51 is essential for viability. In mice, the knockout of the *RAD51* gene causes early embryonic lethality of homozygotes [33]. Additional knockout of the p53 gene postponed the embryonic lethality, but did not prevent it. Mouse Rad51$^{-/-}$ cells fail to proliferate even in culture [33–35]. It was demonstrated using a construct with the tet-repressible promoter in the chicken DT-40 cells that upon depletion of Rad51 protein the cells were no longer capable of undergoing even a single cell cycle and exhibited accumulation of spontaneous chromosomal breaks, indicating an important role of Rad51 in the DSBs repair [35]. Experiments in mice demonstrated that knockouts of several other HR proteins including RAD50, BRCA1, BRCA2, XRCC2, RAD51B, and RAD51D also cause embryonic lethality indicating an essential function of HR for viability in higher eukaryotes [36].

4 The Biochemical Activities of RAD51

RAD51 promotes DNA strand exchange, a basic reaction of HR. DNA strand exchange includes three major steps: (i) RAD51 filament formation, (ii) the search for DNA homology and DNA strand exchange proper, (iii) and DNA branch migration. RAD51 is an ATPase; ATP hydrolysis plays an important role in the function of this protein.

4.1 RAD51 Filament Formation

RAD51 binds to ssDNA forming a helical right-handed nucleoprotein filament in which protein monomers wrap around ssDNA [21, 37]. The filament formation by RAD51 is often referred as presynaptic stage of DNA strand exchange. In vitro, the

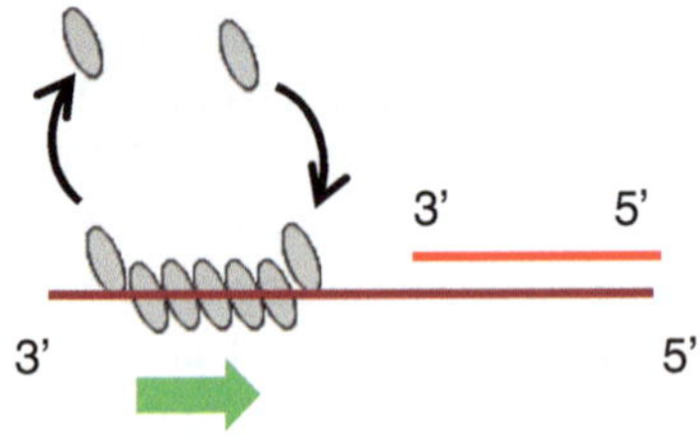

Fig. 2 RAD51 polymerizes on ssDNA with a 3′-5′ polarity. The elongation of the RAD51 filament shows cooperativity in which new monomeric units bind preferably to the existing filament end rather than to free DNA. Disassembly of the filament occurs with the same polarity, but initiates at the opposite filament end. The *green arrow* indicates the polarity of RAD51 polymerization

filament may extend over several thousand of bp of ssDNA, e.g., covering the entire circular ssDNA of bacteriophages M13 (6,407 nt) or ϕX174 (5,386 nt). Filament formation includes the initiation stage, when several monomeric units of RAD51 establish the initial complex on ssDNA and the elongation step, when additional monomeric units are added to the growing end of the filament (Fig. 2). The elongation of the RAD51 filament shows cooperativity in which new monomeric units bind preferably to the existing filament end rather than to free DNA [38–40]. The elongation occurs with a distinct 3′ to 5′ polarity relative to ssDNA, which is opposite to the polarity of RecA filament elongation (5′ to 3′) [41]. Disassembly of the filament occurs with the same polarity, but initiates at the opposite filament end. Thus, the filament has the growing end and the dissociating end. RAD51 can bind both ssDNA and dsDNA; however, in the presence of physiological (or slightly elevated) concentrations of monovalent salt, it shows preference for ssDNA [41, 42]. Some auxiliary proteins, particularly BRCA2, further enforce RAD51 binding specificity for ssDNA [43]. Formation of active (extended) RAD51-ssDNA filament requires the presence of a nucleotide cofactor, ATP or dATP. In the presence of ADP, RAD51 forms condensed filament that is inactive in DNA strand exchange and prone to dissociation [44]. The ATP binding site is found not in the interior of individual subunits, but at the subunit–subunit interface in the filament [37, 45]. The RAD51 nucleotide binding site is composed by the protein domains contributed by two neighboring protein monomers within the filament. Consequently, RAD51 can bind and hydrolyze ATP only after the assembly of the nucleoprotein filament, which occurs in the presence of DNA [46, 47]. Other conditions, like high concentration of monovalent salt, which promote filament formation in the absence of DNA, also stimulate the ATPase activity of Rad51 and its orthologs [48, 49].

4.2 Why Does Rad51 Protein Hydrolyze ATP?

The role of ATP hydrolysis by RAD51 and other proteins of the RAD51/RecA family has been a topic of extensive investigation. Genetic data show that all RecA/Rad51 ATPase mutants show various degree of deficiency in HR and DNA repair in

E. coli, *S. cerevisiae*, and mammalian cells [50–52]. In biochemical studies, it was shown that ATP hydrolysis leads to formation of low affinity nucleoprotein complexes that are prone to dissociation from DNA [44]. This dissociation may lead to recycling of the protein after competition of DNA strand exchange and is critical for DNA branch activity of the protein (see below) [53]. In the case of RecA and most other members of the family, ADP produced during ATP hydrolysis readily dissociates from the protein-complexes is replaced and by free ATP, as long as the pool of free ATP is available [54]. However, human RAD51 is distinct among the members of the RAD51/RecA family so that under normal conditions, e.g., in the presence of Mg^{2+}, the ADP product remains stably associated with human RAD51 even in the presence of high ATP concentrations and the ATP-regeneration system [44]. Thus, ATP hydrolysis leads to self-inactivation of RAD51, turning it into an ADP-bound form that is inactive in DNA strand exchange. It was found that Ca^{2+} by partial inhibition the ATPase activity of RAD51 prevents self-conversion of the protein into an ADP-bound complex [44]. In the presence of Ca^{2+}, Rad51 nucleoprotein filament remains in an ATP-bound form that is active in DNA strand exchange [44, 55, 56]. On the other hand, branch migration activity of RAD51 that depends on ATP hydrolysis is inhibited by Ca^{2+} (see below) [41]. Thus, the ATPase activity of RAD51 provides a powerful pivot for regulation of DNA strand exchange and branch migration activities of this protein. Given that Ca^{2+} concentration rises in response to DNA damage [57–60] and that Ca^{2+} plays an important role at the early stage of meiosis [61], it is tempting to hypothesize that Ca^{2+} is an in vivo modulator of RAD51. However, cellular concentration of free Ca^{2+} seems to be too low to have a direct effect on RAD51 [62]. It is possible, however, that Ca^{2+} may affect RAD51 by acting in combination with posttranslational modifications of RAD51 or with other proteins that interact with RAD51. Alternatively, the in vitro effect of Ca^{2+} may only mimic the effect of other in vivo factors, which may specifically stabilize or destabilize the RAD51-filament by modulating its ATPase activity.

4.3 The Search for DNA Homology and DNA Strand Exchange

The RAD51 nucleoprotein filament possesses a unique activity; it searches for homologous dsDNA sequences and promotes the exchange of identical DNA strands between homologous dsDNA and ssDNA [6, 46, 47]. This step is known as synaptic step of DNA strand exchange. The mechanism of DNA strand exchange is conserved among the members of Rad51/RecA family [63, 64]; it was best studied for RecA [65, 66]. It is thought that RAD51/RecA proteins contain two DNA binding sites: the primary and the secondary [67, 68]. The primary site accommodates ssDNA during the nucleoprotein assembly. The secondary site of the Rad51-ssDNA nucleoprotein filament is responsible for interaction with dsDNA during the search for homology [69]. Mutational, biochemical and fluorescent spectroscopic analyses indicated that RAD51/RecA binding sites are located in the protein region known as the L1 and L2 loops [70–73]. Although the important details of the mechanism of

DNA strand exchange remain to be investigated, the ability to hold two DNA molecules in close proximity clearly plays an essential role in homology recognition and in exchange of strands promoted by the RAD51 filament. The secondary DNA binding site may play an especially important role in this process. The salient feature of the secondary site is that it binds both ss- and dsDNA, but has higher affinity for ssDNA [67]. Because of this specificity the secondary binding site plays a dual role in DNA strand exchange: it binds the incoming dsDNA weakly as part of the homology search process, and then, upon finding homology and promoting local exchange of DNA strands, this site binds tightly to the resulting displaced DNA strand, stabilizing the nascent DNA heteroduplex product [74]. Furthermore, RPA/SSB by removing ssDNA from the secondary site helps to drive the reaction forward. Several lines of evidence indicate that the primary site after DNA strands are exchanged accommodates the newly formed heteroduplex DNA within the postsynaptic filament [75–78].

4.4 DNA Branch Migration

DNA strand exchange results in formation of a heteroduplex product, also known as the joint molecule (JM). At the postsynaptic step, Rad51/RecA extends the joint molecules by a process known as heteroduplex extension or branch migration, in which one DNA strand is progressively exchanged for another [79]. When the extending heteroduplex reaches the ssDNA–dsDNA junction on the invading DNA strand, JMs are converted into a four-stranded Holliday junction (HJ) (Fig. 3). Both RecA and RAD51 can promote four-strand branch migration, although Rad51 does it with a significantly lower rate than RecA [41, 80]. It was demonstrated that branch migration is mechanistically distinct from DNA strand exchange proper [41]. DNA strand exchange requires formation of a stable extended filament, which forms when RAD51 binds ATP, but it does not require ATP hydrolysis [44, 55]. It was shown that non-hydrolyzable ATP analogs, e.g., AMP-PNP, support DNA strand exchange. Similarly, the RAD51 K133R mutant that is unable to hydrolyze ATP can promote DNA strand exchange [81]. In contrast, ATP hydrolysis is absolutely required for branch migration. It was recently demonstrated that cycles of RecA/Rad51 polymerization and dissociation drive branch migration (Fig. 3) [41]. Because dissociation of RAD51 depends on ATP hydrolysis, these results explain the requirement in ATP hydrolysis during branch migration [41, 80]. The factors, like Ca^{2+}, that inhibit ATP hydrolysis, inhibit branch migration as well. At the same time Ca^{2+} greatly stimulates DNA strand exchange [41]. While DNA strand exchange activity is commonly considered as the hallmark activity of RAD51/RecA family, branch migration activity of these proteins may play important supplementary functions, e.g., at the initial stages of recombination in vivo by helping to form and stabilize initially unstable joint molecules. Then following this initial stage of branch migration, more potent specialized branch migration proteins, e.g., RAD54 [82, 83], may gain access to the joint molecules and promote their branch migration with a higher rate (Fig. 3).

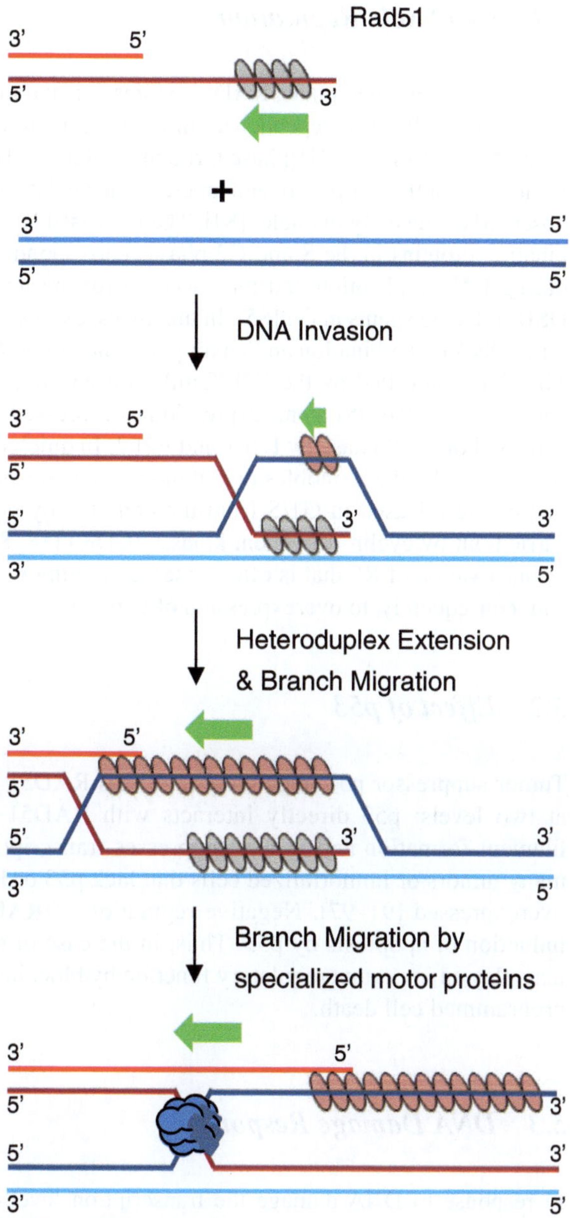

Fig. 3 A hypothetical role of RAD51 branch migration activity during DSB repair. DNA with DSB is excised by 5′ end resection to produce DNA with a 3′ ssDNA tail. With an exposed 3′ tail, RAD51 can promote DNA invasion to form a D-loop and then extend and stabilize this D-loop through branch migration driven by protein polymerization (shown by the *green arrow*). Branch migration is then continued by specialized enzymes, e.g., RAD54 (*blue shape*)

5 Regulation of Rad51 Activity

The activity of Rad51 is regulated through the cell cycle and DNA damage response signaling. The regulatory mechanisms involve transcriptional and posttranslational regulation. In addition, the activity of RAD51 can be either enhanced or inhibited through interaction with several auxiliary proteins.

5.1 Cell Cycle Regulation

In both yeast and mammals, RAD51 expression and intracellular localization show remarkable cell-cycle dependence. In mouse cells RAD51 transcription shows the highest level from late G1 phase through M phase [84]. While G0 phase RAD51 is located in both cytoplasm and nuclei, during late G1-S-G2 phase Rad51 was observed exclusively in nuclei [84]. The increased level of Rad51 and other recombination proteins in the S and G2 phases corresponds to the important role of HR during DNA replication and may account for the increased resistance of cells to DSB-inducing agents in G2 [85]. In mammals, expression of the RAD51 and several other DNA recombination and repair genes including RAD54, BARD1, MSH2, and MLH1 is controlled by the E2F family transcription factors [86, 87]. During G0 and early G1 RAD51 gene expression is repressed by E2F4, and subsequently activated in G1/S phase by E2F1 and E2F2. In quiescent cells, E2F1 and E2F2 are sequestered by the retinoblastoma tumor suppressor protein (Rb); the transcription factors are released in G1/S from the complex by phosphorylation of Rb that is carried out by cyclin-dependent kinases (CDK) [88, 89]. Hyperactivation of CDK or inactivation of Rb that is often observed in tumor cells leads to E2F1 activation and, consequently, to overexpression of RAD51.

5.2 Effect of p53

Tumor suppressor p53 negatively regulates RAD51 [90]. The regulation occurs at two levels: p53 directly interacts with RAD51 and inhibits nucleoprotein filament formation and it also suppresses transcription of the RAD51 gene. In many tumors or immortalized cells that lack p53 or Rb, RAD51 was found to be overexpressed [91–97]. Negative regulation of RAD51 occurs in parallel with induction of apoptosis by p53. Thus, in the case of extensive DNA damage p53 may play an important regulatory function by blocking DNA repair and promoting programmed cell death.

5.3 DNA Damage Response

In response to DNA damage the transcription level of the RecA gene, bacterial homologue of RAD51, increases dramatically [98, 99]. In contrast, no or little increase in the Rad51 expression in response to DNA damage was reported. Instead, RAD51 undergoing multiple posttranslational modifications that significantly affect its ability to interact with its protein partners and translocate to the nucleus and form distinct structures known as “foci” [100, 101]. Since Rad51 does not have a Nuclear Localization Signal (NLS) sequence, its nuclear entry likely requires the interaction with other proteins containing functional NLS sequences [102]. RAD51 foci likely identify the sites where DSB repair is carried out by RAD51 and several other recombination proteins including RAD52 and RAD54, the ssDNA-binding protein RPA, and the tumor suppressor BRCA2 [101, 103–106].

5.3.1 BRCA2-Dependent Mechanisms of Regulation

Several lines of evidence indicate that interaction between RAD51 and BRCA2 is especially important for HR and DNA repair [107]. BRCA2 was shown to stimulate loading of RAD51 on ssDNA helping to displace RPA from ssDNA [108–110]. Mutations in BRCA2 severely disrupt RAD51 foci formation, induce spontaneous chromosomal instability, and elevated sensitivity to ionizing radiation during S and G2 [111, 112]. The interaction between RAD51 and BRCA2 is thought to be limited to S and G2 phases of the cell cycle by CDK-dependent phosphorylation of the C terminus BRCA2 site that is involved in RAD51 interaction [113]. The regulatory circuit that governs BRCA2 phosphorylation is thought to involve ATM, p53 and CHK2 which promote CDK inactivation and cell cycle arrest after DNA damage. ATM transduces the DNA damage signal to p53, leading to the transcriptional activation of p21 that inhibits CDK activity [114]. ATM also phosphorylates and activates CHK2, which in turn phosphorylates CDC25 phosphatase, leading to its degradation or cytoplasmic sequestration [115]. Because CDC25 is required for CDK activation, loss of CDC25 activity helps to maintain the phosphorylated/inactivated state of CDK [116]. Inactivation of CDK reduces its ability to phosphorylate S3291 of BRCA2, thus stimulating interactions between RAD51 and the C-terminal region of BRCA2.

5.3.2 BRCA2-Independent Mechanisms of RAD51 Regulation

Rad51 recruitment to DNA damage sites can also be mediated through BRCA2-independent mechanisms. Recently, it was reported that Polo-like kinase 1 (Plk1) phosphorylates RAD51 at serine 14 (S14) during the cell cycle and in response to DNA damage. S14 phosphorylation licenses subsequent Rad51 phosphorylation at threonine 13 (T13) by casein kinase 2 (CK2), which in turn triggers direct binding to the Nijmegen breakage syndrome gene product, Nbs1 [117]. Nbs1, together with its binding partners Mre11 and Rad50, is efficiently recruited to DNA damaged sites at the early stages of DNA damage response [118, 119].

Further regulation of HR is provided by RAD51 phosphorylation mediated by CHK1 [120]. It was demonstrated that Chk1 interacts with RAD51. It was also shown that RAD51 is phosphorylated in a Chk1-dependent manner on Thr 309 which is located in a Chk1 consensus phosphorylation site. Chk1-depleted cells failed to form RAD51 nuclear foci after exposure to hydroxyurea, and cells expressing a phosphorylation-deficient mutant $RAD51^{T309A}$ were hypersensitive to hydroxyurea.

Human RAD51 interacts with DNA-damage-activated kinase cAbl, which phosphorylates RAD51 on Tyr54 and Tyr315 [121, 122]. This phosphorylation occurs in a sequential manner, Tyr315 first followed by Tyr54. Tyr315 is close to the subunit–subunit interacting site and its phosphorylation is related to the dissociation of Rad51 polymer [123]. RAD51 can also be phosphorylated by the chimeric BCR/Abl kinase resulting from the chromosomal translocation that produces Philadelphia chromosome and is linked to the frequent incidence of chronic myelogenous leukemia [124]. It was shown that cells lacking cAbl and the related Arg protein [125] or both do not have defined defects in double-strand DNA break repair and in fact appear to be somewhat radioresistant relative to $abl^{+/+}$ epithelial cells [126]. These data indicate that activated cAbl inhibits DNA repair, while p53 mediates growth arrest and perhaps apoptosis [127]. The precise role played by cAbl and, specifically, the effect of RAD51 phosphorylation in the DNA damage response remains to be elucidated.

RAD51 competes with RPA for ssDNA sites during nucleoprotein filament assembly. Sumoylation of RPA1 was recently suggested to promote HR by facilitating the recruitment of RAD51 [128]. In contrast, conjugation of RAD51 with UBL1 (ubiquitin-like 1) promoted by Ubc9/UBE21 may inhibit DSB repair and decrease cell resistance to ionizing radiation [129].

5.4 *Effect of RAD54 and Other Stimulatory Proteins*

Rad51 interacts with Rad54 functionally and physically; interactions between these two proteins are extensive and critical to the function of HR in eukaryotes [3, 130, 131]. Rad54 is a motor protein that translocates on dsDNA and promotes chromatin remodeling and branch migration of Holliday junctions in an ATPase-dependent manner [83, 132]. Physical interactions between Rad51 and Rad54 proteins are species-specific and conserved from archaea to humans [132]. In *S. cerevisiae*, overexpression of Rad54 can suppress certain repair phenotypes of rad51 mutants [51, 130]; and the rate and extent of Rad51 recruitment to the HO-induced DSB is significantly reduced in the absence of Rad54 [133]. In mouse ES cells, IR-induced Rad51 foci co-localize with Rad54 foci [104]. Moreover, Rad51 foci formation shows dependence on Rad54.

Yeast Rad54 strongly stimulates the DNA strand exchange activity of Rad51 [134]. This stimulation is evolutionarily conserved: archaeal, Drosophila, and human Rad54 orthologs stimulate DNA strand exchange activity of their cognate Rad51 [81, 135–137]. The stimulation depends on the ATPase activity of Rad54 [134], indicating that Rad54 translocation on dsDNA plays a role in stimulation of DNA strand exchange activity. The mechanism of DNA strand exchange stimulation involves formation of a complex between Rad54 and the Rad51 nucleoprotein filament [138–140]. In this complex, translocation of potential target DNA by Rad54 is linked to the DNA homology search process promoted by the Rad51 nucleoprotein filament. It was proposed that the translocation activity of Rad54 may both provide a more efficient delivery of dsDNA to the site of the homology search within the filament and cause transient disruption of dsDNA base pairs making them available for interaction with the ssDNA bound within the Rad51 nucleoprotein filament.

While ATPase-dependent dsDNA translocation by Rad54 may play a critical role in stimulation of DNA strand exchange, Rad54 employs additional mechanisms of Rad51 stimulation. By forming a complex with the Rad51 nucleoprotein filament, Rad54 stabilizes the filament and increases the Rad51 ability to compete with RPA for ssDNA binding [133, 141]. The filament stabilization function of Rad54 does not depend on its ATPase activity [141].

Interaction between Rad51 and Rad54 has synergistic effect on the activities of both proteins. While Rad54 stimulates the DNA strand exchange activity of Rad51, in its turn, Rad51 stimulates the dsDNA-dependent ATP hydrolysis of Rad54 [138, 139], increases the processivity of Rad54 DNA translocation along DNA [136], stimulates chromatin remodeling [137, 142–144], and DNA branch migration activity of Rad54 protein [145].

In addition to RAD54, several other proteins including BLM [146], RAD51AP [147, 148], and Hop2-Mnd1 [149, 150] were reported to stimulate DNA strand exchange activity of human RAD51.

6 RAD51 and Tumorigenesis

6.1 RAD51 Mutants Found in Cancers

As it was discussed in previous sections, RAD51 is overexpressed in many cancers or immortalized cells. In addition, mutations or polymorphism in the *RAD51* gene have been identified in several human tumors, including breast cancer and head and neck squamous cell carcinoma [151–159] suggesting the involvement of the human RAD51 protein in tumor suppression mechanisms. Most of the *RAD51* mutations in tumor cells were found in its noncoding region, but a missense RAD51 mutation, in which Arg150 is replaced by Gln (R150Q), was identified in patients with bilateral breast cancer [151]. In vitro, this mutant protein showed altered DNA binding [160].

6.2 RAD51 as a Target for Anticancer Therapies

High levels of RAD51 are associated with elevated rates of DNA recombination as well as enhanced resistance to DNA-damaging chemotherapies and/or ionizing radiation in several experimental tumor systems [91, 161, 162]. In contrast, mutations that reduce RAD51 activity or expression decrease cell survival. Complete inactivation of RAD51 causes cell lethality [33]. Consequently, RAD51 was recognized as an important target for anticancer therapies. Given the elevated levels of

DNA damage in cancer cells compared to normal cells, a further increase of DNA damage by inhibiting HR components, in combination with chemotherapeutic drugs, could lead to cancer cell death. Indeed, recent studies have shown that ATM, DNA-PK, and CHK1 inhibitors have preferential toxicity toward cancer cells following treatment with genotoxic agents [163]. In addition, it was proposed that overexpression of RAD51 observed in most cancer cells represents a compensatory mechanism for the loss of alternative repair pathways in cancer cells occurring due to intrinsic instability of their genome [164]. In the absence of these alternative DNA repair pathways, inhibition of RAD51 could make cancer cells more sensitive to cytotoxic agents than normal cells [164, 165].

6.2.1 Antisense Strategies

Antisense strategies have been successfully used to attenuate Rad51-mediated radioresistance in in vitro and in vivo studies [166, 167]. In more recent studies, inhibition of RAD51 expression by small interfering RNA (siRNA) enhanced the effect of cisplatin treatment of cancer (HeLa) cells [164]. Importantly, cancer cells were found to be more sensitive to combination treatment with siRNA and cisplatin than non-transformed cells.

6.2.2 Small Molecules Inhibitors of RAD51

Anti-cancer therapies based on specific inhibitors that target DNA repair proteins are developing at a fast pace. The use of small molecule inhibitors offers significant advantages over both siRNA inhibition and antibody microinjection, which include good delivery properties, good in vivo stability, a low probability of inducing an immune response, and low cost. Recently, it was shown that specific inhibitors of poly-ADP ribose polymerase 1 (PARP1) are effective in killing cells that carry mutant BRCA1 and BRCA2 proteins. PARP1 is a DNA damage sensor protein that is particularly important for the repair of DNA single-strand breaks [168–171]. Inactivation of PARP1 by specific inhibitors results in persistent DNA single-strand breaks. When the replication fork encounters these DNA single-strand breaks, they are converted to DNA DSBs which are repaired through the HR pathway. Combination of specific PARP1 inhibitors with mutations in the *BRCA1* and *BRCA2* genes, which inactivate HR, was proven to be deadly for cancer cells carrying these mutations. Several PARP1 inhibitors are currently undergoing clinical trials against familial breast cancer in which BRCA1 or BRCA2 are mutated. However, resistance to PARP1 inhibitors has already been observed, as cells partially restore HR. In addition, while BRCA1 and BRCA2 mutations are common in hereditary cancers, they are relatively rare among sporadic cancers, which represent 80–90% of all breast cancer cases. Therefore, developing specific inhibitors against key proteins of HR, like RAD51, may represent alternative or supplementary strategy, as these inhibitors may be used in combination with both cytotoxic agents and PARP1 inhibitors.

Several small-molecule inhibitors of RAD51 have been described. Screening a small library of 185 compounds identified 4,4′-diisothiocyanostilbene-2,2′-disulfonic acid (DIDS) that inhibited the RAD51-mediated strand exchange [172]. A surface plasmon resonance analysis revealed that DIDS directly binds to RAD51. A gel mobility shift assay showed that DIDS inhibited the DNA-binding activity of RAD51. In another study, it was found fortuitously that sodium metatungstate ($Na_6H_2W_{12}O_{40}$) inhibited the ATPase and DNA strand exchange activities of the archaeal Rad51 from Methanococcus voltae [173]. The tungsten cluster appears to be bound between the DNA-binding loops L1 and L2 anchoring the protein in its inactive conformation. However, in both these studies the inhibitors lack specificity for RAD51; these inhibitors can only be used for in vitro studies of Rad51 activities.

Recently, two specific RAD51 inhibitors were identified. One of them, named RI-1 (3-chloro-1-(3,4-dichlorophenyl)-4-(4-morpholinyl)-1H-pyrrole-2,5-dione), was obtained from a high-throughput screen of a library of 10,000 small-molecule compounds by searching for an inhibitor of RAD51 binding to ssDNA [174].

RI-1 inactivates RAD51 by directly binding to a protein surface that serves as an interface between protein subunits in RAD51 filaments. Cell-based experiments demonstrated that RI-1 specifically inhibits HR and sensitizes human cancer cells to mitomycin C (MMC).

Another compound, named B02 (*E/Z*)-3-benzyl-2-(2-(pyridin-3-yl)vinyl) quinazolin-4(3H)-one, was identified by screening of a 202,556-compound library for a specific inhibitor of the RAD51 DNA strand exchange activity [175]. B02 acts by binding to RAd51 and disrupting RAD51 binding to DNA and formation of the nucleoprotein filament. Importantly, B02 shows a substantial inhibitory effect on HR and DNA repair in human and mouse cells. The results show that B02 inhibits DSB-induced HR and increases cell sensitivity to DNA interstrand cross-linking agents, cisplatin and mitomycin C (MMC). In combination with the PARP1 inhibitor, AZD2281 (olaparib) [176], B02 synergistically increased cell sensitivity to the alkylating agent MMS. Overall, these recent studies [174, 175, 177] show that specific RAD51 inhibitors may be instrumental for the analysis of RAD51 activities and cellular functions and for development of combination anticancer therapies.

References

1. Wyman C, Kanaar R (2006) DNA double-strand break repair: all's well that ends well. Annu Rev Genet 40:363–383
2. Moynahan ME, Jasin M (2010) Mitotic homologous recombination maintains genomic stability and suppresses tumorigenesis. Nat Rev Mol Cell Biol 11(3):196–207
3. Krogh BO, Symington LS (2004) Recombination proteins in yeast. Annu Rev Genet 38: 233–271
4. Pâques F, Haber JE (1999) Multiple pathways of recombination induced by double-strand breaks in *Saccharomyces cerevisiae*. Microbiol Mol Biol Rev 63(2):349–404
5. Mimitou EP, Symington LS (2011) DNA end resection—unraveling the tail. DNA Repair (Amst) 10(3):344–348

6. Kowalczykowski SC (2008) Structural biology: snapshots of DNA repair. Nature 453(7194):463–466
7. Sung P, Krejci L, Van Komen S, Sehorn MG (2003) Rad51 recombinase and recombination mediators. J Biol Chem 278(44):42729–42732
8. Haber JE (2006) Chromosome breakage and repair. Genetics 173(3):1181–1185
9. Holloman WK, Radding CM (1976) Recombination promoted by superhelical DNA and the recA gene of *Escherichia coli*. Proc Natl Acad Sci USA 73(11):3910–3914
10. West SC (2009) The search for a human Holliday junction resolvase. Biochem Soc Trans 37(Pt 3):519–526
11. West SC (2003) Molecular views of recombination proteins and their control. Nat Rev Mol Cell Biol 4(6):435–445
12. Neale MJ, Keeney S (2006) Clarifying the mechanics of DNA strand exchange in meiotic recombination. Nature 442(7099):153–158
13. Game JC, Mortimer RK (1974) A genetic study of X-ray sensitive mutants in yeast. Mutat Res 24:281
14. Game JC (2000) The Saccharomyces repair genes at the end of the century. Mutat Res 451(1–2):277–293
15. Suslova NG (1969) Isolation of *Saccharomyces cerevisiae* mutants sensitive to ionizing radiation. Radiobiologiya 9(6):933–936
16. Zacharov IA, Suslova NG, Fedorova IV (1970) The genetic basis of radiosensetivity in yeast. Genetika 6(5):76–83
17. Symington LS (2002) Role of RAD52 epistasis group genes in homologous recombination and double-strand break repair. Microbiol Mol Biol Rev 66(4):630–670, table of contents
18. Game JC (1993) DNA double-strand breaks and the RAD50-RAD57 genes in Saccharomyces. Semin Cancer Biol 4(2):73–83
19. Shinohara A, Ogawa H, Ogawa T (1992) Rad51 protein involved in repair and recombination in *S. cerevisiae* is a RecA-like protein. Cell 69(3):457–470
20. Aboussekhra A, Chanet R, Adjiri A, Fabre F (1992) Semidominant suppressors of Srs2 helicase mutations of *Saccharomyces cerevisiae* map in the RAD51 gene, whose sequence predicts a protein with similarities to procaryotic RecA proteins. Mol Cell Biol 12(7):3224–3234
21. Ogawa T, Yu X, Shinohara A, Egelman EH (1993) Similarity of the yeast RAD51 filament to the bacterial RecA filament. Science 259(5103):1896–1899
22. Lin Z, Kong H, Nei M, Ma H (2006) Origins and evolution of the recA/RAD51 gene family: evidence for ancient gene duplication and endosymbiotic gene transfer. Proc Natl Acad Sci USA 103(27):10328–10333
23. Sung P (1997) Yeast Rad55 and Rad57 proteins form a heterodimer that functions with replication protein A to promote DNA strand exchange by Rad51 recombinase. Genes Dev 11(9):1111–1121
24. Sonoda E, Zhao GY, Kohzaki M, Dhar PK, Kikuchi K, Redon C, Pilch DR, Bonner WM, Nakano A, Watanabe M, Nakayama T, Takeda S, Takami Y (2007) Collaborative roles of gammaH2AX and the Rad51 paralog Xrcc3 in homologous recombinational repair. DNA Repair (Amst) 6(3):280–292
25. Suwaki N, Klare K, Tarsounas M (2011) RAD51 paralogs: roles in DNA damage signalling, recombinational repair and tumorigenesis. Semin Cell Dev Biol 22(8):898–905
26. Miyagawa K, Tsuruga T, Kinomura A, Usui K, Katsura M, Tashiro S, Mishima H, Tanaka K (2002) A role for RAD54B in homologous recombination in human cells. EMBO J 21(1–2):175–180
27. Wesoly J, Agarwal S, Sigurdsson S, Bussen W, Van Komen S, Qin J, van Steeg H, van Benthem J, Wassenaar E, Baarends WM, Ghazvini M, Tafel AA, Heath H, Galjart N, Essers J, Grootegoed JA, Arnheim N, Bezzubova O, Buerstedde JM, Sung P, Kanaar R (2006) Differential contributions of mammalian Rad54 paralogs to recombination, DNA damage repair, and meiosis. Mol Cell Biol 26(3):976–989
28. Paull TT, Lee JH (2005) The Mre11/Rad50/Nbs1 complex and its role as a DNA double-strand break sensor for ATM. Cell Cycle 4(6):737–740

29. Brugmans L, Verkaik NS, Kunen M, van Drunen E, Williams BR, Petrini JH, Kanaar R, Essers J, van Gent DC (2009) NBS1 cooperates with homologous recombination to counteract chromosome breakage during replication. DNA Repair (Amst) 8(12):1363–1370
30. Wrighton KH (2010) DNA repair: a new fan of the Fanconi anaemia pathway. Nat Rev Mol Cell Biol 11(9):603
31. Chu WK, Hickson ID (2009) RecQ helicases: multifunctional genome caretakers. Nat Rev Cancer 9(9):644–654
32. Clark AJ (1973) Recombination deficient mutants of *E. coli* and other bacteria. Annu Rev Genet 7:67–86
33. Lim DS, Hasty P (1996) A mutation in mouse rad51 results in an early embryonic lethal that is suppressed by a mutation in p53. Mol Cell Biol 16(12):7133–7143
34. Tsuzuki T, Fujii Y, Sakumi K, Tominaga Y, Nakao K, Sekiguchi M, Matsushiro A, Yoshimura Y, Morita T (1996) Targeted disruption of the Rad51 gene leads to lethality in embryonic mice. Proc Natl Acad Sci USA 93(13):6236–6240
35. Sonoda E, Sasaki MS, Buerstedde JM, Bezzubova O, Shinohara A, Ogawa H, Takata M, Yamaguchi-Iwai Y, Takeda S (1998) Rad51-deficient vertebrate cells accumulate chromosomal breaks prior to cell death. EMBO J 17(2):598–608
36. Brugmans L, Kanaar R, Essers J (2007) Analysis of DNA double-strand break repair pathways in mice. Mutat Res 614(1–2):95–108
37. Reymer A, Frykholm K, Morimatsu K, Takahashi M, Norden B (2009) Structure of human Rad51 protein filament from molecular modeling and site-specific linear dichroism spectroscopy. Proc Natl Acad Sci USA 106(32):13248–13253
38. Takahashi M, Strazielle C, Pouyet J, Daune M (1986) Co-operativity value of DNA-RecA protein interaction. Influence of the protein quaternary structure on the binding analysis. J Mol Biol 189:711–714
39. Menetski JP, Kowalczykowski SC (1985) Interaction of recA protein with single-stranded DNA. Quantitative aspects of binding affinity modulation by nucleotide cofactors. J Mol Biol 181(2):281–295
40. De Zutter JK, Knight KL (1999) The hRad51 and RecA proteins show significant differences in cooperative binding to single-stranded DNA. J Mol Biol 293(4):769–780
41. Rossi MJ, Mazina OM, Bugreev DV, Mazin AV (2011) The RecA/RAD51 protein drives migration of Holliday junctions via polymerization on DNA. Proc Natl Acad Sci USA 108(16):6432–6437
42. Liu Y, Stasiak AZ, Masson JY, McIlwraith MJ, Stasiak A, West SC (2004) Conformational changes modulate the activity of human RAD51 protein. J Mol Biol 337(4):817–827
43. Carreira A, Hilario J, Amitani I, Baskin RJ, Shivji MK, Venkitaraman AR, Kowalczykowski SC (2009) The BRC repeats of BRCA2 modulate the DNA-binding selectivity of RAD51. Cell 136(6):1032–1043
44. Bugreev DV, Mazin AV (2004) Ca^{2+} activates human homologous recombination protein Rad51 by modulating its ATPase activity. Proc Natl Acad Sci USA 101(27):9988–9993
45. Yu X, Jacobs SA, West SC, Ogawa T, Egelman EH (2001) Domain structure and dynamics in the helical filaments formed by RecA and Rad51 on DNA. Proc Natl Acad Sci USA 98(15):8419–8424
46. Baumann P, Benson FE, West SC (1996) Human Rad51 protein promotes ATP-dependent homologous pairing and strand transfer reactions in vitro. Cell 87(4):757–766
47. Gupta RC, Bazemore LR, Golub EI, Radding CM (1997) Activities of human recombination protein Rad51. Proc Natl Acad Sci USA 94(2):463–468
48. Wu Y, He Y, Moya IA, Qian X, Luo Y (2004) Crystal structure of archaeal recombinase RADA: a snapshot of its extended conformation. Mol Cell 15(3):423–435
49. Wu Y, Qian X, He Y, Moya IA, Luo Y (2005) Crystal structure of an ATPase-active form of Rad51 homolog from *Methanococcus voltae*. Insights into potassium dependence. J Biol Chem 280(1):722–728
50. Konola JT, Logan KM, Knight KL (1994) Functional characterization of residues in the P-loop motif of the RecA protein ATP binding site. J Mol Biol 237(1):20–34

51. Morgan EA, Shah N, Symington LS (2002) The requirement for ATP hydrolysis by *Saccharomyces cerevisiae* Rad51 is bypassed by mating-type heterozygosity or RAD54 in high copy. Mol Cell Biol 22(18):6336–6343
52. Stark JM, Hu P, Pierce AJ, Moynahan ME, Ellis N, Jasin M (2002) ATP hydrolysis by mammalian RAD51 has a key role during homology-directed DNA repair. J Biol Chem 277(23):20185–20194
53. Kowalczykowski SC, Krupp RA (1995) DNA-strand exchange promoted by RecA protein in the absence of ATP: implications for the mechanism of energy transduction in protein-promoted nucleic acid transactions. Proc Natl Acad Sci USA 92(8):3478–3482
54. Menetski JP, Varghese A, Kowalczykowski SC (1988) Properties of the high-affinity single-stranded DNA binding state of the *Escherichia coli* recA protein. Biochemistry 27(4):1205–1212
55. Ristic D, Modesti M, van der Heijden T, van Noort J, Dekker C, Kanaar R, Wyman C (2005) Human Rad51 filaments on double- and single-stranded DNA: correlating regular and irregular forms with recombination function. Nucleic Acids Res 33(10):3292–3302
56. Hilario J, Amitani I, Baskin RJ, Kowalczykowski SC (2009) Direct imaging of human Rad51 nucleoprotein dynamics on individual DNA molecules. Proc Natl Acad Sci USA 106(2):361–368
57. Schieven GL, Kirihara JM, Gilliland LK, Uckun FM, Ledbetter JA (1993) Ultraviolet radiation rapidly induces tyrosine phosphorylation and calcium signaling in lymphocytes. Mol Biol Cell 4(5):523–530
58. Gafter U, Malachi T, Ori Y, Breitbart H (1997) The role of calcium in human lymphocyte DNA repair ability. J Lab Clin Med 130(1):33–41
59. Sakai H, Ito E, Cai RX, Yoshioka T, Kubota Y, Hashimoto K, Fujishima A (1994) Intracellular Ca^{2+} concentration change of T24 cell under irradiation in the presence of TiO_2 ultrafine particles. Biochim Biophys Acta 1201(2):259–265
60. Negre-Salvayre A, Salvayre R (1992) UV-treated lipoproteins as a model system for the study of the biological effects of lipid peroxides on cultured cells. 4. Calcium is involved in the cytotoxicity of UV-treated LDL on lymphoid cell lines. Biochim Biophys Acta 1123(2):207–215
61. Whitaker M (2006) Calcium at fertilization and in early development. Physiol Rev 86(1):25–88
62. Keith CH, Maxfield FR, Shelanski ML (1985) Intracellular free calcium levels are reduced in mitotic Pt K2 epithelial cells. Proc Natl Acad Sci USA 82(3):800–804
63. Sung P (1994) Catalysis of ATP-dependent homologous DNA pairing and strand exchange by yeast RAD51 protein. Science 265(5176):1241–1243
64. Seitz EM, Brockman JP, Sandler SJ, Clark AJ, Kowalczykowski SC (1998) RadA protein is an archaeal RecA protein homolog that catalyzes DNA strand exchange. Genes Dev 12(9):1248–1253
65. Galletto R, Kowalczykowski SC (2007) RecA. Curr Biol 17(11):R395–R397
66. Roca AI, Cox MM (1997) RecA protein: structure, function, and role in recombinational DNA repair. Prog Nucleic Acid Res Mol Biol 56:129–223
67. Mazin AV, Kowalczykowski SC (1996) The specificity of the secondary DNA binding site of RecA protein defines its role in DNA strand exchange. Proc Natl Acad Sci USA 93(20):10673–10678
68. Mazin AV, Kowalczykowski SC (1999) A novel property of the RecA nucleoprotein filament: activation of double-stranded DNA for strand exchange in trans. Genes Dev 13(15):2005–2016
69. Howard-Flanders P, West SC, Stasiak A (1984) Role of RecA protein spiral filaments in genetic recombination. Nature (London) 309(5965):215–219
70. Matsuo Y, Sakane I, Takizawa Y, Takahashi M, Kurumizaka H (2006) Roles of the human Rad51 L1 and L2 loops in DNA binding. FEBS J 273(14):3148–3159
71. Rehrauer WM, Kowalczykowski SC (1996) The DNA binding site(s) of the *Escherichia coli* RecA protein. J Biol Chem 271(20):11996–12002
72. Malkov VA, Camerini-Otero RD (1995) Photocross-links between single-stranded DNA and *Escherichia coli* RecA protein map to loops L1 (amino acid residues 157–164) and L2 (amino acid residues 195–209). J Biol Chem 270(50):30230–30233

73. Wang Y, Adzuma K (1996) Differential proximity probing of two DNA binding sites in the *Escherichia coli* recA protein using photo-cross-linking methods. Biochemistry 35(11): 3563–3571
74. Mazin AV, Kowalczykowski SC (1998) The function of the secondary DNA-binding site of RecA protein during DNA strand exchange. EMBO J 17(4):1161–1168
75. Chow SA, Honigberg SM, Bainton RJ, Radding CM (1986) Patterns of nuclease protection during strand exchange. recA protein forms heteroduplex DNA by binding to strands of the same polarity. J Biol Chem 261(15):6961–6971
76. Pugh BF, Cox MM (1987) recA protein binding to the heteroduplex product of DNA strand exchange. J Biol Chem 262:1337–1343
77. Rosselli W, Stasiak A (1990) Energetics of RecA-mediated recombination reactions: Without ATP hydrolysis RecA can mediate polar strand exchange but is unable to recycle. J Mol Biol 216:335–352
78. Ullsperger CJ, Cox MM (1995) Quantitative RecA protein binding to the hybrid duplex product of DNA strand exchange. Biochemistry 34(34):10859–10866
79. Cox MM, Lehman IR (1981) RecA protein of *Escherichia coli* promotes branch migration, a kinetically distinct phase of DNA strand exchange. Proc Natl Acad Sci USA 78: 3433–3437
80. Murayama Y, Kurokawa Y, Mayanagi K, Iwasaki H (2008) Formation and branch migration of Holliday junctions mediated by eukaryotic recombinases. Nature 451(7181): 1018–1021
81. Sigurdsson S, Van Komen S, Petukhova G, Sung P (2002) Homologous DNA pairing by human recombination factors Rad51 and Rad54. J Biol Chem 277(45):42790–42794
82. Mazina OM, Rossi MJ, Deakyne JS, Huang F, Mazin AV (2012) Polarity and bypass of DNA heterology during branch migration of Holliday junctions by human RAD54, BLM, and RECQ1. J Biol Chem 287(15):11820–11832
83. Bugreev DV, Mazina OM, Mazin AV (2006) Rad54 protein promotes branch migration of Holliday junctions. Nature 442(7102):590–593
84. Yamamoto A, Taki T, Yagi H, Habu T, Yoshida K, Yoshimura Y, Yamamoto K, Matsushiro A, Nishimune Y, Morita T (1996) Cell cycle-dependent expression of the mouse Rad51 gene in proliferating cells. Mol Gen Genet 251(1):1–12
85. Takata M, Sasaki MS, Sonoda E, Morrison C, Hashimoto M, Utsumi H, Yamaguchi-Iwai Y, Shinohara A, Takeda S (1998) Homologous recombination and non-homologous end-joining pathways of DNA double-strand break repair have overlapping roles in the maintenance of chromosomal integrity in vertebrate cells. EMBO J 17(18):5497–5508
86. Ren B, Cam H, Takahashi Y, Volkert T, Terragni J, Young RA, Dynlacht BD (2002) E2F integrates cell cycle progression with DNA repair, replication, and G(2)/M checkpoints. Genes Dev 16(2):245–256
87. Ishida S, Huang E, Zuzan H, Spang R, Leone G, West M, Nevins JR (2001) Role for E2F in control of both DNA replication and mitotic functions as revealed from DNA microarray analysis. Mol Cell Biol 21(14):4684–4699
88. Trimarchi JM, Lees JA (2002) Sibling rivalry in the E2F family. Nat Rev Mol Cell Biol 3(1): 11–20
89. Dynlacht BD (2008) Live or let die: E2F1 and PI3K pathways intersect to make life or death decisions. Cancer Cell 13(1):1–2
90. Lazaro-Trueba I, Arias C, Silva A (2006) Double bolt regulation of Rad51 by p53: a role for transcriptional repression. Cell Cycle 5(10):1062–1065
91. Raderschall E, Stout K, Freier S, Suckow V, Schweiger S, Haaf T (2002) Elevated levels of Rad51 recombination protein in tumor cells. Cancer Res 62(1):219–225
92. Xia SJ, Shammas MA, Shmookler RJ (1997) Reis, Elevated recombination in immortal human cells is mediated by HsRAD51 recombinase. Mol Cell Biol 17(12):7151–7158
93. Maacke H, Opitz S, Jost K, Hamdorf W, Henning W, Kruger S, Feller AC, Lopens A, Diedrich K, Schwinger E, Sturzbecher HW (2000) Over-expression of wild-type rad51 correlates with histological grading of invasive ductal breast cancer [In Process Citation]. Int J Cancer 88(6):907–913

94. Maacke H, Jost K, Opitz S, Miska S, Yuan Y, Hasselbach L, Luttges J, Kalthoff H, Sturzbecher HW (2000) DNA repair and recombination factor Rad51 is over-expressed in human pancreatic adenocarcinoma. Oncogene 19(23):2791–2795
95. Hannay JA, Liu J, Zhu QS, Bolshakov SV, Li L, Pisters PW, Lazar AJ, Yu D, Pollock RE, Lev D (2007) Rad51 overexpression contributes to chemoresistance in human soft tissue sarcoma cells: a role for p53/activator protein 2 transcriptional regulation. Mol Cancer Ther 6(5):1650–1660
96. Thacker J (2005) The RAD51 gene family, genetic instability and cancer. Cancer Lett 219(2): 125–135
97. Hine CM, Seluanov A, Gorbunova V (2008) Use of the Rad51 promoter for targeted anti-cancer therapy. Proc Natl Acad Sci USA 105(52):20810–20815
98. Witkin EM (1991) RecA protein in the SOS response: milestones and mysteries. Biochimie 73:133–141
99. Frank EG, Ennis DG, Gonzalez M, Levine AS, Woodgate R (1996) Regulation of SOS mutagenesis by proteolysis. Proc Natl Acad Sci USA 93(19):10291–10296
100. Essers J, Houtsmuller AB, van Veelen L, Paulusma C, Nigg AL, Pastink A, Vermeulen W, Hoeijmakers JH, Kanaar R (2002) Nuclear dynamics of RAD52 group homologous recombination proteins in response to DNA damage. EMBO J 21(8):2030–2037
101. Haaf T, Golub EI, Reddy G, Radding CM, Ward DC (1995) Nuclear foci of mammalian Rad51 recombination protein in somatic cells after DNA damage and its localization in synaptonemal complexes. Proc Natl Acad Sci USA 92(6):2298–2302
102. Gildemeister OS, Sage JM, Knight KL (2009) Cellular redistribution of Rad51 in response to DNA damage: novel role for Rad51C. J Biol Chem 284(46):31945–31952
103. Barlow AL, Benson FE, West SC, Hulten MA (1997) Distribution of the Rad51 recombinase in human and mouse spermatocytes. EMBO J 16(17):5207–5215
104. Tan TL, Essers J, Citterio E, Swagemakers SM, de Wit J, Benson FE, Hoeijmakers JH, Kanaar R (1999) Mouse Rad54 affects DNA conformation and DNA-damage-induced Rad51 foci formation. Curr Biol 9(6):325–328
105. Scully R, Chen J, Plug A, Xiao Y, Weaver D, Feunteun J, Ashley T, Livingston DM (1997) Association of BRCA1 with Rad51 in mitotic and meiotic cells. Cell 88(2):265–275
106. Wong AKC, Pero R, Ormonde PA, Tavtigian SV, Bartel PL (1997) RAD51 interacts with the evolutionarily conserved BRC motifs in the human breast cancer susceptibility gene brca2. J Biol Chem 272(51):31941–31944
107. Thorslund T, West SC (2007) BRCA2: a universal recombinase regulator. Oncogene 26(56): 7720–7730
108. Jensen RB, Carreira A, Kowalczykowski SC (2010) Purified human BRCA2 stimulates RAD51-mediated recombination. Nature 467(7316):678–683
109. Liu J, Doty T, Gibson B, Heyer WD (2010) Human BRCA2 protein promotes RAD51 filament formation on RPA-covered single-stranded DNA. Nat Struct Mol Biol 17(10):1260–1262
110. Yang H, Li Q, Fan J, Holloman WK, Pavletich NP (2005) The BRCA2 homologue Brh2 nucleates RAD51 filament formation at a dsDNA-ssDNA junction. Nature 433(7026):653–657
111. Connor F, Bertwistle D, Mee PJ, Ross GM, Swift S, Grigorieva E, Tybulewicz VL, Ashworth A (1997) Tumorigenesis and a DNA repair defect in mice with a truncating Brca2 mutation. Nat Genet 17(4):423–430
112. Tutt A, Connor F, Bertwistle D, Kerr P, Peacock J, Ross G, Ashworth A (2003) Cell cycle and genetic background dependence of the effect of loss of BRCA2 on ionizing radiation sensitivity. Oncogene 22(19):2926–2931
113. Esashi F, Christ N, Gannon J, Liu Y, Hunt T, Jasin M, West SC (2005) CDK-dependent phosphorylation of BRCA2 as a regulatory mechanism for recombinational repair. Nature 434(7033):598–604
114. Brugarolas J, Chandrasekaran C, Gordon JI, Beach D, Jacks T, Hannon GJ (1995) Radiation-induced cell cycle arrest compromised by p21 deficiency. Nature 377(6549):552–557
115. Falck J, Mailand N, Syljuasen RG, Bartek J, Lukas J (2001) The ATM-Chk2-Cdc25A checkpoint pathway guards against radioresistant DNA synthesis. Nature 410(6830):842–847

116. Jinno S, Suto K, Nagata A, Igarashi M, Kanaoka Y, Nojima H, Okayama H (1994) Cdc25A is a novel phosphatase functioning early in the cell cycle. EMBO J 13(7):1549–1556
117. Yata K, Lloyd J, Maslen S, Bleuyard JY, Skehel M, Smerdon SJ, Esashi F (2012) Plk1 and CK2 act in concert to regulate Rad51 during DNA double strand break repair. Mol Cell 45(3):371–383
118. Bekker-Jensen S, Lukas C, Kitagawa R, Melander F, Kastan MB, Bartek J, Lukas J (2006) Spatial organization of the mammalian genome surveillance machinery in response to DNA strand breaks. J Cell Biol 173(2):195–206
119. Chapman JR, Jackson SP (2008) Phospho-dependent interactions between NBS1 and MDC1 mediate chromatin retention of the MRN complex at sites of DNA damage. EMBO Rep 9(8): 795–801
120. Sorensen CS, Hansen LT, Dziegielewski J, Syljuasen RG, Lundin C, Bartek J, Helleday T (2005) The cell-cycle checkpoint kinase Chk1 is required for mammalian homologous recombination repair. Nat Cell Biol 7(2):195–201
121. Yuan ZM, Huang Y, Ishiko T, Nakada S, Utsugisawa T, Kharbanda S, Wang R, Sung P, Shinohara A, Weichselbaum R, Kufe D (1998) Regulation of Rad51 function by c-Abl in response to DNA damage. J Biol Chem 273(7):3799–3802
122. Popova M, Shimizu H, Yamamoto K, Lebechec M, Takahashi M, Fleury F (2009) Detection of c-Abl kinase-promoted phosphorylation of Rad51 by specific antibodies reveals that Y54 phosphorylation is dependent on that of Y315. FEBS Lett 583(12):1867–1872
123. Conilleau S, Takizawa Y, Tachiwana H, Fleury F, Kurumizaka H, Takahashi M (2004) Location of tyrosine 315, a target for phosphorylation by cAbl tyrosine kinase, at the edge of the subunit-subunit interface of the human Rad51 filament. J Mol Biol 339(4):797–804
124. Slupianek A, Jozwiakowski SK, Gurdek E, Skorski T (2009) BCR/ABL kinase interacts with and phosphorylates the RAD51 paralog, RAD51B. Leukemia 23(12):2308–2310
125. Kruh GD, Perego R, Miki T, Aaronson SA (1990) The complete coding sequence of arg defines the Abelson subfamily of cytoplasmic tyrosine kinases. Proc Natl Acad Sci USA 87(15):5802–5806
126. Yuan ZM, Huang Y, Ishiko T, Kharbanda S, Weichselbaum R, Kufe D (1997) Regulation of DNA damage-induced apoptosis by the c-Abl tyrosine kinase. Proc Natl Acad Sci USA 94(4):1437–1440
127. Van Etten RA (1999) Cycling, stressed-out and nervous: cellular functions of c-Abl. Trends Cell Biol 9(5):179–186
128. Dou H, Huang C, Singh M, Carpenter PB, Yeh ET (2010) Regulation of DNA repair through deSUMOylation and SUMOylation of replication protein A complex. Mol Cell 39(3): 333–345
129. Li W, Hesabi B, Babbo A, Pacione C, Liu J, Chen DJ, Nickoloff JA, Shen Z (2000) Regulation of double-strand break-induced mammalian homologous recombination by UBL1, a RAD51-interacting protein. Nucleic Acids Res 28(5):1145–1153
130. Clever B, Interthal H, Schmuckli-Maurer J, King J, Sigrist M, Heyer WD (1997) Recombinational repair in yeast: functional interactions between Rad51 and Rad54 proteins. EMBO J 16(9):2535–2544
131. Golub EI, Kovalenko OV, Gupta RC, Ward DC, Radding CM (1997) Interaction of human recombination proteins Rad51 and Rad54. Nucleic Acids Res 25(20):4106–4110
132. Mazin AV, Mazina OM, Bugreev DV, Rossi MJ (2010) Rad54, the motor of homologous recombination. DNA Repair (Amst) 9(3):286–302
133. Wolner B, van Komen S, Sung P, Peterson CL (2003) Recruitment of the recombinational repair machinery to a DNA double-strand break in yeast. Mol Cell 12(1):221–232
134. Petukhova G, Stratton S, Sung P (1998) Catalysis of homologous DNA pairing by yeast Rad51 and Rad54 proteins. Nature 393(6680):91–94
135. Haseltine CA, Kowalczykowski SC (2009) An archaeal Rad54 protein remodels DNA and stimulates DNA strand exchange by RadA. Nucleic Acids Res 37(8):2757–2770
136. Mazina OM, Mazin AV (2004) Human Rad54 protein stimulates DNA strand exchange activity of hRad51 protein in the presence of Ca^{2+}. J Biol Chem 279(50):52042–52051

137. Alexiadis V, Kadonaga JT (2002) Strand pairing by Rad54 and Rad51 is enhanced by chromatin. Genes Dev 16(21):2767–2771
138. Van Komen S, Petukhova G, Sigurdsson S, Stratton S, Sung P (2000) Superhelicity-driven homologous DNA pairing by yeast recombination factors Rad51 and Rad54. Mol Cell 6(3):563–572
139. Mazin AV, Bornarth CJ, Solinger JA, Heyer WD, Kowalczykowski SC (2000) Rad54 protein is targeted to pairing loci by the Rad51 nucleoprotein filament. Mol Cell 6(3):583–592
140. Solinger JA, Lutz G, Sugiyama T, Kowalczykowski SC, Heyer WD (2001) Rad54 protein stimulates heteroduplex DNA formation in the synaptic phase of DNA strand exchange via specific interactions with the presynaptic Rad51 nucleoprotein filament. J Mol Biol 307(5):1207–1221
141. Mazin AV, Alexeev AA, Kowalczykowski SC (2003) A novel function of Rad54 protein. Stabilization of the Rad51 nucleoprotein filament. J Biol Chem 278(16):14029–14036
142. Jaskelioff M, Van Komen S, Krebs JE, Sung P, Peterson CL (2003) Rad54p is a chromatin remodeling enzyme required for heteroduplex DNA joint formation with chromatin. J Biol Chem 278(11):9212–9218
143. Alexeev A, Mazin A, Kowalczykowski SC (2003) Rad54 protein possesses chromatin-remodeling activity stimulated by the Rad51-ssDNA nucleoprotein filament. Nat Struct Biol 10(3):182–186
144. Zhang Z, Fan HY, Goldman JA, Kingston RE (2007) Homology-driven chromatin remodeling by human RAD54. Nat Struct Mol Biol 14(5):397–405
145. Rossi MJ, Mazin AV (2008) Rad51 protein stimulates the branch migration activity of Rad54 protein. J Biol Chem 283(36):24698–24706
146. Bugreev DV, Mazina OM, Mazin AV (2009) Bloom syndrome helicase stimulates RAD51 DNA strand exchange activity through a novel mechanism. J Biol Chem 284(39):26349–26359
147. Wiese C, Dray E, Groesser T, San Filippo J, Shi I, Collins DW, Tsai MS, Williams GJ, Rydberg B, Sung P, Schild D (2007) Promotion of homologous recombination and genomic stability by RAD51AP1 via RAD51 recombinase enhancement. Mol Cell 28(3):482–490
148. Modesti M, Budzowska M, Baldeyron C, Demmers JA, Ghirlando R, Kanaar R (2007) RAD51AP1 is a structure-specific DNA binding protein that stimulates joint molecule formation during RAD51-mediated homologous recombination. Mol Cell 28(3):468–481
149. Pezza RJ, Voloshin ON, Vanevski F, Camerini-Otero RD (2007) Hop2/Mnd1 acts on two critical steps in Dmc1-promoted homologous pairing. Genes Dev 21(14):1758–1766
150. Chi P, San Filippo J, Sehorn MG, Petukhova GV, Sung P (2007) Bipartite stimulatory action of the Hop2-Mnd1 complex on the Rad51 recombinase. Genes Dev 21(14):1747–1757
151. Kato M, Yano K, Matsuo F, Saito H, Katagiri T, Kurumizaka H, Yoshimoto M, Kasumi F, Akiyama F, Sakamoto G, Nagawa H, Nakamura Y, Miki Y (2000) Identification of Rad51 alteration in patients with bilateral breast cancer. J Hum Genet 45(3):133–137
152. Levy-Lahad E, Lahad A, Eisenberg S, Dagan E, Paperna T, Kasinetz L, Catane R, Kaufman B, Beller U, Renbaum P, Gershoni-Baruch R (2001) A single nucleotide polymorphism in the RAD51 gene modifies cancer risk in BRCA2 but not BRCA1 carriers. Proc Natl Acad Sci USA 98(6):3232–3236
153. Wang WW, Spurdle AB, Kolachana P, Bove B, Modan B, Ebbers SM, Suthers G, Tucker MA, Kaufman DJ, Doody MM, Tarone RE, Daly M, Levavi H, Pierce H, Chetrit A, Yechezkel GH, Chenevix-Trench G, Offit K, Godwin AK, Struewing JP (2001) A single nucleotide polymorphism in the 5′ untranslated region of RAD51 and risk of cancer among BRCA1/2 mutation carriers. Cancer Epidemiol Biomarkers Prev 10(9):955–960
154. Wang F, Zou YF, Feng XL (2011) Association of XRCC3 C722T and RAD51 G135C polymorphisms with head and neck cancer susceptibility in a Polish population. Exp Mol Pathol 90(3):318, author reply 319
155. Sliwinski T, Walczak A, Przybylowska K, Rusin P, Pietruszewska W, Zielinska-Blizniewska H, Olszewski J, Morawiec-Sztandera A, Jendrzejczyk S, Mlynarski W, Majsterek I (2010) Polymorphisms of the XRCC3 C722T and the RAD51 G135C genes and the risk of head and neck cancer in a Polish population. Exp Mol Pathol 89(3):358–366

156. Blasiak J, Przybylowska K, Czechowska A, Zadrozny M, Pertynski T, Rykala J, Kolacinska A, Morawiec Z, Drzewoski J (2003) Analysis of the G/C polymorphism in the 5′-untranslated region of the RAD51 gene in breast cancer. Acta Biochim Pol 50(1):249–253
157. Jakubowska A, Gronwald J, Menkiszak J, Gorski B, Huzarski T, Byrski T, Edler L, Lubinski J, Scott RJ, Hamann U (2007) The RAD51 135 G>C polymorphism modifies breast cancer and ovarian cancer risk in Polish BRCA1 mutation carriers. Cancer Epidemiol Biomarkers Prev 16(2):270–275
158. Kadouri L, Kote-Jarai Z, Hubert A, Durocher F, Abeliovich D, Glaser B, Hamburger T, Eeles RA, Peretz T (2004) A single-nucleotide polymorphism in the RAD51 gene modifies breast cancer risk in BRCA2 carriers, but not in BRCA1 carriers or noncarriers. Br J Cancer 90(10):2002–2005
159. Hasselbach L, Haase S, Fischer D, Kolberg HC, Sturzbecher HW (2005) Characterisation of the promoter region of the human DNA-repair gene Rad51. Eur J Gynaecol Oncol 26(6):589–598
160. Ishida T, Takizawa Y, Sakane I, Kurumizaka H (2007) Altered DNA binding by the human Rad51-R150Q mutant found in breast cancer patients. Biol Pharm Bull 30(8):1374–1378
161. Vispe S, Cazaux C, Lesca C, Defais M (1998) Overexpression of Rad51 protein stimulates homologous recombination and increases resistance of mammalian cells to ionizing radiation. Nucleic Acids Res 26(12):2859–2864
162. Xu ZY, Loignon M, Han FY, Panasci L, Aloyz R (2005) Xrcc3 induces cisplatin resistance by stimulation of Rad51-related recombinational repair, S-phase checkpoint activation, and reduced apoptosis. J Pharmacol Exp Ther 314(2):495–505
163. Bolderson E, Richard DJ, Zhou BB, Khanna KK (2009) Recent advances in cancer therapy targeting proteins involved in DNA double-strand break repair. Clin Cancer Res 15(20):6314–6320
164. Ito M, Yamamoto S, Nimura K, Hiraoka K, Tamai K, Kaneda Y (2005) Rad51 siRNA delivered by HVJ envelope vector enhances the anti-cancer effect of cisplatin. J Gene Med 7(8):1044–1052
165. Luo J, Emanuele MJ, Li D, Creighton CJ, Schlabach MR, Westbrook TF, Wong KK, Elledge SJ (2009) A genome-wide RNAi screen identifies multiple synthetic lethal interactions with the Ras oncogene. Cell 137(5):835–848
166. Taki T, Ohnishi T, Yamamoto A, Hiraga S, Arita N, Izumoto S, Hayakawa T, Morita T (1996) Antisense inhibition of the RAD51 enhances radiosensitivity. Biochem Biophys Res Commun 223(2):434–438
167. Ohnishi T, Taki T, Hiraga S, Arita N, Morita T (1998) In vitro and in vivo potentiation of radiosensitivity of malignant gliomas by antisense inhibition of the RAD51 gene. Biochem Biophys Res Commun 245(2):319–324
168. Bryant HE, Schultz N, Thomas HD, Parker KM, Flower D, Lopez E, Kyle S, Meuth M, Curtin NJ, Helleday T (2005) Specific killing of BRCA2-deficient tumours with inhibitors of poly(ADP-ribose) polymerase. Nature 434(7035):913–917
169. Helleday T, Lo J, van Gent DC, Engelward BP (2007) DNA double-strand break repair: from mechanistic understanding to cancer treatment. DNA Repair (Amst) 6(7):923–935
170. Farmer H, McCabe N, Lord CJ, Tutt AN, Johnson DA, Richardson TB, Santarosa M, Dillon KJ, Hickson I, Knights C, Martin NM, Jackson SP, Smith GC, Ashworth A (2005) Targeting the DNA repair defect in BRCA mutant cells as a therapeutic strategy. Nature 434(7035):917–921
171. Helleday T (2010) Homologous recombination in cancer development, treatment and development of drug resistance. Carcinogenesis 31(6):955–960
172. Ishida T, Takizawa Y, Kainuma T, Inoue J, Mikawa T, Shibata T, Suzuki H, Tashiro S, Kurumizaka H (2009) DIDS, a chemical compound that inhibits RAD51-mediated homologous pairing and strand exchange. Nucleic Acids Res 37(10):3367–3376
173. Li Y, He Y, Luo Y (2009) Crystal structure of an archaeal Rad51 homologue in complex with a metatungstate inhibitor. Biochemistry 48(29):6805–6810
174. Budke B, Logan HL, Kalin JH, Zelivianskaia AS, Cameron McGuire W, Miller LL, Stark JM, Kozikowski AP, Bishop DK, Connell PP. (2012) RI-1: a chemical inhibitor of RAD51 that disrupts homologous recombination in human cells. Nucleic Acids Res 40(15):7347–7357

175. Huang F, Motlekar NA, Burgwin CM, Napper AD, Diamond SL, Mazin AV (2011) Identification of specific inhibitors of human RAD51 recombinase using high-throughput screening. ACS Chem Biol 6(6):628–635
176. Menear KA, Adcock C, Boulter R, Cockcroft XL, Copsey L, Cranston A, Dillon KJ, Drzewiecki J, Garman S, Gomez S, Javaid H, Kerrigan F, Knights C, Lau A, Loh VM Jr, Matthews IT, Moore S, O'Connor MJ, Smith GC, Martin NM (2008) 4-[3-(4-cyclopropanecarbonylpiperazine-1-carbonyl)-4-fluorobenzyl]-2 H-phthalazin- 1-one: a novel bioavailable inhibitor of poly(ADP-ribose) polymerase-1. J Med Chem 51(20):6581–6591
177. Huang F, Mazina OM, Zentner IJ, Cocklin S, Mazin AV (2012) Inhibition of homologous recombination in human cells by targeting RAD51 recombinase. J Med Chem 55(7): 3011–3020

Index

L. Panasci et al. (eds.), *Advances in DNA Repair in Cancer Therapy*,
Cancer Drug Discovery and Development 72, DOI 10.1007/978-1-4614-4741-2,
© Springer Science+Business Media, LLC 2013

MIX
Papier aus verantwortungsvollen Quellen
Paper from responsible sources
FSC® C105338

If you have any concerns about our products,
you can contact us on
ProductSafety@springernature.com

In case Publisher is established outside the EU,
the EU authorized representative is:
Springer Nature Customer Service Center GmbH
Europaplatz 3, 69115 Heidelberg, Germany

Printed by Libri Plureos GmbH
in Hamburg, Germany